Weeds and Weed Management on Arable Land

An Ecological Approach

Weeds and Weed Management on Arable Land

An Ecological Approach

Sigurd Håkansson

Department of Ecology and Crop Production Science
Swedish University of Agricultural Sciences
Uppsala
Sweden

CABI *Publishing*

CABI *Publishing* is a division of CAB *International*

CABI Publishing
CAB International
Wallingford
Oxon OX10 8DE
UK

CABI Publishing
44 Brattle Street
4th Floor
Cambridge, MA 02138
USA

Tel: +44 (0)1491 832111
Fax: +44 (0)1491 833508
E-mail: cabi@cabi.org
Website: www.cabi-publishing.org

Tel: +1 617 395 4056
Fax: +1 617 354 6875
E-mail: cabi-nao@cabi.org

A catalogue record for this book is available from the British Library, London, UK.

Library of Congress Cataloging-in-Publication Data

Hakansson, Sigurd.
  Weeds and weed management on arable land : an ecological approach /
Sigurd Håkansson.
      p. cm.
Includes bibliographical references (p.).
  ISBN 0-85199-651-5 (alk. paper)
  1. Weeds. 2. Weeds--Ecology. 3. Weeds--Control. 4. Weeds--Sweden.
5. Weeds--Ecology--Sweden. 6. Weeds--Control--Sweden. I. Title.
  SB611 .H35 2003
  632′.5--dc21

        2002010487

ISBN 0 85199 651 5

Typeset by AMA DataSet Ltd, UK.
Printed and bound in the UK by Biddles Ltd, Guildford and King's Lynn.

Contents

Preface

Background and Focus of the Book

This book may in the first instance be of interest to teachers, postgraduate students and researchers in weed science, but will, I hope, also prove useful to undergraduate students who are specializing in this and related subjects. Discussions and illustrations in the book also relate to issues that are of interest in other fields of agriculture, horticulture and plant ecology. There are also concepts in the book that I hope will attract other categories of readers interested in biology and ecology.

As a researcher and teacher in weed science and crop production ecology for almost 40 years, I have had the opportunity to follow developments in these subjects during a period of great change. In weed research, the interest has gradually shifted from a strong focus on chemical weed control in the 1950s and 1960s to an increasing accent on integrated weed management during the 1980s and 1990s.

Investigations with a broader biological and ecological approach to weed management received proportionally little financial support in most countries from the 1950s to the late 1970s. This also applies to Sweden, although such investigations were awarded *relatively* more money in Sweden than in many other countries.

In 1954, I had the opportunity to start research on wild garlic (*Allium vineale* L.) occurring as a weed on arable land in South-eastern Sweden. In discreet opposition to demands from above for experiments focusing on chemical control, I was able to give my investigation a broader ecological approach. My youthful ambition was to present a case study illustrating the minimum knowledge needed for understanding the possibilities and difficulties of a plant growing and persisting as a weed in different cropping systems – which is, at the same time, the minimum knowledge needed for rational combinations of direct and indirect control of the weed.

When I became a university lecturer in the 1960s, the period of early, almost unreserved enthusiasm over modern herbicides had not yet passed. A discussion on weed management was largely synonymous with a discussion on chemical weed control. Like the bodies funding research, students were influenced by the prevailing circumstances, most of them expecting weed science courses to be focused on chemical weed control.

In spite of this, I persisted with presenting weed science courses with a wider, more ecological approach. The courses were on the whole accepted and appreciated by the majority of my students, although protests also arose against 'wasting time on economically unimportant issues'. However, general opinions gradually changed in favour of broader ecological approaches to weed management. The proportions between different subject areas in my courses may have varied more or less from year to year, but the overall structure of the courses has remained unchanged over the years. This structure is reflected in the present book.

The strong focus on chemical control in weed research and weed management practices for about three decades resulted in an overuse of herbicides, which indeed deserves criticism. However, my personal opinion is that this criticism must not lead to a complete rejection of chemical weed control. Pluralism is always fruitful; at the same time as the development of cropping systems without any use of chemical weed control should be supported, a cautious and restricted use of herbicides will certainly do well in many cropping systems seriously structured to minimize negative environmental impacts. Verdicts on adverse side effects of herbicides should be based on comprehensive herbicide tests, which are now being increasingly performed and reported. Like other cultural measures in a cropping system, chemical weed control should be based on broad biological and ecological knowledge and considerations.

Many herbicides are positive tools in the control of vegetation since they allow a reduction in tillage where soil erosion is a problem. This is particularly important in warm areas with high precipitation, on sloping ground and in dry, windy areas, but intensive tillage and passes with heavy machinery have negative impacts everywhere. Soil compaction is one of the persistent problems in modern agriculture and intensified soil tillage can never be a satisfactory means of replacing chemical weed control. Minimum and zero tillage are being practised or attempted in many agricultural areas for the purpose of minimizing soil tillage disadvantages.

A one-sided focus on the short-term effects of weed control measures, either chemical or mechanical, should thus be avoided. In harmony with intentions that are generally recognized today, different control and cultural measures should be combined into systems of integrated weed management. I hope that the facts and viewpoints in my book will contribute to a balanced development of such systems.

Guide to Central Topics Covered in Chapters 1–13

Chapter 1

All plants occurring in an arable field in a given situation without the grower's intention are termed 'weeds' in that situation; thus also 'volunteer' crop plants.

Chapters 2 and 3

Classification of plant species into life forms on the basis of ecologically significant traits is discussed. For species occurring as arable weeds, the classification here considers traits with primary influence on the appearance of the species in crop stands differing as to times and means of establishment and harvesting, and with various geometry, annual developmental rhythm and duration (1 or more years). The importance of dormancy of weed seeds, resulting in soil seed banks, is discussed. Seasonal variation in the germination readiness of seeds in the weed seed banks is illustrated (e.g. Fig. 1) and its influence on the appearance of weeds in different situations is dealt with. Perennial weeds are classified according to properties of vegetative structures determining their establishment and persistence as arable weeds.

Chapters 4 and 5

Arable weeds (non-parasitic) can be looked upon as plants in early stages of secondary vegetation succession (Tables 1–4). Relationships between plants with different life forms and their ability to establish as weeds in different crops (considering associated cultural measures) are therefore found to present essentially similar patterns in Scandinavian, European and global perspectives (Tables 3–15).

The occurrence of plants with C_4 photosynthesis as arable weeds is illustrated (Tables 12, 14 and 15) and differences are discussed. Particular attention is devoted to the geographical distribution of rhizomatous C_4 species. The distribution is discussed comparing the occurrence of the C_3 species *Elymus repens*.

A weed with better possibilities to establish, grow and reproduce in a given crop than in the others in a crop sequence when its growing plants are not subjected to direct control may often be more strongly restricted in that crop than in the others when its growing plants are effectively killed by direct control and its reproduction thus prevented. This may apply even if plants of the weed are effectively controlled when emerged in the other crops. Though important and easily explained, this is too often overlooked in general discussions on the growth and reproduction of weeds in crop sequences.

Chapter 6

The early growth of plants from seeds is illustrated, considering influences of soil conditions, seedbed preparation and means of sowing. Both crops and weeds are regarded (Tables 18–21; Figs 3–12). The conditions for their germination (or bud activation) and their emergence and further early growth largely determine their competitive position in the future growth of the crop–weed stand (Tables 22–24).

Chapters 7 and 8

The concept of plant competition and factors affecting the competitive outcome in plant stands of short duration are analysed regarding both crop–weed stands and pure crop stands. Phenotypic plasticity enables plants to adapt to different environments, including competitive conditions. Differing plasticity is one of the reasons for differences in the relative competitiveness of plants. Variation in properties ('quality') of harvested crop products is, to a great extent, caused by the plastic responses of the crop plants to differences in their environment, including their competitive situation.

Competition in a plant stand always modifies, sometimes very drastically, the response of plants to external factors. This also means that the relative competitive abilities of genetically identical plants may be strongly affected by changes in the external environment. This applies to their competitive ability in terms of both effects on other plants and responses to competition from them (e.g. Tables 31, 32 and 36; Fig. 34). The ranking order of crops or weeds as to their relative competitiveness may thus differ strongly with the environment.

Weakening of plants caused by mechanical or chemical means, diseases, etc. leads to an increased suppressive effect of competition from surrounding plants, when these plants are not weakened or are less weakened (Tables 26, 27, 34, 35 and 37). The importance of interactive effects of weed control measures and crop–weed competition is emphasized here and further stressed in Chapters 10 and 11.

Chapter 9

Methods, difficulties and pitfalls in measuring results of competition in plant stands of short duration are dealt with. 'Reference models' are described. They have been derived both to deduce reasonable characters of various relationships between production and density of different plants in mixed stands, and to test the usefulness of various ratios or indices for characterizing the relative competitiveness of plants and to specify restrictions in their usefulness. Thus, for instance, UPR, which is a central ratio in my own work, has been evaluated in this way. Reference models are based on production-density equations of pure stands of identical plants. The pure stands are then assumed to represent mixtures of different plant categories (A, B, etc.) uniformly distributed in the stands at varying densities (see Eqns 4–6, p. 138).

Chapter 10

Soil tillage is studied considering both indirect effects on weed populations and direct effects on growing weed plants. Life history and annual developmental rhythm largely determine the response of the weeds to different kinds of tillage at different times of the year. Principles are discussed and illustrated on the basis of experiments. Effects of tillage in interaction or combination with effects of other measures and situations in a cropping system are dealt with. Under the motto 'no more tillage than necessary', various forms of 'reduced tillage' are reviewed.

Chapter 11

Chemical weed control measures are discussed as elements in cropping systems, considering intended and unintended effects. By weighing desired and adverse effects, the justification for chemical weed control is debated. Considering both positive and negative effects, comparisons with tillage are of great interest.

Chapter 12

A number of management methods, applied or hypothetical, are surveyed regarding experienced or possible effects. These methods include mechanical or physical measures, the use of cover crops and mulches, timing and methods of harvesting, breeding for increased competitiveness of crops and biological weed control.

Chapter 13

Issues requiring knowledge for the guidance of rational integrated weed management in a cropping system are listed.

The importance of these issues is illustrated in case studies on the appearance of a number of plants with different traits as arable weeds in Sweden (e.g. Tables 42–44; Figs 66–71). Differences in the occurrence of these plants in diverse agricultural situations are described and changes over periods of years are discussed in relation to changes in the cropping systems. Difficulties in interpreting causalities expose gaps in current knowledge.

Acknowledgements

This book would not have been written without positive encouragement and support from my friends at the Department of Ecology and Crop Production Science in Uppsala. This Department, one of the major departments of the Swedish University of Agricultural Sciences, has changed names and fields of responsibility several times during the past five decades, but weed science and related crop production ecology have always been central fields. My successor, Professor Håkan Fogelfors, has always encouraged me and strongly supported my writing this book. I am sincerely indebted to him and to his co-workers for providing me with excellent opportunities to continue working at the Department, not only according to the traditional rights of a professor emeritus but also as a member of the team.

There are a number of people to whom I am indebted and whom I would like to thank here. First I want to mention Karl-Gustav Ursberg, who has been my skilled technical assistant for more than 30 years and who has more recently given me invaluable help with many computer problems. He has, among other things, scanned all my graphs and given them a uniform style. I am also very grateful to all those at the University who have read different chapters in my manuscript and given me valuable criticism. I mention them in alphabetical order: Dr Lars Andersson, Dr Tommy Arvidsson, Dr Bengt Bodin, Dr Ullalena Boström, Professor Sten Ebbersten, Professor Inge Håkansson, Dr Erik Hallgren, Dr Margareta Hansson and Dr Lars Ohlander.

From colleagues in neighbouring countries, I have had many valuable viewpoints and strong encouragement and support in writing my book. I express my sincere gratitude to all of them: Dr Angelija Buciene and Dr Virginijus Feiza in Lithuania, Professor Haldor Fykse in Norway, Dr Jukka Salonen in Finland and Professor Jens C. Streibig in Denmark.

As Swedish is my mother tongue, my text would have had too much of a Swedish flavour without competent revision. I am very indebted to Dr Mary McAfee, Wiltshire, England, for her many improvements of my language.

The Swedish Research Council for Environment, Agricultural Sciences and Spatial Planning has financially supported the publishing of this book, which I gratefully acknowledge. I also acknowledge with sincere gratitude direct and indirect financial support from the Department of Ecology and Crop Production Science and the Royal Swedish Academy of Agriculture and Forestry.

Last but not least, I want to thank my wife Kerstin for having supported my writing by her tolerance and constant encouragement. I also want to thank my daughters and their families for stimulating me by showing keen interest in the progress of the project. I dedicate this book to the three generations of my family.

Sigurd Håkansson
Uppsala, October, 2002

1

Introduction

_____

Comments on the book

This book deals with the appearance and management of plants with different traits occurring as weeds on arable land. It describes and discusses matters and relationships that are important as a basis for understanding the varying occurrence of weeds in different crops and cropping systems and, at the same time, understanding the response of different weeds to specified management measures. This understanding is particularly crucial in planning systems of weed management measures over long periods of time; in other words, in planning 'integrated weed management'.

Principles are stressed before details. Although the matters discussed are often illustrated by experimental results representing conditions in temperate humid areas, the illustrations exemplify principles of general interest. Some illustrations directly treat global questions (e.g. Tables 14 and 15). Most of the illustrations are based on previously published material, although not always material published in English. Some previously unpublished illustrations, based on experiments devised for the author's university lectures, are also presented. The need for increased knowledge and understanding of certain relationships is stressed and experimental and measurement methods are discussed in many cases.

The book focuses in particular on issues that have been emphasized in the author's lecturing and experimental work.

New approaches to certain issues may make the book a useful complement to other books on weeds and weed managements, such as those by King (1966), Altieri and Liebman (1987), Radosevich et al. (1996) and Liebman et al. (2001). In addition, tables and diagrams in this book are largely based on material representing areas that are very seldom discussed in international textbooks. Principles emphasized in research from these areas underlie many approaches to weed management problems in tropical agriculture (e.g. Alström, 1990; Åfors, 1994).

Any procedure in a cropping system has some effect on weed populations, either directly on growing weed plants or indirectly through influences on weed seeds or other propagules in the soil, etc. Definitive distinctions can therefore not be made between weed management procedures and other cropping procedures. This should, of course, go without saying in discussions on integrated weed management, when long-term effects of procedures and conditions in a cropping system are to be considered. When, for instance, seedbed preparation and sowing are discussed in the following, effects on crops and weeds are evaluated at the same time.

The concept of weeds

The term 'weed' is given many meanings in the literature, although it usually refers to plants occurring in situations where they are unwanted. The word is usually understood as a term limited to herbaceous plants, but it sometimes also includes shrubs and trees. According to *The Shorter Oxford English Dictionary*, the word usually represents 'a herbaceous plant not valued for use or beauty, growing wild or rank, and regarded as cumbering the ground or hindering the growth of superior vegetation'. Outside the scope of this book, 'weed' is a word frequently used figuratively, characterizing creatures of various kinds regarded as unprofitable, sickly, troublesome or noxious.

As this book focuses on arable land, 'weeds' are discussed within the definition of plants occurring on arable land without the grower's intention. The term 'weed' therefore includes volunteer plants of cultivated species established in arable fields in situations where they are not intended.

Even when restricted to plants on arable land, numerous definitions of 'weed' are found in the literature. However, they largely fit into the descriptions above. (For examples of definitions, see Rademacher, 1948; Salisbury, 1961; Holzner, 1982; Radosevich *et al.*, 1996.)

Weeds on arable land are frequently called *arable weeds*. This brief and convenient designation has been widely used in the literature and is also used here. Arable weeds may be synonymously called *agrestal weeds* (Holzner, 1982).

A plant species, as such, is not a weed in the meaning given above. Weeds are those plants of the species that occur in situations where they are unwanted. This should be kept in mind when weed plants representing different species are named '*weed species*' for the sake of brevity, in this book as in other literature. Characteristic features and various origins of arable weeds are well described by Baker (1965).

Within many species, certain biotypes have become cultivated plants whereas other biotypes occur as weeds. An example is *Daucus carota*. Even genetically rather similar plants within one and the same species occur as troublesome weeds under certain conditions with respect to climate, soil and/or stages of agricultural development, but are, or have been, grown as cultivated plants under other conditions. *Avena strigosa* is an example.

Volunteer plants of cultivated species may originate from seeds or vegetative plant parts (e.g. potato tubers) lost from previous crops or sown with contaminated seed, etc. Plants of this kind may cause great harm, particularly when they resemble the present crop plants to the extent that they can neither be controlled nor separated from the harvested product. When they represent a previously grown cultivar of the same species as the present crop, volunteers with unwanted properties may lower the quality of the harvested product.

Arable land includes agricultural and horticultural fields where tillage is a regular measure. The present book is structured primarily to consider agricultural conditions, but the fundamentals dealt with here are naturally valid for parallel conditions in horticulture.

Although weeds are, on the whole, understood to be unwanted plants, all plants other than the intended crop plants are usually termed weeds. This even applies to plants causing little harm. It should be emphasized that many plants termed weeds can in fact be favourable from various aspects when they occur at low densities.

Extensive experience indicates that only a minority of the plant species inhabiting a geographical area as wild plants are able to become persistent weeds in intensively cultivated agricultural or horticultural fields. It will be seen in the following that there is no simple biological definition characterizing an arable weed with universal validity. Not unexpectedly, the significance of a specified plant trait is conditional. Traits such as life form, annual rhythm of development and growth, dormancy pattern of seeds and/or other propagules, competitive ability of plants, etc., are discussed in the following with respect to their influences on the ability of a plant to become a weed in a given situation.

Some definitions and descriptions

A few words and concepts are listed in alphabetical order below, and defined and commented on regarding their use in this book.

- *Control*: see 'Weed control'.
- *Cropping system*: The concept comprises a crop rotation or a period of years embracing a typical crop sequence with associated measures for the establishment, growth and harvesting of the crops. It thus also includes measures for managing weeds and pests. Typical cropping systems in Scandinavia and Finland are discussed on pp. 20–21.
- *Fallow*: The word stands for an arable field temporarily set aside from plant production in order to control weeds, to make the soil richer by growing plants for green manuring, etc.
- *Latin names of plants*: Unless otherwise stated, names are used according to *Flora Europaea* (Tutin *et al.*, 1964–1980).
- *Ley*: Particularly in Scandinavian literature, the word is used in the meaning of an arable field with a (fodder) crop consisting of one or more grasses, mostly mixed with some leguminous plant, largely red clover (seldom clover or lucerne solely). From the 19th century on, perennial (usually grass–clover) leys with a duration of 2–3 (1–5) years have been regularly alternated with annual crops, mainly cereals, sometimes also other crops and/or tilled fallows, etc., in more or less fixed rotations. The leys are usually under-sown in cereals. From the following year, the 'first ley year', it is cut for fodder. In recent decades, however, an increasing number of farms have discontinued livestock farming and growing of leys for fodder.
- *Management*: See under 'Weed management'.
- *Names of plants*: See under 'Latin names of plants'.
- *Rhizomes*: Underground shoot (stem) branches, growing plagiotropically to various lengths until they change to orthotropic growth, forming aerial shoots.
- *Scientific names of plants*: See under 'Latin names of plants'.
- *Seed*: The word is used here in a broad sense, comprising an entire dispersal unit containing a seed. It does not only mean a seed in its strictly organographic meaning.
- *Seed bank*: Here understood as the soil seed bank, representing seeds surviving in the soil for longer or shorter periods of years due to dormancy.
- *Stolons*: Shoot (stem) branches creeping on the soil surface, rooting from their nodes.
- *Weed*: see under the heading 'The concept of weeds' p. 1.
- *Weed control*: Activities and modifications of measures or conditions in the cropping system intended to reduce weed populations.
- *Weed management*: A general term for activities, procedures and modifications of conditions in the cropping system that are intended to influence weed populations. The term thus includes 'weed control'. The word 'management' stresses the intention of regulating weed populations to appropriate levels, considering both short-term economic and long-term ecological aspects. 'Weed management' may thus also include measures with the aim of preserving weed populations at some low or moderate level.

2

Classification of Plants Based on Traits of Ecological Significance

Plants have been classified in many ways and with various aims. We have the taxonomic groupings with taxa on different levels: families, genera, species, subspecies and varieties. However, even species that are closely related in a taxonomic classification may differ greatly in characteristics of ecological importance. A woody plant, a bush or a tree, may be found in the same genus as a herb, which may, in turn, be a short-lived annual or a long-lived perennial plant. Plants diverging in those ways differ strongly in respect of their ability to establish and build up persistent populations in different environments. Plants have therefore also been classified with regard to traits of ecological significance, such as 'life form', 'growth form' and 'lifespan' (for an overview of definitions and classification proposals, see Krumbiegel, 1998). The following discussions are restricted to *vascular plants*.

Life Form, Growth Form, Lifespan

The term 'life form' and its well-known definition by Raunkiær (1934) are discussed below as an introduction to the classification of arable weeds into 'life forms' such as this term is understood in this book.

'Growth form' is a term frequently used for characterizing genetically fixed morphological structures of vegetative plant parts. The term considers structures in a wider sense than 'life form' according to the above. It has reference to the structure recognized as the basis behind modifications into different 'growth types', which describe types of morphological adaptations of plant individuals to the environment, enabled by the phenotypic plasticity of the individuals. Krumbiegel (1998) has discerned 20 growth forms among annual plants. Perennial weeds have been grouped by Korsmo (e.g. 1930) on the basis of the traits of their perennating vegetative structures, without distinguishing them in terms of 'life form' and 'growth form' as these terms are defined above.

The 'lifespan' of plants, from their early growth through germination or sprouting of individuals, may be defined here as the length of time of their survival by vegetative structures adapted to enduring winters or dry seasons and their subsequent growth. For perennial plants, the lifespan of a given genet (individual or clone: e.g. Harper, 1977) may be related either to separate individuals or to clones. The lifespan of a clone may be very long even in cases when the life length of individuals or individual perennating structures is short, often as short as about 1 year.

Different 'growth types' of the same genotype developing under different environmental conditions may be easily confused with genetically determined 'growth forms'. Some weed species appear with morphologically dissimilar plants. In many cases, we have an insufficient knowledge of

©CAB *International* 2003. *Weeds and Weed Management on Arable Land: an Ecological Approach* (S. Håkansson)

what is due to genetic variation (growth forms) or to modifications in response to the environment (growth types). It is also often difficult or inconvenient to distinguish the terms 'life form' and 'growth form' as defined above. Even a 'growth type', representing a common adaptation to environmental conditions in a larger area, is difficult to distinguish from 'growth form' without thorough studies of genetic conditions. The term 'life form' is therefore often used in a wide sense, comprising 'growth form'. Of necessity, 'life type' frequently must also be considered a 'life form'. Thus, 'life form' is used in that wide sense in this book.

However, the classification according to Raunkiær's (1934) life-form system will first be presented briefly, because it is often used or referred to in literature on weeds.

Raunkiær's Life-form System

Vascular plants are classified by Raunkiær (1934) into life forms with regard to their ways of surviving unfavourable seasons, i.e. winters in temperate climates and dry seasons in warmer areas. The classification is based on the position of surviving meristematic tissues in vegetative *buds, including shoot apices*, or in *seeds*. The meristems in these structures are the initiators of new plant units or individuals growing in the subsequent vegetation period. The following main groups of plants are distinguished.

- *Phanerophytes.* Vegetative buds are situated on stem structures reaching high levels above ground; trees and bushes and, in warm climates, sometimes also tall plants with non-woody stems.
- *Chamaephytes.* Vegetative buds are positioned above, but near, the ground surface; plants with woody or non-woody stems.
- *Hemicryptophytes.* Vegetative buds are situated at the soil surface, from slightly above to slightly below; plants with taproots, with above-ground or

shallow below-ground runners, with surviving stem bases or shallow rootstocks, etc.

- *Cryptophytes.* Vegetative buds are situated on underground structures. Geophytes and helophytes are distinguished as subgroups. *Geophytes* are terrestrial plants with more deep-growing rhizomes or creeping roots, with stem or root tubers, bulbs, etc. *Helophytes* are plants in swamps, at shores, etc., with their perennating buds situated in waterlogged soil or in water.
- *Therophytes.* The only meristems normally surviving from one growing season to the following are the seed embryos; annual plants, whose vegetative parts do not survive adverse periods (winters or dry periods) between two growing seasons.
- *Comments regarding parallel or alternative survival through vegetative and generative units.* Most phanerogams that survive adverse periods between growing seasons by vegetative organs can do this also by seeds. Individuals of facultative winter annuals (see below), which, after germination in late summer or autumn, survive the winter as young plants, appear as hemicryptophytes, whereas those developed from seeds in the spring perform as therophytes.

The plant categories distinguished in Raunkiær's system naturally differ in their ability to establish and persist under different abiotic (edaphic, climatic) and biotic (e.g. competitive) conditions (e.g. Crawley, 1986). They therefore also have different abilities to appear in different successional stages of plant communities (see below).

Raunkiær's classification system can be used for comparing the life-form composition of weeds in various crops and/or agroecosystems. For this purpose, however, a classification where the term 'life form' is used in a wider sense seems better suited and more convenient and is therefore used in this book, as described below.

Classification Using the Term 'Life Form' in a Wide Sense

As stated above, it is frequently difficult, or inconvenient, to distinguish between the terms 'life form' and 'growth form'. 'Life form' has therefore sometimes been used for both of them (cf. Krumbiegel, 1998). Many authors (including Håkansson, 1995a,b,c) have used the term in a wider meaning, including both 'growth form' and 'lifespan'. Consequently, 'annuals', 'biennials', 'perennials', etc. are 'life forms' in the meaning in which 'life form' is used here.

Overview

Before more details of the classification of arable weeds are presented, terms for a superordinate grouping level are defined below. Synonyms to the terms chosen as first-hand terms are added in parentheses.

1. *Monocarpic plants (hapaxanthic plants, semelparous plants).* Plant individuals normally die entirely following the formation of generative reproduction organs, seeds in phanerogams. Seed formation is started and completed within one growing season.

- *Annuals (monocyclic plants).* Phanerogams with seed production. In temperate areas with real winters, annuals are traditionally subdivided into two categories:
  - *Summer annuals.* Plant species with individuals that normally complete their growth and development, including seed production, within one growing season.
  - *Winter annuals.* Plant species with individuals that can survive winter, mostly in early stages of development, after germination in late summer or autumn, and then continue growth and set seed in the following growing season. Most annuals of this category are 'facultative winter annuals', germinating more or less frequently also in the spring, in which case the plants have a summer annual performance.
- *Biennials (bicyclic plants).* Plant individuals reach their generative phase, set seed and die in their second growing season.
- *Monocarpic perennials (pluriennials, polycyclic plants).* Plant individuals need more than two growing seasons to reach their generative phase.

2. *Polycarpic plants (pollacanthic plants, iteroparous plants).* Genets (individuals or clones) survive winters or dry seasons by means of vegetative structures (perennating organs) and can repeat vegetative development and growth and form organs for generative reproduction (seeds in phanerogams) in more than 1 year.

- *Perennials (= polycarpic perennials)*
  - *Stationary perennials.* Slight, or slow, horizontal extension of individuals or clones through shoots or roots. These are also known as *simple perennials*.
  - *Creeping perennials.* Horizontal extension of individuals and clones through plagiotropic ('creeping') shoots or roots. These are also known as *running* or *wandering perennials*.

'Perennials' are usually understood to be herbaceous plants, whereas woody plants (trees and bushes) are considered additional groups, although they could be seen as groups among the perennials. Unless otherwise stressed, discussions in this book concern herbaceous vascular plants.

Seed banks and germination of weed seeds in the soil in different seasons

Long survival of dormant seeds enables a plant population to build up a bank of seeds of different ages in the soil. Germination of these seeds in portions over a period of years is a prerequisite for the persistence of populations of annual wild plants and weeds. Unless seeds are imported from outside, these populations would otherwise be eliminated in years when environmental conditions obstruct plant growth and reproduction. Some perennials largely rely on continuous vegetative survival (by 'bud banks'), but sexual reproduction by seeds is usually of importance, particularly for their long-term survival, and, in fact, most perennial weeds build up seed banks in the soil. *Tussilago farfara* is an example of an exception. Its seeds germinate or die in a few weeks following maturation.

There is extensive literature on the dormancy of seeds, the build-up of soil seed banks, the conditions under which seed dormancy can be broken and the prerequisites of germination (further, see Chapter 6). The seasonal variation in the germination among seeds in soil seed banks is briefly commented on here.

In temperate areas, these seeds exhibit a more or less distinct seasonal variation between dormant and non-dormant states, largely determined by the seasonal temperature variation. There is thus a seasonal variation in the intensity of germination and seedling emergence. Different plants perform differently. Figure 1 illustrates variations typical of summer- and winter-annual weeds, looked upon as entire groups, in South and Central Sweden. Attention should be paid to differences exhibited when the soil is undisturbed and disturbed by shallow tillage, respectively. Possible flushes of germination and seedling emergence following heavy rain cannot be seen in Fig. 1, as the curves there average different years, fields and species.

Classification of weeds

The following classification is largely based on works by Korsmo (e.g. 1930, 1954) in Norway, though modified in some details (Håkansson, 1975a, 1982, 1983a, 1992).

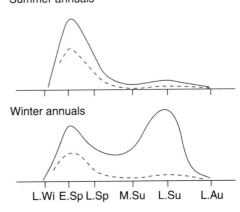

Fig. 1. Seasonal variation in seedling emergence from weed seeds in arable soil in South and Central Sweden. Average appearance of summer- and winter-annual weeds common in Sweden, as found in two investigations (Håkansson, 1983a, 1992). *Solid lines*: Emergence in a short period (2 or 3 weeks) following shallow soil tillage (harrowing or simulated harrowing) carried out at different points in time as the only soil operation in the study year. *Broken lines*: Corresponding emergence from soil not tilled in that year. L.Wi, late winter; E.Sp, early spring; L.Sp, late spring; M.Su, mid-summer; L.Su, late summer; L.Au, late autumn. There were differences between species, sites and years, but the general pattern was similar. A few species, considered summer annuals in Sweden, e.g. *Chamomilla suaveolens*, *Spergula arvensis* and *Urtica urens*, germinated more readily than other summer annuals after soil tillage in summer and early autumn. They are excluded here. Possible 'secondary emergence flushes' are averaged out in the curves.

Only *phanerogams* occur among the monocarpic plants, whereas vascular cryptogams, such as species of *Equisetum*, are represented among the polycarpic perennials.

A plant species sometimes appears in more than one life form. The appearance may differ within a species, even in a small geographical area. However, *for agronomic purposes, it seems reasonable to classify a species within a restricted climatic area on the basis of its most typical appearance there*. Plant species presented in this chapter as examples of different life forms are classified on the basis of their typical, or most common, appearance in North-European countries.

Annuals

The plants normally reproduce solely by seeds. In temperate climates with cold winters, summer and winter annuals are distinguished.

1. *Summer annuals.* Seeds in their soil seed banks germinate mainly in spring or early summer (Fig. 1). Plants then emerge in the early part of the growing season, flower and set seed in the year of emergence. When seeds germinate in the autumn – which also happens – the resulting plants normally die in the winter. Soil tillage stimulates germination, the stimulation being quantitatively most pronounced in the spring. Later in the growing season, the stimulation can also be relatively strong, but lower in absolute terms.

Examples of weeds: *Avena fatua, Chenopodium album, Polygonum persicaria, Sinapis arvensis.*

2. *Winter annuals (facultative).* Typical of most winter annual weeds in Northern Europe is the fact that seeds in the seed banks may germinate in any season under suitable temperature and moisture conditions. Germination is strongly favoured by soil tillage, especially in summer and autumn. In these seasons, extensive germination mainly occurs after tillage (Fig. 1; Håkansson, 1983a, 1992). It may also be triggered by heavy rain (Roberts and Potter, 1980). Plants having emerged in the early part of the growing season flower and set seed in the same year, like those of summer annual species. Plants from seeds that have germinated later in the growing season survive winter to a large extent in young stages. In typical cases, they flower and set seed in the following growing season. Annual species flowering both in their first and in their second year of life, depending on the time of germination, are here called *facultative* winter annuals. Milberg *et al.* (2000) characterize plants with a flexible time of germination as *germination generalists.*

Certain winter annuals (e.g. *Apera spicaventi*) germinate predominantly in late summer or autumn, although to a lesser extent also in spring (Avholm and Wallgren, 1976). No major weeds in Scandinavia are *obligate* winter annuals by the definition that all plants develop from seeds germinated in the later part of the summer or in the autumn and therefore do not set seed until the next growing season. (On biotype diversity, see pp. 12–13.)

As tillage strongly triggers seed germination of winter-annual weeds in late summer and early autumn (Fig. 1), it should be stressed that discussions on the occurrence of these weeds in autumn-sown crops apply to situations where seedbed preparation by tillage precedes autumn sowing – unless otherwise stated. To what extent the occurrence of winter-annual weeds will be changed by exclusion of this tillage is not well understood. One question is to what extent such exclusion will be compensated for, or even counteracted, by the consequences of more seeds remaining in the superficial soil, if deep tillage by ploughing is excluded in the cropping system.

The strong effect of tillage on germination in late summer and autumn applies to all winter annuals observed in the Swedish experiments. The main reasons for this effect may vary with species and with the conditions in the soil closely surrounding the seeds. Seed age and previous environmental influences are of great importance. Depending on all these factors, light will be more or less influential. In any case, a suddenly improved gas exchange caused by tillage will facilitate germination (cf. Håkansson, 1983a). An example of the influence of seed age is that newly shed, fresh seeds of *Matricaria perforata* germinate much better in the presence of light than in darkness, whereas the response of older seeds to light is weaker (Kolk, 1962). It has also been often observed that fresh seeds of *M. perforata* germinate much more frequently on the soil surface than after shallow burial by tillage.

Based on certain observations of my own, it may be suspected that the emergence of winter-annual plants in spring seedbeds

(Fig. 1) to some extent – or in some cases – is a continuation of germination started in the autumn (see discussion on p. 34).

Koch (1969) grouped annual weeds with regard to their germination seasonality on the basis of observations in Germany. He points out that there is often a great variation within the same species even within a small area. None the less, his grouping seems to be reasonably valid for temperate climates characteristic of Central and Northern Europe and is presented as follows.

- Germination mainly in early spring: e.g. *Galeopsis speciosa*, *G. tetrahit*, *Polygonum aviculare*, *Bilderdykia convolvulus*.
- Germination mainly in spring: e.g. *Chenopodium album*, *Polygonum lapathifolium*, *P. persicaria*, *Urtica urens* (often rather late).
- Germination mainly in late spring: e.g. *Amaranthus retroflexus*, *Galinsoga parviflora*, *G. ciliata*, *Setaria viridis*, *Solanum nigrum*, *Sonchus asper*, *S. oleraceus*.
- Germination mainly in autumn and spring: e.g. *Alopecurus myosuroides*, *Chamomilla recutita*, *Fumaria officinalis*, *Galium aparine*, *Matricaria perforata*, *Viola arvensis*.
- Germination throughout the year: e.g. *Capsella bursa-pastoris*, *Lamium amplexicaule*, *L. purpureum*, *Myosotis arvensis*, *Poa annua*, *Senecio vulgaris*, *Spergula arvensis*, *Stellaria media*, *Thlaspi arvense*, *Veronica persica*.
- Germination mainly in autumn: e.g. *Apera spica-venti*, *Centaurea cyanus*, *Lapsana communis*, *Veronica hederifolia*.
- Germination mainly in winter: e.g. *Ranunculus arvensis*.

Biennials, monocarpic perennials (pluriennials)

Like annuals, these plants normally reproduce only by seed. Although their seeds mainly germinate in spring or early summer, they rarely flower and set seed in their first year, unlike winter annuals that have emerged in the spring. After an individual has flowered and set seed, in the second year or later, it normally dies entirely. The plants mostly remain in the vegetative phase for one or more years, strengthening their vegetative structures and accumulating food reserves before entering the generative phase. Plants classified as biennials usually flower in the second year in their typical habitats. Examples are *Cirsium vulgare*, *Arctium* spp. and *Daucus carota*. However, depending on both genetic diversity and environmental conditions, their vegetative phase may last over 2 years or more, and they may even flower in the first year under conditions extremely favourable to their growth.

Many *monocarpic perennials* (or *pluriennials*) usually grow vegetatively for several years until they flower. *Cirsium palustre* and *Angelica silvestris* are examples of species that may need a period of 5–7 years of strengthening their vegetative structures before flowering (Sjörs, 1971). In the genus *Agave* there are species whose individuals build up a very strong leaf rosette for 20–30 years before they flower and set seed; after that they die completely within a few months.

Stationary perennials

The individuals of a stationary perennial plant extend only slightly or slowly from the spot where they originally established. With increasing age and size, their perennating parts may become fragmented due to attacks of herbivores or fungi or to death of tissues with age or for other reasons. Genets such as grass tufts may therefore gradually become more or less partitioned, forming crowds of separate units. A big grass tussock may thus consist of several autonomous individuals. If not mechanically dispersed by soil tillage, these individuals crowd in very dense clones. Their wintering buds are usually situated from slightly above to slightly below the ground surface.

In most cases, the stationary perennials can be characterized as *hemicryptophytes* in Raunkiær's life-form system. They comprise species with very different morphological habits. These may be distributed into two main subgroups depending on whether their perennating regenerative organs are: (1) short stems or (2) taproots.

1. *Perennating organs are short stems or stem parts, branched or unbranched, from which adventitious roots and aerial shoots develop.* New orthotropic aerial shoots annually develop from meristems (buds) on the perennating stems. Adventitious roots develop from these stem parts and mostly also from lower parts of the new orthotropic shoots. In many cases, for instance among grasses, the primary roots rapidly become replaced or complemented by adventitious roots from the stem bases. The size of the perennating system and the number of aerial shoots usually increase from year to year. Examples:

- Tufted grasses and sedges. Perennating stems are the basal parts of aerial shoots, vertical or almost vertical: e.g. *Agrostis capillaris, Arrhenatherum elatius, Deschampsia flexuosa, Holcus lanatus.*
- Dicotyledonous plants with short vertical or almost vertical perennating stems close to the soil surface: e.g. *Leontodon autumnalis, Plantago major, Ranunculus acris.*
- Dicotyledonous plants with a usually oblique or horizontal perennating stem close to the soil surface, in some species connected with persistent primary roots developing as more or less branched taproots: e.g. *Alchemilla* spp., *Artemisia vulgaris, Centaurea jacea, Leucanthemum vulgare, Plantago lanceolata, Senecio jacobaea.*

2. *Taproots, more or less branched.* In undisturbed plants, new aerial shoots normally develop from buds situated near the soil surface on the upper part of the taproot. This part is a stem, becoming more or less vertically split or branched with age. Examples of species: *Anchusa officinalis, Bunias orientalis, Centaurea scabiosa, Rumex crispus, R. longifolius,* *R. obtusifolius, Taraxacum officinale, Symphytum officinale.*

The regenerative ability of the various parts of a fragmented taproot differs between species. In *Taraxacum officinale,* for instance, all parts of the taproot and its branches exceeding a few millimetres in thickness are regenerative (Korsmo, 1930; Kvist and Håkansson, 1985). Healy (1953) and Hudson (1955) report that, in *Rumex crispus* and *R. obtusifolius,* only the upper 5–7 cm of the taproot can develop new shoots. Fykse (1986) reports similar observations in these species and in *R. longifolius.* However, Cavers and Harper (1964) noted plant development from lower parts of the taproot in *Rumex* in early spring, otherwise not. In laboratory experiments with *R. crispus,* new shoots only developed from buds on the upper stem part of the taproot, the 'crown', but in the spring, shoots developing from lower parts of the taproot were seen in the field, on plants with dead crowns (Kvist and Håkansson, 1985).

Creeping perennials

The plants extend horizontally by means of creeping vegetative organs. These are either plagiotropic lateral stems or plagiotropic thickened roots. The plagiotropic stems are either: (1) above-ground prostrate stems, rooting from their nodes, sometimes morphologically specialized *stolons,* or (2) underground stems, *rhizomes.* (3) Creeping roots, originally thin, become regenerative after secondary growth in thickness.

By means of the creeping stems or roots, new plants gradually develop at various distances from the position of an original plant. Fragmentation, caused by ageing, fungi, animals, soil cultivation, etc., results in clones with increasing numbers of separate individuals. Plants may disperse over wide areas by means of cultivation implements or other agents.

1. *Dispersal and regeneration by stolons or prostrate above-ground stems.* Stolons are plagiotropic shoots developed from buds at lower nodes of orthotropic shoots. The stolons may branch both plagiotropically and orthotropically (forming assimilating green shoots). It seems that they do not grow out until orthotropic primary shoots have reached a stage when they produce a surplus of photosynthates (S. Håkansson, unpublished data). The stems of the stolons often markedly differ from those of the orthotropic shoots. They have often long distinct internodes. In many species, on the other hand, longer or shorter parts of the basal stems of the aerial shoots lie on the ground but are otherwise morphologically rather similar to the vertical stems.

New individuals arise through the development of roots and shoots from the nodes of the creeping stems. They become separate individuals with the death of the joining stolon internodes. Lower parts of these individuals become perennating organs, enabling a continued growth and vegetative spread and multiplication. Stolon internodes usually do not survive the winter. Examples of North-European species with stolons or prostrate stems are *Agrostis stolonifera, Glechoma hederacea, Poa trivialis, Potentilla anserina, P. reptans, Prunella vulgaris, Ranunculus repens, Veronica serpyllifolia.*

2. *Dispersal and regeneration by underground plagiotropic stems, rhizomes.* The following description of rhizomes applies more directly to those with distinct nodes, well separated by comparatively long, slender internodes. These are typical of rhizomatous grasses such as *Elymus repens* (Fig. 50) but are also represented in many other plant families. Principles also apply to corresponding underground stem structures with short and stout internodes (according to the definition of a 'rhizome' in the above). Rhizomes originally grow out from basal underground stem bases of aerial shoots when these have developed a foliage large enough for producing a surplus of photosynthates (Håkansson, 1967, 1982; Håkansson and Wallgren, 1976). Lateral buds develop from superficial tissues in the axils of scale leaves at the nodes of the rhizomes. Rhizome branches of various orders may develop within a season and penetrate the soil to different depths.

New aerial shoots develop both by formation of orthotropic shoots from nodes near the soil surface of earlier aerial shoots (tillering) and/or by formation of orthotropic shoots from the rhizomes. In the latter case, the new aerial shoots are mostly a result of rhizome apices changing from plagiotropic to orthotropic growth. In undisturbed plants, they usually only develop to a minor extent from lateral rhizome buds. When these are activated, they mostly grow plagiotropically forming rhizome branches. However, the apices of these sometimes rather soon bend upwards forming new orthotropic shoots.

Not only the plagiotropic rhizome parts are regenerative but also the underground vertical stem bases of aerial shoots. Fragments of these have proved to develop new plants (Håkansson, 1969a). The proportion of buds developing aerial shoots is enhanced with an increased degree of fragmentation, until the fragments have become too small to contain enough food reserves and/or too many of them lack viable buds.

- Rhizomes sensitive to soil cultivation.
  - Rhizome system shallow: e.g. *Achillea millefolium, A. ptarmica, Cerastium arvense, Galium mollugo, Lamium album, Urtica dioica.*
- Rhizomes tolerant to soil cultivation.
  - Rhizome system shallow; branches with spool-shaped swellings: e.g. *Mentha arvensis, Stachys palustris.*
  - Rhizome system shallow; branches without spool-shaped swellings: e.g. *Agrostis gigantea, Elymus repens, Holcus mollis, Polygonum amphibium.*
  - Rhizome system reaching greater depths: e.g. *Equisetum arvense* (lateral tubers frequently formed at the nodes), *Phragmites australis, Tussilago farfara.*

Comments: 'Rhizome system shallow' means that rhizomes normally grow within the topsoil layer; the great majority of them mostly in the upper 10-cm soil layer at undisturbed growth. For species with rhizome branches reaching greater depths, the rhizome system may penetrate the soil from shallow layers to depths far below the topsoil layer. The maximum depth depends on species and soil properties. The degree of tolerance of the rhizome structures to soil cultivation is far from being strictly correlated with the morphological robustness (when rhizomes representing different species are compared).

3. *Dispersal and regeneration by plagiotropic thickened roots.* Perennating thickened roots develop among slender roots as a result of secondary growth in thickness. They grow at various inclinations, horizontal to vertical. After a period of growth in thickness, new aerial shoots develop from buds in the thickened roots. The horizontal roots enable an effective dispersal of the plant, even without intervention by soil cultivation implements or other agents. Aerial shoots develop from buds at irregular distances from each other on the root. These buds are differentiated from tissues below cortex. In undisturbed plants, only a few buds grow out forming aerial shoots, whereas most of them stop growing in early stages of development owing to correlative inhibition.

Fragmented thickened roots are regenerative. The proportion of buds developing aerial shoots is enhanced with an increased degree of fragmentation, until the fragments have become too small to contain enough food reserves. New roots develop both from the thickened roots and from the underground stem parts of the shoots. Fragments of the stem bases of aerial shoots are regenerative (if the shoots are not very young). They develop shoots and new roots from tissues at nodes (e.g. Korsmo, 1930, 1954; Håkansson, 1969d).

- Perennating root system shallow: e.g. *Rumex acetosella*, *Sonchus arvensis*.
- Perennating root system reaching greater depths: e.g. *Cirsium arvense*, *Convolvulus arvensis* (in combination with rhizomatous structures).

Perennials with other or modified regenerative structures

Perennials with regenerative structures other than those presented above also occur as arable weeds. *Allium vineale* can be mentioned as an example. This species, perennating and multiplying by *bulbs*, can become a noxious arable weed under special conditions, e.g. locally in Sweden (Håkansson, 1963a). Some species have tubers or bulbs that can be regarded as parts of rhizomes. Examples are *Cyperus rotundus* with *tubers*, i.e. swollen regenerative stem parts. These alternate with thin internodes in rhizome branches and/or they terminate the branches (see Tables 12 and 14). *Oxalis latifolia* forms bulb-like organs at the tips of its rhizome branches (e.g. Holm *et al.*, 1997). The two latter species are important weeds in a global perspective, particularly *C. rotundus*, but none of them in cold-temperate regions. Species such as *Mentha arvensis* and *Stachys palustris* form spool-shaped swellings near the tips of rhizome branches.

Structures with tubers or bulbs as special formations on rhizome branches may be regarded as variants of rhizomatous structures; the plants classified here among the rhizomatous perennials. Besides the main perennating structures of a plant, there are often additional structures with a more or less pronounced regenerative capacity. As understood from the above, underground stem parts are thus more or less regenerative besides the thickened roots in plants such as *Cirsium arvense* and *Sonchus arvensis* and, particularly, *Convolvulus arvensis*.

Classification Problems with Species Distributed over Wide Geographical Areas

As previously stated, the life-form classification of a plant species is often conditional. A species may comprise genotypes representing diverse life forms occurring within the same or in different areas. It may also differ phenotypically in response to the environment. For example, its ability to survive winters as a winter annual or as a perennial differs, not only between climatic areas but also from year to year in the same field.

The grouping of annual plants into summer and winter annuals seems to be particularly conditional. In species classified as winter annuals under Swedish and North-European conditions, spring germination is followed by a summer-annual performance of the established plants, whereas plants established after autumn germination largely survive the winter as young plants. Spring and autumn germination may sometimes be represented by different biotypes, both in the same area and in different geographical regions. If a population of weed species includes such biotypes and these are not easily distinguished by morphological characters, it may be practically reasonable, or inevitable, to consider it a taxonomic unit characterized as a facultative winter annual.

Assume now that a geographically widely-distributed annual species is characterized as a (facultative) winter annual in northern Europe. This characterization may be irrelevant, or have a different meaning, in warmer areas. It may be useful throughout Europe but have somewhat different meanings in the North and the South. In southern Europe, the performance of a species as a summer or a (facultative) winter annual will mainly depend on its germination seasonal-

ity and, in a Mediterranean climate, on the resistance of its growing plants to summer drought. It may only slightly, or not at all, depend on its ability to survive winters, which are moist and mild. A plant being a typical summer annual in a North-European climate does not germinate until spring in that climate, though the dormancy of many seeds in its soil seed bank is broken as early as in late autumn after a period of low temperatures. There, however, the temperatures in late autumn and winter are too low for germination. In southern Europe, even seeds of a plant with (possibly) identical genetic characters might germinate in the winter. The temperatures are low enough to break dormancy, but, at the same time, frequently high enough to allow germination.

In the hot tropics, a grouping of annual plants into summer and winter annuals is inapplicable. However, there might be differences in the germination seasonality among the annuals motivating divisions into other subgroups. Thorough investigations are needed to clarify this issue (cf. Baskin and Baskin, 1998).

With the classification problems now being discussed still in mind, relationships between different groups of weeds and their occurrence in crops of different types are examined in Chapter 5. The question here is to what extent relationships valid in wider geographical perspectives can be defined. First, a simple grouping of crops based on their duration is presented.

3

Annual and Perennial Crops

The arable crops are grouped here considering their lifespan, annual life cycle and degree of stand closure. A grouping into annual and perennial crops is illustrated with special regard to crops in temperate areas. The annual crops are subdivided with respect to the season in which they are sown or planted. The grouping is indicated in Tables 3 and 5. At the same time, these tables present grading of the relative abundance of various weeds in different categories of crop based on broad experiences from Sweden. The crop–weed relations seen there are analysed in Chapter 5. However, it should be stressed here that the grading applies to cropping systems with 'ordinary' soil tillage, where sowing or planting is immediately preceded by soil cultivation. In a non-tillage situation with direct sowing, the conditions may differ very strongly.

In its position as a crop, a species or variety may differ completely from its position in the life-form system presented above. Thus, for instance, the potato plant is basically a perennial, but is a spring-planted annual crop. Sugar beet is a biennial plant, but is an annual spring-sown crop when grown for sugar production. Spring-sown varieties of small-grain cereals, rape and turnip rape are both summer-annual plants and spring-sown annual crops. Autumn-sown varieties of these species may on the whole be characterized as biennials. They should not be considered winter annuals if they do not normally flower before winter

when sown in the spring. However, there are differences, between both species and varieties of winter crops, in their disposition to flowering as annuals. Low spring temperatures sometimes enhance this disposition.

1. *Annual crops*, e.g. cereals, oilseed crops, peas, beans, sugar beet, potatoes and various vegetables. In accordance with the above, they are subdivided into *spring crops* (spring-sown or spring-planted annual crops) and *winter crops* (autumn-sown annual crops). The spring crops are here subdivided into *crops with stands that close late* in the growing season, resulting in a weak competitive effect on weeds from early stages of development, e.g. row crops, such as potatoes, sugar beet and most vegetables, and *crops whose stands close early*, such as small-grain cereals and (largely) oilseed crops, which become comparatively competitive from early stages.
2. *Perennial crops.* In the Nordic countries, the principal perennial crops in a crop sequence on arable land are leys for cutting, mostly leys with grass plus clover or grass only, more seldom clover in pure stands and rarely with lucerne alone or in mixtures. These leys, the plants of which are mostly undersown in spring cereals, usually have a duration of 2–4 years. Pastures for grazing mostly have a much longer duration even where established on arable land. The weed flora in leys alters with the age of the ley (Table 4) following principles of vegetation successions on non-tilled land, with

modifications caused by the annual cutting. When surveying the weed occurrence in leys (Tables 3 and 5), *young leys* (chiefly in the meaning of first-year leys) and *older leys* are distinguished in order to briefly characterize typical weed flora changes with the ageing of the ley.

Conditions in perennial crops or plantations, such as orchards and vineyards, are discussed later on (Tables 12, 13). These crops represent much more heterogeneous crop categories than the grass–clover or grass leys. In extreme cases, the entire fields around trees or bushes are, each year, either tilled or treated chemically to manage weeds, or covered with grass and/or other vegetation, which may be cut as needed. The 'average field' with such crops could be described as a field with mosaics or strips resembling areas with annual crops (or black fallows) and areas with dense stands of crops such as leys for cutting. The two types of area in the mosaics or strips thus favour different life forms of weeds.

4

Weed Communities Looked Upon as Early Stages in Secondary Vegetation Succession

'Vegetation succession' is a concept encompassing those changes in the composition of plant communities that normally take place with the progress of time. Following the classic work of Clements (1916), further studies have illustrated and analysed important patterns in vegetation changes. Crawley (1986) describes vegetation successions as 'community dynamics occurring on a time scale of the order of the life-spans of the dominant plants (in contrast to much slower, evolutionary changes, occurring over hundreds or thousands of generations, or the much more rapid seasonal or annual fluctuations in species' abundances)'. Primary and secondary successions are distinguished in the literature (for reviews, see Tilman, 1986, 1990; Brown and Southwood, 1987; Miles, 1987; Mortimer, 1987; Bazzaz, 1990; Hansson and Fogelfors, 1998).

'Primary succession' concerns changes in the composition of plant communities starting with the establishment of plants on bare ground exposed after, for instance, glacial recession and volcanic eruption.

'Secondary succession' refers to changes in the composition of communities of plants repopulating an area of bare soil after destructive disturbance of a previous vegetation cover. *The following solely concerns secondary successions.*

The first plants repopulating bare ground predominantly originate from seeds and vegetative plant parts remaining in the soil. Mostly, they also come from dispersal units entering from outside, but usually to a minor extent. Densities and proportions of the taxa and life forms of these first plants thus largely depend on the previous vegetation and its treatments. Plants that are superior competitors in a short-term perspective become dominant, e.g. plants with high growth rates and/or originating from propagules with large amounts of food reserves. For reasons discussed below, more short-lived plants occur among the dominants in the first year than in later successional stages (Table 1). Rapidly growing monocarpic plants (mostly annuals) and polycarpic plants (perennials) capable of fast and strong vegetative renewal after disturbance are usually represented among these dominants. See, for instance, Numata (1982) and Crawley (1986).

The changes in the vegetation on abandoned arable land are prominent examples of secondary vegetation successions. These changes, summed up in Table 1, strikingly illustrate that arable weed communities can be characterized as early stages in such successions. An understanding of the dynamics in these changes means an understanding of relations between weed communities on arable land and many plant communities in the surrounding landscape.

Secondary vegetation successions, as briefly characterized in Table 1, largely result from plant competition. It is easily understood that late emergence of a plant in relation to neighbouring plants is a competitive disadvantage. Thus, it is hard, or impossible, for new plants to establish in an already closed canopy of plants. This particularly applies to seedlings (Table 2).

Seedlings frequently use up their food reserves and die in the shade without reaching stages of positive net photosynthesis

Table 1. Secondary vegetation succession: development from the start on bare soil following disturbance, e.g. by soil tillage. Vegetation changes typical of the successional course on abandoned arable land in humid climate.[a]

Time	Vegetation succession (biomass proportions)	Comments
Year 1: First growing season after disturbance	More short-lived plants (largely annuals on abandoned arable land) among the dominants than in subsequent years	Weed communities in annual crops correspond to the first stage of succession. Weed control can be seen as an obstacle to the early successional course
Year 2 and onwards	Persistent perennials increasing in proportion to short-lived plants	Comparable to changes in leys (grassland) sown in arable fields
After varying time	Woody plants increasingly established, influencing the herbaceous vegetation more and more	Landscape gradually less open in areas with abandoned arable fields. Afforestation, now or earlier, is a way of governing and speeding up forest establishment. To keep landscape open, spontaneous growth of trees must be actively prevented
Finally	Forests of various types	More stable plant communities – although the composition of species is not constant

[a]The vegetation on headlands surrounding arable fields, on verges and ditch-banks, etc., where the soil is not tilled, but cutting sometimes occurs, has many similarities to leys. Although varying greatly, such vegetation largely represents intermediate stages in secondary succession. Perennial plants usually dominate the vegetation. Perennial arable weeds often occur, and many of them, particularly creeping plants, can easily invade adjacent arable fields from there. If preventive measures are not taken, woody vegetation often forms curtains of trees on such ground. Translated from Håkansson (1995d).

Table 2. Number and biomass of above-ground shoots of weed plants established from seeds, either in pure stands or mixed with ley plants sown at different times in relation to the weeds.[a]

Ley plants: sown 0, 18 or 45 d.b.s. (= days before sowing of weeds)	Weed plants 53 days after their sowing	
	No. (m^{-2})	Dry weight (g m^{-2})
No ley plants sown	878	337
Stands even, not cut		
0 d.b.s.	707	236
18 d.b.s.	637	9.5
45 d.b.s.	41	0.04
Stands uneven and cut		
45 d.b.s. (A)[a]	182	0.28
45 d.b.s. (B)[a]	171	0.32

[a]Greenhouse experiment in boxes, 80 × 80 cm^2, at Uppsala 1973. Seeds of the annuals *Chenopodium album*, *Chamomilla recutita*, *Thlaspi arvense* and *Stellaria media*, and the perennial *Rumex crispus* were mixed in soil that was spread as thin superficial layers, afterwards watered repeatedly. This was done either on previously unsown soil or on soil sown with seeds of ley species (*Phleum pratense*, *Festuca pratensis* and *Trifolium pratense*, 8 + 8 + 4 kg ha^{-1} of seed in 'even' or 'uneven' stands). Ley plants sown in even stands were not cut. Uneven stands had either four gaps of 15 × 15 cm^2 (A), or one gap of 30 × 30 cm^2 (B) per box. They were cut at a height of 7 cm 27 days after weed sowing. (From Håkansson, 1979.)

(Table 20), or they die due to drought or mineral deficiency resulting from the water and nutrient uptake by the older plants.

In the experiment that lies behind Table 2, concerning establishment of weed seedlings in stands of ley plants (description with the table), water stress was eliminated by repeated watering. In spite of this, most weed plants originating from seeds sown 45 days after the ley plants had been sown were dead, and the others were largely dying, on the 53rd day after their sowing, particularly in even and uncut ley stands. The 'uneven' ley stands were at least as even as 'good ley stands' in ordinary field situations. More plants per unit area were alive in the gaps than in the other parts of the ley stand, but the plant weights did not differ noticeably. Cutting, which was done in the uneven ley stands (at a height of 7 cm 27 days after the sowing of weeds), had probably influenced the weed seedlings more than the arranged gaps. In case (B) as well as in case (A), the gaps were obviously too small to favour the seedlings appreciably (cf. Fig. 2).

In free vegetation development, perennials with long-lived individuals or clones become increasingly dominant because, each year, they can start growth from vigorous vegetative units. Annuals then meet with difficulties since they start growth as seedlings, which are competitively weak relative to these perennial plants. Perennials with short-lived genets soon meet with difficulties too, as they frequently have to re-establish as seedlings.

The biomass in communities of undisturbed plants in late successional stages is usually dominated by a few species. At the same time, these communities can have a high species diversity. Even annuals may establish, but mainly in spots with destroyed or weakened perennials. Establishment of seedlings, representing perennials as well as annuals, is favoured, or maybe only possible, in temporary gaps or patches with weakened plants. New perennial species, although weakly competitive in juvenile stages, can establish in this way and then increase their proportion in the plant community, if they are sufficiently

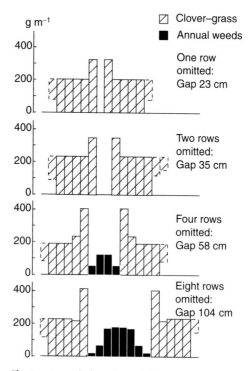

Fig. 2. Annual plants have difficulties in producing vigorous growth in already well-established plant stands. This was illustrated in a field experiment with a first-year ley stand of red clover and grass, where gaps were arranged. Spontaneously emerging annual weeds (mainly winter annuals) could grow vigorously only in gaps exceeding a certain width. The ley plants were sown in N–S aligned rows spaced at 11.5 cm. Gaps were arranged through omission of ley plant rows. Bars show fresh weights of aerial plant shoots, the bases of which were situated within strips of a width of 11.5 cm and a length of 1 m; actual or omitted ley plant rows were situated in the centres of the strips. Weights are means from strips with the same position relative to gaps, on both sides of these. Weights were determined at the normal time of hay-making, in mid-June. Weeds weighed less than 1 g in strips with ley plants. In gaps with only one or two rows omitted, they weighed 1.3 and 2.2 g, respectively. *Matricaria perforata* and *Chamomilla recutita* represented 70–80% of weed biomasses, *Capsella bursa-pastoris* and *Myosotis arvensis* 20–25%. (From Håkansson and Jendeberg, 1972.)

competitive in more advanced stages of plant development.

Grazing in permanent pastures reduces the competition and dominance of certain species. Moderate grazing therefore mostly facilitates diversity.

In humid climates, successions lead to the formation of forests, providing that ground and temperature conditions enable stands of trees to establish and persist (Table 1). Here, arable fields are largely situated on previous forest land and free successional changes in the vegetation of abandoned arable fields sooner or later result in some kind of forest.

However, the establishment of trees is often a slow process. Seedlings of trees often have difficulties in establishing among perennial grasses and herbs, which may rapidly form dense covers (e.g. Olsson, 1987). It therefore often takes a long time until trees establish spontaneously, unless these dense covers are broken or weakened by some disturbance. It is thus often in temporary gaps or in spots with weakened vegetation that seedlings resulting in the first trees establish. Trees of certain species, e.g. aspen, spread vegetatively by suckers from creeping roots of older individuals.

5

Weeds with Diverse Life Forms in Various Types of Crops

Overview in a North-European Perspective

Comments on crops and cropping systems in the Nordic countries

The types of rotations or sequences in which crops are alternated in the fields vary extremely between agricultural areas and farms and have changed with time. An idea of crop sequences in the Nordic countries may be obtained from the following acreage proportions of the three main types of crop in the last five decades of the 20th century: *annual spring crops 30–60%, annual autumn-sown crops 5–35%, leys for cutting (hay, silage) 20–40%.*

Among annual crops, cereals have dominated. Spring-sown cereals have been barley, oats or wheat in various proportions and autumn-sown cereals have been wheat and, to different extents, rye and triticale. Oilseed crops have been sown to various extents. Particularly in Denmark and South Sweden, sugar beet has represented considerable acreages, in certain areas also potatoes. Further north, row crops are mainly potatoes in minor acreages.

Leys for cutting are largely composed of grasses (mainly timothy, *Phleum pratense*) and clovers (mainly red clover, *Trifolium pratense*) undersown in spring cereals in the year preceding the 'first year of ley' and are mostly maintained for 2, up to 4, years (but sometimes longer, particularly in northern Scandinavia and Finland in the earlier decades under discussion). The leys are usually broken by ploughing, sometimes after shallower, 'sward-fragmenting' tillage or, in recent decades, a herbicide treatment.

The acreage of leys in the areas as a whole has decreased since the early 1950s, largely resulting from the loss of leys on farms giving up stock keeping. Behind the statistics of mean acreages of leys there are thus stock-keeping farms with 30–50% leys in alternation with annual crops, mainly cereals, and farms with solely annual crops. The proportions of different kinds of farms vary between countries and districts.

The percentage acreage of leys for cutting has generally been higher in Finland and Norway and lower in Denmark than in Sweden, but changes over time have been considerable in all areas. The percentage acreage of autumn-sown cereals has, on the whole, been lower in Denmark, and much lower in Finland and Norway, than in Sweden.

In the first two or three decades after 1950, levels of fertilizer use increased dramatically. In the last two decades they have stabilized and even decreased in some situations. In the first years following 1950, the use of herbicides increased rapidly, mainly in cereals. At the same time, an increasing number of herbicides was introduced, enabling the control of most weeds of importance. In the 1970s, herbicides were used in most annual crops.

Combine harvesting gradually replaced other forms of cereal harvesting after its introduction in the late 1940s. As combine harvesting is done later than harvesting with older methods, more weed seeds have matured and, for many species, more of them are shed or dropped to the ground and more left with the stubble, which is longer than at previous harvesting by binder. Particularly where cereals have replaced leys, the increased addition of seeds of cereal-adapted weeds to the soil seed banks would have caused enhanced problems with many weed species if not counteracted by chemical control. Regular use of herbicides in cereals thus has largely hidden the negative influences of modern harvesting techniques and, on farms without livestock, increased proportions of cereals in the crop sequences.

Relationships of weeds to crops largely depend on times and types of tillage

Arable land is in most cases characterized by soil tillage recurring with various frequencies over a period of years. *The following description applies to cropping systems where the soil is tilled before a new crop is sown or planted.* Thus, where only annual crops are grown in a crop sequence, some kind of tillage is practised every year. The general outlines presented in this section are illustrated by the results of field surveys in subsequent sections.

Where frequent tillage kills or strongly destroys assimilating weed shoots, the long-term persistence of weed populations depends on the establishment of new plants from seeds in the soil seed bank or on the regeneration from surviving vegetative structures. The selective effect of tillage on plant species or biotypes over periods of many years varies naturally with type, frequency and seasonal distribution of the soil operations in the cropping system. In combination with the different abilities of diverse weeds to grow and reproduce in competition with the crops grown, this selection determines the composition of the banks of seeds and other regenerative structures of weeds in the soil. In industrialized

countries, weeds important in 'full-tillage' cropping systems very rarely represent species relying on regular renewal of their populations by dispersal from outside (cf. Holzner, 1982; Holzner *et al.*, 1982).

The influence of tillage is drastically apparent in the case study on *Allium vineale* (pp. 223–233), a plant proven to be very sensitive to tillage in the spring (Fig. 67) and for that reason being of minor importance as a weed in annual crops established after spring tillage. The frequency of years with spring tillage was judged to have been a dominant factor regulating its potential to persist in a cropping system up to the mid-1950s, i.e. until chemical weed control became increasingly influential. Based on experiments and field surveys, the maximum frequency of years with spring tillage allowing its persistence seems to have been 30–40% under the prevailing farming conditions (Fig. 70).

In Table 3, the relative potential of plants with diverse life forms to grow and reproduce in various crops is graded. This grading (originally made for the Swedish extension service: Håkansson, 1977, 1989) applies to the main agricultural areas in Scandinavia and Finland but, in principle, to wider temperate areas. Table 4 illustrates the occurrence of different categories of weeds in Swedish grass–clover leys of different ages. Table 5 presents the grading of a number of weed species important in Sweden. Like Table 3, it weighs up knowledge based on field observations of myself and others, interviews and literature information (e.g. Ferdinandsen, 1918; Bolin, 1922; Korsmo, 1930; Hultén, 1950; Granström and Almgård, 1955; Mikkelsen and Laursen, 1966; Thörn, 1967; Mukula *et al.*, 1969; Hallgren, 1972, 1993; Fiveland, 1975; Gummesson, 1975; Raatikainen and Raatikainen, 1975; Raatikainen *et al.*, 1978; Streibig, 1979; Haas and Streibig, 1982; Hanf, 1982).

Table 5 grades the potential of a number of plant taxa (species or groups of closely related species of interest as weeds in Sweden) to grow and reproduce in diverse crops. The grading by Roman numerals is valid solely for comparisons within taxa.

Table 3. Relative potential of plants with diverse life forms to grow and reproduce as weeds in different kinds of crops in the absence of active control by chemical or mechanical means. Characteristics of plant groups seen as entireties.[a] (Cf. Håkansson, 1995d.)

| | Annual crops | | | Perennial crops: mainly leys in crop rotations | |
| | Potato, sugar beet, vegetables | Cereals, oilseed crops | | | |
Life form (including lifespan) of weed plants		Spring-sown	Autumn-sown	Young leys	Older leys
1. Annuals					
1.1. Summer annuals	III	III	I–II	I	–
1.2. Winter annuals (facultative)	III	II	III	II	I
2. Biennials	–	–	–	–	–
3. Perennials					
3.1. Stationary perennials	I	–	–	II	III
3.2. Creeping perennials					
3.2.1. Above-ground prostrate shoots, stolons	I–II	–	–	II	III
3.2.2. Underground plagiotropic shoots, rhizomes					
3.2.2.1. Sensitive to soil cultivation	I	–	–	I	III
3.2.2.2. Tolerant to soil cultivation	III	III	II–III	III	II–III
3.2.3. Plagiotropic thickened roots	III	III	II–III	II–III	I–II

[a]Grading: III, maximal; II, somewhat reduced; I, limited; –, minimal.
Note: Grading values are comparable only horizontally. They do not inform on quantitative abundance or importance.

Table 4. Occurrence of plants of annual weeds and aerial shoots of perennial weeds in leys for cutting.[a] Averages of early-summer counts in leys of different ages (in number of years from sowing: I, II, or III + older) by Hallgren (1972). Leys were grass–clover leys on clay soils in the surroundings of Uppsala.

	Number m^{-2}		
	Year I	Year II	Year III + older
Summer annuals	4	3	0.2
Winter annuals	62	28	6
Annuals, total	66	31	6
Perennials	6	35	104

[a]The most common annual weeds were *Chamomilla recutita*, *Viola arvensis*, *Matricaria perforata* and *Capsella bursa-pastoris*. Among perennial weeds, *Elymus repens*, *Achillea millefolium* and *Taraxacum officinale* were most abundant. The strong increase in the *number* of shoots of perennials with increased ley age resulted here from creeping plants. Their biomass did not increase in proportion to the shoot number. Stationary perennials (largely *T. officinale*) increased by plant size in the first place.

Though occurring in varying frequencies and abundances, the taxa listed are of similar interest in neighbouring countries, many of them far outside the Nordic areas.

Representing the first stage of succession according to Table 1, the weeds in the annual crops in a field that has been regularly tilled for a period of many years, are frequently dominated by annuals with strong seed banks and plants with a high growth rate. However, perennials tolerant to (or sometimes even favoured by) tillage normally also occur to various extents in annual crops. They can even appear more abundantly than annual weeds in an annual crop if they have been greatly favoured in preceding years. Perennial weeds of importance in annual crops are mainly creeping perennials with rhizomes or plagiotropic roots capable of a rapid plant regeneration after mechanical disturbance. Examples are *Elymus repens* and *Equisetum arvense* with rhizomes and *Cirsium arvense* and *Sonchus arvensis* with creeping roots.

Weeds with diverse annual life cycles respond differently to tillage at specified

Table 5. Distribution[a] and average abundance[b] of plant taxa (species, or groups of species within a genus) as weeds in arable fields in Sweden and grading[c] of their relative potential to grow and reproduce in different kinds of crops in the absence of active control by chemical or mechanical means. Taxa grouped with respect to their life form. Only plants of concern as weeds in modern Swedish agriculture are presented. (Cf. Table 3.)

Life form (including lifespan) and taxon of weed plants[d]	Distribution in the country and average abundance in arable fields		Annual crops	Cereals, oilseed crops		Perennial crops: mainly leys in crop rotations	
			Potato, sugar beet, vegetables	Spring-sown	Autumn-sown	Young leys	Older leys
1. Annuals							
1.1. Summer annuals							
Avena fatua	SC	1	III	III	II	I	–
Bilderdykia convolvulus	A	4	III	III	II	I	–
Chenopodium album	A	4	III	III	I	I	–
Chrysanthemum segetum	S	1	III	III	I	–	–
Erysimum cheiranthoides	A	2	III	II	II	I	–
Fumaria officinalis	SC	3	III	II	–	–	–
Galeopsis spp. (a)	A	4	III	III	III	I–II	I
Polygonum aviculare	A	3	III	II	III	I–II	I
Polygonum spp. (b)	A	3–4	III	III	I	I	–
Raphanus raphanistrum	SC	1	III	III	I	I	–
Sinapis arvensis	SC	1–2	III	III	II	I	–
Solanum nigrum	SC	1	III	I	–	–	–
Sonchus spp. (c)	SC	1	III	I	–	–	–
Spergula arvensis	A	3	III	II	I	I	–
Urtica urens	SC	1	III	I	–	–	–
1.2. Winter annuals (facultative)							
Apera spica-venti	S(C)	1	I	I	III	II	–
Buglossoides arvensis	SC	1	I	I	III	I	–
Capsella bursa-pastoris	A	3	III	II	III	II	I
Centaurea cyanus	SC	1	II	II	III	II	–
Chamomilla recutita	SC	1	II	I	III	II	–
Galium aparine (d)	SC	3	III	II	III	I	–
Geranium spp. (e)	SC	1	II	I	III	III	III
Lamium purpureum (f)	A	3	III	III	III	II	II
Lapsana communis	SC	2	III	III	III	III	II
Matricaria perforata (g)	SC	4	II	II	III	III	I
Myosotis arvensis	A	4	III	II	III	III	II
Papaver spp. (h)	SC	1	II	I	III	II	–
Poa annua	A	4	III	II	III	III	III
Senecio vulgaris	A	1	III	II	II	I	–
Stellaria media	A	4	III	III	III	III	II
Thlaspi arvense	A	3	III	III	III	III	I
Veronica spp. (i)	SC	3	III	II	III	II	–
Viola arvensis (j)	A	4	III	III	III	II	I
2. Biennials			–	–	–	–	–
3. Perennials							
3.1. Stationary perennials							
Tufted grasses							
Agrostis capillaris	A	3	–	–	–	I	III
Deschampsia cespitosa	A	4	–	–	–	I	III

continued

Table 5. *Continued.*

Life form (including lifespan) and taxon of weed plants[d]	Distribution in the country and average abundance in arable fields		Annual crops			Perennial crops: mainly leys in crop rotations	
			Potato, sugar beet, vegetables	Cereals, oilseed crops		Young leys	Older leys
				Spring-sown	Autumn-sown		
Dicotelydonous herbs with short subterranean stems							
Alchemilla spp.	A	1	–	–	–	I	III
Artemisia vulgaris	SC	1	–	–	–	–	III
Plantago lanceolata	SC	2	I	–	–	II	III
Plantago major	A	3	I	–	–	II	III
Ranunculus acris	A	1	–	–	–	–	III
Herbs with taproot							
Barbarea vulgaris	A	2	–	–	–	II	III
Bunias orientalis	C	1	–	–	–	–	III
Rumex crispus	A	2	I	–	–	I	III
Taraxacum officinale coll.	A	4	II	I	I	II	III
3.2. Creeping perennials							
3.2.1. Above-ground prostrate shoots, stolons							
Agrostis stolonifera	SC	1	I	–	–	I	III
Potentilla anserina	SC	1	I	–	–	I	III
Prunella vulgaris	A	2	I	–	–	I	III
Ranunculus repens	A	2	II	I	I	II	III
3.2.2. Underground plagiotropic shoots, rhizomes							
3.2.2.1. Sensitive to soil cultivation							
Achillea millefolium	A	1	I	–	–	I	III
Galium mollugo	A	1	I	–	–	I	III
Urtica dioica	A	1	I	–	–	I	III
3.2.2.2. Tolerant to soil cultivation							
Rhizomes shallow, with spool-shaped swellings							
Mentha arvensis	A	1	III	III	II	II	I
Rhizomes shallow, without spool-shaped swellings							
Agrostis gigantea	A	1	III	III	III	III	III
Elymus repens	A	4	III	III	III	III	III
Holcus mollis	S	1	III	III	III	III	III
Rhizomes reaching greater depths							
Equisetum arvense	A	2	III	III	II	III	II
Phragmites australis	A	1	III	III	III	III	II
Tussilago farfara	A	1	III	III	III	III	III
3.2.3. Plagiotropic thickened roots							
Thickened roots shallow							
Rumex acetosella	A	1	III	III	III	III	II
Sonchus arvensis	A	2	III	III	II	II	I
Thickened roots reaching greater depths							
Cirsium arvense	A	2	III	III	II	II	I
Convolvulus arvensis (k)	SC	1	III	III	III	III	II
Perennials with other structures							
Allium vineale (with bulbs)	S	1	–	–	III	III	III

[a]Distribution in Sweden: A, all Sweden; C, Central Sweden; S, Southern Sweden.
[b]Average abundance at present, denoted 1–4: 1, frequent in very limited areas of Sweden, or under very special conditions, or occurring as a more scattered weed over larger areas; 4, a frequent and, on average, abundant weed in arable fields in all, or most of the important agricultural areas in the country.
[c]III, maximal; II, somewhat reduced; I, limited; –, minimal.
Note: Grading values are comparable only horizontally. They do not inform on quantitative abundance or importance.
[d](a) G. bifida, G. speciosa and G. tetrahit; (b) P. lapathifolium (in A) and P. persicaria (mainly in S); (c) S. asper, most often, and S. oleraceus; (d) information will often include L. spurium; (e) G. dissectum, G. molle, G. pusillum and G. robertianum; (f) information will sometimes include L. amplexicaule and L. hybridum; (g) information will sometimes include Chamomilla recutita and C. suaveolens; (h) P. argemone, P. dubium and P. rhoeas; (i) V. agrestis, V. arvensis and V. hederifolia; (j) information will sometimes include V. tricolor; (k) comparatively large numbers of regenerative underground stems also develop.

times of the year. This largely explains differences in the weed flora composition in crops established in different seasons. The diverging germination seasonality of typical summer and winter annuals (Fig. 1) explains dissimilarities in their appearance as weeds in spring- and autumn-sown crops (Table 3). Individual weed species differ more or less from the average germination patterns illustrated in Fig. 1, which is one reason for differences in their occurrence (Table 5). At least the short-term influence of tillage on perennial weeds with a strong capacity for vegetative regeneration depends more on the time of tillage in relation to their vegetative life cycles than to the time in relation to their seed germination seasonalities.

Annual weeds grow less abundantly in grass–clover leys than in annual crops due to the competitive conditions in the leys. Winter annuals develop more frequently than summer annuals in leys. They grow particularly in first-year leys, and with vigorous plants not least in patches with weakened or lacking ley plants (Fig. 2, Tables 2 and 4; Hallgren, 1976). Winter-annual weeds are, to a great extent, plants established in the previous year, not least those in first-year leys. Summer annuals establish best after germination in early spring (e.g. Galeopsis spp. with early germinating seeds). Both categories of annuals normally establish to decreasing extents with increasing age of a ley (Tables 3 and 4).

The vegetative structures of many creeping perennials, as well as those of most stationary perennials, are sensitive to tillage.

When regrowing after tillage, these are to a great extent weakly competitive and many of them die. In competitive annual crops that have been established after normal tillage, most of these perennials therefore less often develop vigorous plants from vegetative structures (Table 3). Examples of such perennials are presented in Table 5.

Plants from seeds in the seed banks of these perennials often emerge frequently in annual crops, but they usually produce weak growth and do little harm in competitive crops, such as cereals. They survive to a considerable extent even in competitive crops, but are largely killed by tillage preceding the following crop. If the annual crop is less competitive (e.g. sugar beet, potatoes, vegetables), seedlings of species such as *Taraxacum officinale*, *Rumex crispus*, *Plantago major*, *P. lanceolata* and *Ranunculus repens* can become very vigorous plants, causing great weed problems. In addition, plants originating from vegetative structures that have been weakened but not killed by previous tillage can produce a very vigorous growth if not controlled.

Of particular interest is the situation when seedlings of these perennials establish among ley plants undersown in cereals. They largely survive in the cereal crop, pass the winter and grow in the subsequent ley in parallel with the ley plants. Many of those perennials that are sensitive to tillage tolerate ordinary cutting and competition in leys for fodder. They can therefore become important components in the ley vegetation, some of them reducing the quality or quantity of the fodder.

Typical ley weeds increase their bio-mass in the ley stand with increasing age of the ley (Tables 3 and 5). Creeping perennials can spread and multiply vegetatively in the ley. Among stationary perennials, the biomass increase is largely a result of an increasing size of plants that established as seedlings together with the ley plant seed-lings in the year before the 'first-year ley'. This applies to species with taproots, e.g. *Taraxacum officinale* and *Rumex crispus*, and to tufted grasses such as *Agrostis capillaris* and *Deschampsia cespitosa*. In field experiments, *T. officinale* has proved to increase not only in plant size but also in number of individuals (Åberg, 1956), which indicates a continuous seedling establish-ment. As shown in Fig. 2 and Table 2, seedlings have great difficulties in establish-ing under dense plant canopies. Among the seedlings behind the figures in Table 2, there were not only seedlings of annuals from both small and large seeds but also seedlings of the perennial *Rumex crispus* with compara-tively large seeds. However, seedlings may establish in periods immediately following mowing and in the gaps or spots with weak plants that frequently occur even in stands considered uniform under ordinary field conditions.

Vigorous regrowth from *Taraxacum* and *Rumex* taproots surviving previous tillage is sometimes seen in cereal fields (cf. Kvist and Håkansson, 1985), but mainly after weak tillage or in thin crops. In weakly competitive annual crops, many perennial plants sensitive to tillage can grow much more vigorously after establishment as seed-lings than in cereals. In some agricultural areas, e.g. North America (Alex, 1982), *Taraxacum officinale* seems to be more recognized as a cereal weed than in Sweden. The effective wind dispersal of seeds facilitates invasion and persistence of *T. officinale* on any ground.

Creeping perennials that are tolerant to tillage and therefore important weeds in competitive annual crops grow vigorously or at least persist in leys. Some of them, particularly rhizomatous grasses such as *Elymus repens* are very persistent in the leys. Depending on climate, soil, character

of ley plants, fertilization, cutting practices, etc., these grasses either increase or decrease with increasing age of the ley. *E. repens*, on average, seems to persist at a rather unchanged level (e.g. Hagsand and Thörn, 1960; Raatikainen and Raatikainen, 1975). It seems to decrease over years at low supply of nitrogen, but increase at high supply, which agrees with responses to nitrogen recorded by Tilman (1990).

The two tall-growing weeds with creep-ing roots, *Cirsium arvense* and *Sonchus arvensis*, almost invariably decrease with ageing of the ley. However, they usually persist in old leys too (Raatikainen and Raatikainen, 1975). Like *E. repens*, they can therefore rapidly increase in an annual crop after the breaking up of an old ley. Because of their tolerance to, or even stimulation by, regular tillage, creeping perennials of this type are important weeds in annual as well as perennial crops in a crop sequence (Table 3), though different species have different preferences (Table 5).

Perennials sensitive to tillage, and therefore mainly ley weeds, cause only slight problems in cereals and other compet-itive annual crops following the breaking of leys. Even when plants of that category have produced a vigorous growth in a ley, they are usually strongly checked when the ley is broken by ploughing. However, the ability of the plants to recover vigorously after this ploughing also depends on the degree of competition in the following crop.

The cultivated plants in leys for cutting are mostly stationary perennials. They are tufted grasses (e.g. *Phleum pratense* and *Festuca pratensis*) and legumes with tap-roots (e.g. *Trifolium pratense* and *Medicago sativa*), which are sensitive to tillage, like most weeds with these life forms. When the rhizomatous grass *Poa pratensis* is grown in a ley, it is sometimes seen as a weed in subse-quent annual crops, but is less persistent than *Elymus repens*. The stoloniferous clover *Trifolium repens* sometimes survives tillage on ground that is sufficiently moist to suit its growth.

On the whole, however, cultivated ley plants, as well as typical ley weeds, mostly cause only minor problems in annual crops, and when they do, they mainly cause physical problems in the seedbed of the first crop following a ley. Thus, even in situations when perennials sensitive to tillage are sufficiently killed by the ley-breaking operations, both weeds and cultivated ley plants with strong vegetative structures may be insufficiently fragmented and cause unfavourable seedbed conditions and an uneven establishment of the new crop.

Changes in tillage and other measures selectively affecting weeds

In comparison with the tillage implements available today, those used in older farming had weaker effects, allowing many plants suppressed by modern tillage to persist as important arable weeds. Lower fertility levels and other soil conditions restricting crop growth, inferior sowing techniques, etc., entailed competitive situations that allowed more plants to occur as abundant weeds than situations in modern agriculture. The weed flora in earlier times thus comprised many species enduring conditions in weakly competitive cereal stands, but these species are overrun by competition from the cereals in high-input agriculture. As understood from an extensive Swedish literature from the 18th to the early 20th century (reviewed by Fogelfors, 1979), more plant species were noxious cereal weeds in earlier times than today. The number of economically important weed species has decreased much more than the total number of weed species occurring in a given agricultural area (e.g. Andreasen *et al.*, 1996).

A number of those stationary and creeping perennials that are easily checked by tillage today were often troublesome in cereal fields in older times when tillage was less effective. A few examples are *Artemisia vulgaris*, *Rumex crispus*, *Taraxacum officinale* and *Ranunculus repens*. They represent plant categories that *will again become troublesome cereal weeds at reduced tillage if not controlled by chemicals or other measures.*

The abundance of species and biotypes representing different life forms in a given field at a given time is always determined by many factors. Climate, soil and cropping system factors interact. Tillage, fertilizer use, particularly the nitrogen supply, liming and drainage are important.

These factors have various direct effects on the weeds. However, their effects, especially their selective effects, are strongly modified by competition, particularly in those crops whose stands close rapidly, such as cereals. The competitive effects, in turn, depend to a great extent on factors that can be determined, at least partly, by the grower. Such factors are the nutritional and water conditions as well as the crop species and varieties, the plant density and spatial distribution and the depth of sowing. The influence of these factors is evaluated later on in this book in various connections.

As previously mentioned, methods and times of harvest have a considerable influence on the reproduction and spread of weeds. This influence is selective. Combine harvesting takes place later and leaves a longer stubble than did harvesting with older methods. Different weeds respond differently, depending on their annual developmental rhythm, on their life form and morphology and, particularly, on the position and type of their reproductive organs. Without counteractions, an increased addition of seeds to the seed bank will increase the populations of many weeds. Changes in harvest times also change the times when mechanical measures against perennials such as *Elymus repens* can be undertaken. Chemical control has levelled out the differences between diverse weeds that would otherwise have resulted from the changed harvest times and methods. It has, on the whole, also reduced the occurrence of weeds, even those whose reproduction and spread are favoured by the changes in harvesting in recent decades (see below).

It is apparent from the above that thorough modern tillage strikingly restricts the opportunities of many perennials to become noxious weeds (several creeping as well as stationary perennials). It should also be stressed again that these perennials have greater possibilities for recovering after tillage in a weak crop than in a crop exerting strong competition.

Increased fertilization with nitrogen facilitates a rapid stand closure of a crop. It seems that plants of any life form that have declined most strongly as weeds in the past century include those that grow slowly in young stages and are, at the same time, light-demanding but less nutrient-dependent (e.g. Mittnacht *et al.*, 1979; Haas and Streibig, 1982; Mahn, 1984; Erviö and Salonen, 1987). *Equisetum arvense* appears to be such a plant; it was shown by Andersson (1997) that nitrogen application at normal levels in high-input agriculture suppresses this species in competition with cereals.

Among the more important weeds in modern high-input agriculture are plants that grow rapidly, produce tall aerial shoots and, in addition, respond as positively to high fertilizer levels as do the competing crops. Examples are *Chenopodium album* (Table 32) and *Elymus repens* (Thörn, 1967; Ellenberg, 1974; Tilman, 1990; Hansson and Fogelfors, 1998). These species are only two examples of a great number of plants with very broad adaptability, largely due to great phenotypic plasticity. When emerging sufficiently early in a crop with tall-growing, shading plants, shoots of these plants grow taller and their assimilating leaves are placed higher up than in a low-growing or less shading crop. Plants of *C. album* emerging late in an already dense stand may, on the other hand, survive with very short shoots, thereby using minimum amounts of assimilates for stem formation and allowing the major part of their assimilates to be used for the formation of assimilating leaves. Even very small individuals are then able to produce a surplus of resources sufficient for setting a few seeds. Shoots from rhizomes of *E. repens* developing in already dense stands of a crop mainly allocate their resources to leaves with lower contents of dry matter relative to their assimilating surfaces than leaves developed in better light. They provide a very minute rhizome production, but if still active when light improves, they may start a rapid tillering and rhizome production.

Prostrate or low-growing plants can also persist as abundant weeds in dense crops if they can stand low light supply. *Stellaria media* is an example (e.g. Fogelfors, 1973). It adapts to a dense crop stand by a more erect growth and development of thinner leaves than in a more open stand. Photosynthesis is thereby facilitated in relation to respiration.

As understood from the above and as further illustrated in Chapter 7 (e.g. Table 27), the effect of herbicides and many other factors is strengthened by plant competition. What we usually call 'herbicide effects', both with regard to their general strength and to their selectivity, are thus in reality a joint effect of phytotoxicity and competition. The reproduction and long-term development of the populations of different weed species and biotypes are selectively influenced by parasites, herbivorous insects and other herbivores. The strength and selective pattern of these influences are certainly considerably modified by the plant stand density and structure, including the species composition, and by the resulting competitive conditions in the plant stand.

With regard to species and biotypes, the weed flora differs strongly between geographical areas. With regard to the life forms of major arable weeds, on the other hand, the weed flora differs less under comparable cropping conditions. This may be seen in the following.

Scandinavia and Finland – Field Surveys and General Observations

The occurrence of weeds with different life forms is discussed below on the basis of experiences illustrated in Table 5 and data from field surveys in the different Nordic countries, summed up in Tables 6–10. Tables 6, 9 and 10 are based on plant frequencies, Table 7 on plant densities and Table 8 on weights per unit area. Definitions and further information on the surveys and calculations are given with the tables. The material in the first place represents 'full-tillage' cropping systems with 'conventional' use of herbicides.

Table 6. Sweden: occurrence of summer and (facultative) winter annual dicotyledonous weeds in spring-sown cereals (barley, oats, wheat) and autumn-sown cereals (wheat and, to a lesser extent, rye). Weeds counted in untreated plots of field experiments with chemical weed control in Southern and Central Sweden. Counts represent different periods of years, i.e. 1953–1973 and 1971–1990, from reports by Gummesson (1975) and Hallgren (1993), respectively. Figures are frequency values denoting the percentage number of experimental fields[a] where a weed taxon occurred with ≥10 individuals m^{-2}.

	Period 1953–1973		Period 1971–1990	
Categories and taxa of weeds[b]	Spring-sown cereals	Autumn-sown cereals	Spring-sown cereals	Autumn-sown cereals
Summer annuals				
Bilderdykia convolvulus	16	10	16	10
Brassica spp.	2	0	1	0
Chenopodium album	46	4	41	8
Erysimum cheiranthoides	5	0	6	0
Fumaria officinalis	8	2	13	2
Galeopsis bifida + speciosa + tetrahit	37	10	32	11
Polygonum aviculare	3	7	4	11
Polygonum lapathifolium + persicaria	26	2	13	2
Sinapis arvensis	7	0	4	0
Sonchus asper + oleraceus	2	0	2	0
Spergula arvensis	26	2	8	0
Sum of frequency values	178	37	140	44
Winter annuals				
Buglossoides arvensis	0	2	0	1
Capsella bursa-pastoris	2	4	4	6
Centaurea cyanus	2	8	0	3
Chamomilla recutita	0	2	0	2
Consolida regalis	0	0	0	1
Galium aparine (+ *spurium*)	6	13	10	16
Lamium purpureum (+ *amplexicaule*)	8	8	13	13
Lapsana communis	5	3	6	2
Matricaria perforata	11	48	9	42
Myosotis arvensis	9	16	14	23
Papaver argemone + dubium + rhoeas	0	3	0	3
Stellaria media	35	44	33	42
Thlaspi arvense	11	6	7	3
Veronica spp.	3	15	3	16
Viola arvensis (+ *tricolor*)	25	23	26	31
Sum of frequency values	117	195	125	204
Proportions of summer- and winter-annual weeds (%)				
Summer annuals	60	16	53	18
Winter annuals	40	84	47	82
Total	100	100	100	100

[a]Total number of experimental fields: in the period 1953–1973, 1083 fields with spring-sown cereals and 768 fields with autumn-sown cereals; in the period 1971–1990, 925 fields with spring-sown cereals and 964 fields with autumn-sown cereals.
[b]Within and between the genera *Brassica* and *Sinapis*, confusion of species will have occurred to some extent in the identification of young plants. *Chenopodium suecicum* and *Atriplex patula* occur to a minor extent and may sometimes have been recorded as *C. album*. Seedlings of annual *Sonchus* species and the perennial *S. arvensis* may also have been mixed up in a few cases.

Table 7.   Finland: average density (number of plants per unit area) of weeds in spring-sown cereals (wheat, oats and barley) and autumn-sown cereals (wheat and rye). Calculations based on values from summer counts in field plots untreated with herbicides in the survey years. Spring-sown cereals in 2088 fields in 1962–1964 (Mukula *et al.*, 1969) and in 267 fields in 1982–1984 (Erviö and Salonen, 1987). Autumn-sown cereals in 540 fields in 1972–1974 (Raatikainen *et al.*, 1978). Fields of the two surveys in spring cereals were rather similarly distributed in the country.

	Weeds in spring-sown cereals				Weeds in autumn-sown cereals (plants m⁻²) in the 1970s
	Plants[a] m⁻²			1980s rel. to 1960s (%)	
Weed categories	1960s	1980s	Mean		
Summer annuals	281	74	(177)	26	122
Winter annuals	168	73	(121)	43	84
Annuals, total	449	147	(298)	33	206
Perennials[a]	113	19	(66)	17	69
All (incl. unidentified)[a]	597	173	(385)	29	286
Proportions of summer- and winter-annual weeds (%)					
Summer annuals	63	50	(59)		59
Winter annuals	37	50	(41)		41
Total	100	100	(100)		100

[a]For perennial weeds: plant number means number of aerial shoots emerged from vegetative structures plus young plants from seeds.

Frequency or density values do not characterize the relative importance or the competitive influences of different weeds, but they grade in broad lines the relative potential of plants or groups of plants to grow as weeds in various crops.

In many cases, determinations are less suitable for perennial weeds than for annuals and comparisons are therefore focused on summer and winter annuals, primarily or entirely. Weight determinations in leys (Table 8) provide a good basis for comparisons including perennials.

Grass weeds were not recorded in the counts presented in Table 6 and some less frequent weeds are not represented in the material underlying this table and Tables 7–11. This will, however, only marginally influence the values presented and the judgements regarding the relative occurrences of summer and winter annuals as weeds in various crops.

According to the percentage values shown in Tables 6, 9 and 10, summer-annual weeds, looked upon as entire groups, occur less frequently in autumn-sown cereals than do winter annuals in spring cereals. This is not a surprising consequence of the differences in their germination seasonality (Table 2) and their ability to survive winters. Calculations using data reported by Hallgren (1993) indicate similar conditions when the occurrence of these weed categories in spring- and autumn-sown oilseed crops is compared. The same applies when calculations are based on plant weights presented by Hallgren (2000).

Table 7 indicates a higher proportion of summer-annual weeds in autumn-sown cereals in Finland than in the other Nordic countries. One reason might be that the average winter is harder in Finland, resulting in a higher mortality among winter-annual weeds in Finland than in the other areas. There are certainly also other reasons. Thus, for example, many of the more abundant summer annuals in Finland are species with a particularly high potential to compete in autumn-sown crops. They are *Galeopsis* spp. (in Finland mainly *G. bifida* and *G. speciosa*) and *Polygonum aviculare*, which largely germinate in early spring and *Bilderdykia convolvulus*, which both germinates early and has climbing shoots. By virtue of its very high frequency and abundance on arable land in general,

Table 8. Finland: vegetation in leys for cutting (hay) of different ages grown in alternation with annual crops on arable land. Leys mostly undersown in spring cereals. Average values calculated on the basis of data from field surveys in 1966–1968, presented by Raatikainen and Raatikainen (1975). Weights determined at times of hay-making (late June to early July) in a total of 1620 fields distributed throughout Finland.

| | Air-dry weight of above-ground shoots of plants in leys of the following age in years: | | | | |
Category of plants	1	2	3	4	5
Weight (kg ha⁻¹)					
Cultivated grasses	2634	2765	2498	2299	1328
Cultivated clovers	465	477	275	181	87
Weeds, total	566	748	941	1127	1435
Total vegetation	3665	3990	3714	3607	2850
Weight (percentage of total vegetation weight)					
Cultivated grasses	71.9	69.3	67.3	63.8	46.6
Cultivated clovers	12.7	12.0	7.4	5.0	3.0
Weeds, total	15.4	18.7	25.3	31.2	50.4
Total vegetation	100.0	100.0	100.0	100.0	100.0
Specified weeds					
Summer annuals	0.71	0.09	0.04	0.02	0.02
Winter annuals	1.38	0.24	0.17	0.10	0.10
Perennials, total	13.33	18.41	25.13	31.14	50.24
Agrostis capillaris	0.99	1.87	3.98	4.87	7.31
Deschampsia cespitosa	0.53	1.63	5.33	7.64	17.42
Taraxacum officinale	0.22	0.64	1.14	1.78	3.82
Ranunculus repens	1.15	1.79	2.13	1.55	2.10
Elymus repens	3.89	4.43	4.08	4.13	4.50
Cirsium arvense	0.29	0.20	0.15	0.12	0.03
Sonchus arvensis	0.21	0.09	0.04	0.02	0.03
Other perennials	6.05	7.76	8.28	11.03	15.03

Table 9. Norway: occurrence of summer- and winter-annual weeds in spring- and autumn-sown cereals. Calculations based on frequency values of different species from early-summer counts in untreated plots of field experiments with chemical weed control in 1983–1997, as reported by Fykse (1999). Frequency values as defined in Table 6.

Weed category	Spring-sown cereals	Autumn-sown cereals
Sums of frequencies of different weeds[a]		
Summer annuals	189	43
Winter annuals	140	191
Proportions of summer- and winter-annual weeds (%)		
Summer annuals	57	18
Winter annuals	43	82
Total	100	100

[a]Sums of frequencies represent 14 summer- and 13 winter-annual weeds from 340 fields with spring-sown and 170 with autumn-sown cereals.

Chenopodium album becomes one of the major weeds also in autumn-sown crops.

Annual plants are less abundant than perennials as weeds in leys (Tables 8 and 10) at least in leys older than 1 year (Table 4). Competition in spring is stronger in a well wintered ley than in autumn-sown cereals. In relation to winter-annual weeds, summer annuals are therefore much more suppressed in leys than in autumn-sown cereals. Accordingly, summer annuals grow to a lesser extent than winter annuals in leys. In that respect, surveys indicate rather similar conditions in Sweden (Table 4; Hagsand and Thörn, 1960; Thörn, 1967), Finland (Table 8) and Denmark (Table 10). With the ageing of leys, annual weeds also decrease strongly, especially summer annuals, whereas perennials increase, looked upon as an entire group (Tables 4 and 8), but there are

Table 10. Denmark: occurrence of summer- and winter-annual weeds in various kinds of crops. Calculations based on frequency values resulting from surveys in fields well distributed over the country in two periods of years (1967–1970 and 1987–1989), as reported by Andreasen *et al.* (1996).[a]

Weed categories and years	Spring-sown crops	Autumn-sown crops	Grass leys
Sums of frequencies of different weeds			
Summer annuals			
1967–1970	212	102	23
1987–1989	77	36	11
Winter annuals			
1967–1970	340	429	178
1987–1989	177	192	110
Percentage proportions of summer- and winter-annual weeds			
Summer annuals			
1967–1970	38	19	11
1987–1989	30	16	9
Winter annuals			
1967–1970	62	81	89
1987–1989	70	84	91
Sums of frequencies in 1987–1989 as percentages of those in 1967–1970			
Summer annuals	36	35	48
Winter annuals	52	45	62

[a]Spring-sown crops: barley and rape; autumn-sown crops: wheat and rye. Grass leys are second-year leys. Frequency sums represent 16 summer- and 27 winter-annual weeds. Number of surveyed sites 139 (1967–1970) and 213 (1987–1989) evenly distributed among crops.

Table 11. Sweden: number of annual dicotyledonous weed plants per unit area in spring- and autumn-sown cereals. Means for 5-year periods, calculated on the basis of data from Hallgren (1993) reporting on early-summer counts in untreated plots of field experiments with chemical weed control.[a]

	Weed plants m^{-2}	
Years	Spring-sown cereals	Autumn-sown cereals
1955–1959	453	202
1960–1964	407	193
1965–1969	219	142
1970–1974	205	138
1975–1979	221	165
1980–1984	201	156
1985–1989	210	140

[a]Figures based on a minimum of 71 and a maximum of 318 experimental fields similarly distributed in Southern and Central Sweden in the different periods of years.

Summer annuals

Chenopodium album is on the whole the most common summer-annual weed in all the four countries. Examples of other species common in varying abundance in different areas are *Galeopsis bifida*, *G. tetrahit* and *G. speciosa*, *Bilderdykia convolvulus*, *Polygonum lapathifolium*, *P. persicaria* and *P. aviculare*.

Chenopodium album exhibits an 'average performance' of a summer-annual weed in the Nordic countries. It is comparatively sensitive to herbicides used in different annual crops and has therefore decreased after 1950, but not nearly as much as, for instance, *Sinapis arvensis*. This species was very abundant in the annual crops in important agricultural areas up to the 1950s, but has decreased strongly since then because it is very sensitive to phenoxy herbicides and many other weed control chemicals used from the 1950s on. *C. album* is still a very common weed with a comparatively heavy soil seed bank, although it has decreased (cf. Jensen and Kjellsson, 1995). This still makes it a very frequent weed in spring-sown annual crops and it is therefore often seen also in autumn-sown crops and leys.

important species differences, as exemplified in Table 8. (The figures there are somewhat biased with regard to the influence of the ley age. Thus, leys older than 3 years, particularly the 5-year leys, tend to be clustered in certain areas or environments, while elsewhere leys are largely broken when they are younger.) Table 8 will, however, well illustrate the principles of the influence of the ley age on different species. The illustrations are in agreement with other observations in the Nordic countries.

Based on the surveys summarized in Tables 6–10 and many other field observations, the occurrence of the more frequent or otherwise interesting weeds in Scandinavia and Finland are commented on.

However, other summer annuals establish more frequently and abundantly in autumn-sown crops and leys, when seen in relation to their occurrence in the spring-sown crops. This applies to the *Galeopsis* spp., to *Bilderdykia convolvulus*, and to *Polygonum aviculare*.

Polygonum aviculare differs from the more typical summer annuals by often establishing more abundantly in autumn-sown than in spring-sown cereals (Table 6). The main reason for that is certainly that it largely germinates very early in the spring. This facilitates its competitive position in an autumn-sown crop. At the same time, its earlier seedlings are frequently killed by seedbed preparation in the spring, which reduces its occurrence in spring crops. Other summer annuals often producing a vigorous growth in autumn-sown cereals are *Bilderdykia convolvulus* and *Galeopsis* spp. The seeds of these species also germinate early to a considerable extent. *B. convolvulus* is also favoured by climbing shoots in its competition in any crop stand. Because these species have larger seeds than *P. aviculare*, their early seedlings may be destroyed to a lesser extent by seedbed preparation in the spring than early seedlings from smaller seeds (e.g. Habel, 1954). This is probably at least one reason why they establish more frequently than *P. aviculare* in spring cereals, when compared with their establishment in winter cereals.

Avena fatua, which is a true summer annual in the Nordic countries, was a common weed in these countries in earlier times. However, it was almost extinct in the early decades of the 20th century, but came back as a feared weed in the 1950s. The records of this species as a weed in Sweden are presented and the reasons for its varying occurrence discussed in more detail on pp. 234–236.

In fields where *Brassica* species are grown as oilseed crops, volunteer plants of these species often appear. These plants are easily controlled by herbicides in cereals, but not so in the oilseed crops themselves, where the seeds of these volunteers often cause quality problems in the harvested product.

The summer annuals *Sonchus asper* (Fig. 23a), *S. oleraceus*, *Solanum nigrum* and *Urtica urens* are strongly suppressed in crops with dense stands, even in spring-sown cereals. However, in areas where the climate or soil suit them, they grow vigorously and require attentive control in less competitive crops, such as sugar beet, potatoes and vegetables. The competitive weakness of these species is often attributed to late spring germination (e.g. Koch, 1969). However, this is probably not the only reason, because seedlings of the species also emerge earlier in the spring. Relatively high demands for light, and perhaps also sensitivity to drought, may make the annual *Sonchus* species weakly competitive in cereals.

Winter annuals (facultative)

Stellaria media, *Matricaria perforata*, *Viola arvensis* and *Myosotis arvensis* are among the most frequent dicotyledonous winter-annual weeds in the countries now considered. Other dicots of considerable importance in large areas are *Capsella bursa-pastoris*, *Galium aparine*, *Lamium amplexicaule*, *L. purpureum*, *Lapsana communis* and *Thlaspi arvense*.

As tillage strongly increases the seed germination of winter annual weeds in late summer and early autumn (Fig. 1), it is important to stress that discussions on the occurrence of these weeds in autumn-sown crops concern situations where mechanical seedbed preparation precedes sowing (cf. pp. 27–28).

Most winter annual weeds in the Nordic countries easily germinate and establish in the spring as well as in later seasons. This applies particularly to species such as *Lamium purpureum*, *Stellaria media*, *Viola arvensis* and *Galium aparine*, which are therefore important weeds both in spring- and autumn-sown cereals. *S. media* is often more abundant in spring crops than any summer-annual weed. Like *Poa annua*, *L. purpureum* and *S. media* can flower till late autumn. Flowering plants can survive winter and even grow in mild winter periods and continue growing from very early spring. These plants could be considered short-lived perennials. *Cerastium fontanum* is a very variable taxon, appearing as an annual or a short-lived perennial. It is common in certain areas, where it is seen in leys as well as in autumn-sown crops, but seldom abundantly in spring crops.

All species mentioned as winter annuals are among the more frequent weeds in young leys. *Matricaria perforata* may be the most conspicuous of them because of its tall shoots and easily seen flowers, but *Poa annua* and *Stellaria media* are more frequent. The latter is favoured by its tolerance to shading (see above); yet it produces a rather weak growth in shading stands when compared with the dense, thick covers it can form in situations of better light.

Poa annua is a ubiquitous grass, which can be seen in any crop, including leys. Because it is variable (annual or short-lived perennial: see below) and can flower, set seed and germinate throughout the growing season, plants of this species are seen in most crops. However, in dense crop stands, it produces weak growth and probably does not influence the crop very much. In thin crops, on the other hand, this grass can develop dense covers, considerably influencing crop growth.

Several species classified as winter annuals due to their 'average performance' may vary in life form. For the sake of simplicity, it appears reasonable to view them as winter annuals. In *Poa annua*, both annual and perennial types are found (e.g. Law *et al.*, 1977; Netland, 1985). *Stellaria media* frequently spreads by prostrate, rooting shoots and survives winter vegetatively as a short-lived perennial. We can also see parts of flowering plants of *Matricaria perforata* surviving winter and developing new flowering shoots from lateral buds in the following summer. By definition, such plants are perennials. After emergence in spring, plants of *M. perforata* sometimes grow vegetatively throughout the growing season and do not flower until the next year. These plants perform as biennials, a trait certainly genetically determined in this case. However, diverse behaviour of the individuals of a species may not necessarily result only from genetic diversity, but also from environmental influences resulting in different 'growth types' as this term was presented on p. 4.

Apera spica-venti mainly occurs on lighter mineral soils in southern Sweden. This species is a less important weed in spring crops (Table 5). One reason for this is probably its restricted spring germination (cf. Koch, 1969; Avholm and Wallgren, 1976). However, it accidentally appears in abundance in spring crops too. This may result from special weather conditions allowing its germination in the spring. However, plants from seeds germinated in the autumn may also develop in spring crops.

To what extent plants establishing in spring crops represent seedlings surviving after autumn germination is an important question that concerns winter annuals in general. As our knowledge of these matters is very poor, thorough studies on the following important questions are of interest: to what extent does autumn and spring germination within a species result from genetic diversity, and when, or to what extent, is the varying time of germination a consequence of different conditions within or around even genetically similar seeds? In the former case, there is a great potential for selection among the plants in favour of either spring or autumn germination as a response to exchange of crops or to other changes in the cropping system.

Biennials (monocarpic perennials)

Biennials are easily killed or suppressed by tillage in their vegetative phase. In sequences of annual crops, their reproduction is therefore strongly blocked. Flowering plants are sometimes seen in leys on arable land, but rarely in abundance. Examples are *Carduus crispus*, *Cirsium vulgare* and *C. palustre*. Plants of the latter species sometimes do not flower until they have grown vegetatively for more than 1 year. Some species are seen in grazed areas outside arable land. Biennials are particularly often seen at sites adjacent to cropped fields, such as roadsides, ditch-banks, courtyards, and so on, i.e. in areas that are not tilled but where the vegetation cover is otherwise frequently disturbed. Other plants that should be mentioned here are *Arctium* spp. and *Daucus carota* (wild forms of carrots). *D. carota* is an example of a wild plant causing problems in the seed production of its corresponding cultivated forms as a result of interbreeding.

Stationary perennials

As weeds on arable land, these perennials grow vigorously mainly in leys cut for fodder and are often named 'ley weeds' (cf. p. 26). Their seedlings have difficulties in forming strong plants in competitive crops, and their perennating structures also produce a weak or slow regrowth after tillage. Therefore they usually do not have sufficient time to produce a vigorous growth in competitive annual crops such as cereals (sown after effective tillage). Many of them may, however, become very problematic weeds in weakly competitive agricultural or horticultural crops. A great number of species frequently occur on roadsides, ditch-banks and similar habitats adjacent to arable areas.

Tables 4 and 8 indicate an increase in the total amount of perennial weeds with the age of a ley. The increasing species are generally stationary perennials, as is illustrated in Table 8 by the tufted grasses *Agrostis capillaris* and *Deschampsia cespitosa*, and the taproot plant *Taraxacum officinale*. This also applies to typical ley weeds among the tillage-sensitive creeping perennials (see Table 8 and additional data in Raatikainen and Raatikainen, 1975).

Creeping perennials with stolons or prostrate above-ground stems

Seedlings of these plants often have less time to produce vigorous growth in annual crops with dense stands. This also applies to plants arising from their vegetative structures. Some species, e.g. *Ranunculus repens*, sometimes grow vigorously even in cereals and similarly competitive crops, particularly on moist soil. *R. repens* has relatively large seeds and, therefore, strong

seedlings. It also regenerates vegetatively better than many other stoloniferous perennials. The grass *Agrostis stolonifera* also forms strong creeping shoots providing a considerable spread and growth in rather competitive crops in certain areas. Where they are well adapted to climate and soil, many of the creeping perennials can rapidly produce dense soil covers in agricultural and horticultural crops with weakly competitive stands.

Although some stoloniferous perennials thus can become troublesome in annual crops, these perennials are foremost ley weeds with a tendency to increase with the age of the ley. Many of them are common plants on roadsides and similar locations adjacent to arable fields.

Creeping perennials with rhizomes sensitive to soil cultivation

The rhizomes of these plants are often morphologically well developed and are efficient perennating structures when mechanically undisturbed, but produce a slow or weak regrowth in competitive crops after burial by tillage. Many species tolerate the joint effects of competition and annual cutting and can therefore increase on ground where they are able to overcome their weaknesses in early growth. They thus regrow efficiently after cutting, which leaves their underground structures undisturbed.

Several species can therefore produce vigorous plants in leys, particularly in older ones. Most species belong to the typical vegetation on non-tilled but otherwise disturbed ground in the vicinity of arable fields, e.g. on headlands, roadsides and ditch-banks. Examples are *Achillea millefolium*, *A. ptarmica*, *Galium mollugo*, *Lamium album* and, particularly on soil rich in nitrogen, *Urtica dioica*.

Creeping perennials with rhizomes tolerant to soil cultivation

The rhizomes of these plants have a strong capacity to regenerate after breakage and burial, enabling them to produce a rapid and vigorous regrowth after tillage (cf.

Håkansson, 1967; Kvist and Håkansson, 1985). Plants with such rhizomes therefore become important weeds in annual crops as well as in leys. Grasses with shallow rhizomes survive ordinary cutting and competition in leys for fodder. *Elymus repens* is the most widespread and, on average, the most abundant rhizomatous weed in the Nordic countries, especially so in their northernmost areas. Its potential to grow in competitive crops is high and it seems to have rather similar growth conditions in cereals and leys for fodder (Tables 5, 8 and 47). In contrast to another rhizomatous grass, *Agrostis gigantea*, it is evidently favoured by high levels of fertilizers (at least by those containing nitrogen) in its competition with cereal and ley plants. Depending on species composition, cutting regime, fertilizer use, etc., its clones increase or decrease in vigour with increasing age of the ley. On average, however, it persists at a rather unchanged abundance with increasing age of a ley (Table 8; Hagsand and Thörn, 1960). (See special descriptions of the two grasses on pp. 236–242.)

The two species with deep-growing rhizomes, *Equisetum arvense* and *Tussilago farfara*, are weeds of importance in both annual crops and leys. However, they have obvious difficulties in withstanding the competition in dense crops at high fertilizer applications. They decrease in abundance with the ageing of a ley as long as the ley stand is still dense. Their occurrence as arable weeds has diminished considerably, particularly in the years from the 1950s through the 1970s, probably due to increased fertilizer use, drainage, etc. *E. arvense* presents weak growth in strongly N-fertilized, shading crops (Andersson, 1997). All the rhizomatous perennials are frequent plants in more or less disturbed habitats outside the arable areas, such as headlands, roadsides and ditch-banks.

Creeping perennials with plagiotropic thickened roots

A number of species with such roots, e.g. *Cirsium arvense* and *Sonchus arvensis*, occur as important weeds in both annual and perennial crops. Rather short fragments of their thickened roots are regenerative. The roots of some species are more or less dormant in late summer and autumn. After tillage in that period, the regeneration of these species is delayed until the following spring, when dormancy has been broken by the low winter temperatures.

In leys cut for fodder, cutting and competition jointly suppress many plants of this category developing tall shoots, e.g. *Cirsium arvense* and *Sonchus arvensis*. Cutting deprives these plants of the major part of their assimilating foliage, and the regrowth draws great amounts of food reserves from their underground structures. According to general field experience and a field survey (Table 8), their clones, particularly those of *S. arvensis*, therefore mostly weaken considerably with increasing age of a ley. However, their vegetative structures are usually strong enough to survive the mechanical breaking of even an old ley, enabling them to initiate a recovery in subsequent annual crops, where they soon may need to be controlled. Where herbicides are used without restrictions, they are easily controlled in the cereal fields.

Rumex acetosella, a variable perennial with plagiotropic thickened roots, also occurs as a weed in both annual crops and leys. This species has decreased strongly as an arable weed in most agricultural areas in the 20th century. It is competitive mainly on poor, acid soils and can withstand dry conditions (Ellenberg, 1974; Hanf, 1982). Under such poor conditions, it may be a weed of importance in many crops and persists in leys for cutting because of its relatively low-growing shoots.

Changes in weed flora after the 1940s

Changes have differed between agricultural areas and cropping systems. The following describes main changes as average events. Typical changes in the Nordic countries have been described by many authors, e.g. Haas and Streibig (1982) and Salonen (1993). As expected, they stress that the introduction and regular use of modern herbicides has been particularly influential. By ordination methods in a multivariate approach, Salonen thus described the very strong influence of continuous herbicide use on the weed species composition in spring cereals.

Tables 6, 7, 10 and 11 describe changes in the occurrence of annual weeds in cereals in various periods from the 1950s on. Table 11 indicates, for Sweden, a strong decline in annual weeds, which was most striking in the first 15–20 years. Changes continued after that, but were less rapid. In Table 6, the period 1971–1990 can be compared with 1953–1973. The sums of frequency values in the table suggest a weak decline among summer annuals in spring cereals, but only a few species clearly declined and others increased.

Tables 7 and 10 indicate a general decrease of annual weeds from the 1960s to the 1980s in Finland and Denmark, where summer annuals decreased by 65–75% and winter annuals by 50–60%. According to Table 11, the years between the two periods under comparison largely represent years with a strong decline in annual weeds in Sweden. It should be noticed here that Table 11, which is based on densities, suggests a stronger general decline than Table 6, based on frequencies, although the two tables represent the same Swedish material. The figures based on densities probably describe the changes better than those based on frequencies.

Factors other than chemical control have also contributed to reducing weed populations. Crop stands have gradually been improved in many ways, thereby increasing their competitive power. However, increased fertilizer use may favour not only the crops, but sometimes also some important weeds, e.g. *Chenopodium album* (Tables 31 and 32). At the same time, certain observations suggest that strong nitrogen fertilization increases the joint suppressive effect of herbicides and competition on the herbicide-sensitive weeds owing to an increased sensitivity of these weeds at rich nitrogen supply. However, this suggestion needs to be studied in more detail.

A rapid decline in annual weeds sensitive to phenoxy compounds and other herbicides in regular use took place from the 1950s through the 1970s. This applies even to the many farms that replaced leys by cereals in these years (cf. pp. 20–21). Unless prevented by chemical control or drastic tillage on fallows, many weeds would have increased in the strongly cereal-dominated crop sequences, as regards both annuals and perennials. For instance, *Cirsium arvense* on the whole decreased drastically, although it is favoured in cereals in the absence of chemical control. *Sonchus arvensis* also decreased, though less drastically.

Due to the introduction of regular chemical control and/or to other cultural changes, many other perennial weeds have declined in the Nordic countries. They represent species with strong regenerative structures, such as *Equisetum arvense* and *Tussilago farfara*, as well as a number of perennials with weaker structures. Many of these perennials may have declined largely due to increased competition from the crops. It is likely that *Sonchus* and *Equisetum* have been disfavoured by the increased nitrogen supply (Table 35; Andersson, 1997).

According to the above, *Elymus repens*, like annuals such as *Chenopodium album*, has probably been favoured by the increased nitrogen supply. *E. repens* has decreased or increased differently in different areas and periods of years, largely due to varying degrees of mechanical control after harvesting of crops and/or in response to effective herbicides (mainly glyphosate) available in the last few decades (cf. pp. 236–242).

According to the literature presented with Tables 6–11 and related information from the Nordic countries, changes in the abundance of annual species may be briefly described as follows.

A striking decline first took place in Denmark, i.e. in areas with the most intensive agriculture. Most of the major annual weeds have thus decreased there, or remained at stable levels. This even applies to some of those that have increased in parts of the other countries. Such species are *Galium aparine, Lamium purpureum/*

amplexicaule, *Stellaria media* and *Viola arvensis*.

Gummesson (1975) described changes in the occurrence of annual weeds in cereal fields in Sweden in the period 1952–1973. Important weeds that had changed in frequency in spring and winter cereals were ranked in order of the degree of changes as follows.

Weeds that had decreased were: in spring cereals (i) *Sinapis arvensis*, (ii) *Spergula arvensis*, (iii) *Polygonum lapathifolium/persicaria*, and (iv) *Chenopodium album*; in winter cereals (i) *Centaurea cyanus*, (ii) *Galium aparine* and (iii) *Matricaria perforata*.

Weeds that had increased were: in spring cereals (i) *Lamium purpureum/amplexicaule*, (ii) *Fumaria officinalis*, (iii) *Myosotis arvensis*, (iv) *Bilderdykia convolvulus*, and (v) *Viola arvensis*; in winter cereals (i) *Bilderdykia convolvulus*, (ii) *Polygonum aviculare*, (iii) *Myosotis arvensis* and (iv) *Viola arvensis*.

Examples of species that have decreased in Sweden after the 1970s are *Polygonum lapathifolium*, *P. persicaria*, *Sinapis arvensis*, *Spergula arvensis* and *Centaurea cyanus*. Species that have increased are *Lamium purpureum* and *Myosotis arvensis*, and more or less *Fumaria officinalis*, *Polygonum aviculare* and *Viola arvensis* as well.

Chenopodium album and *Stellaria media* changed only slightly in the earlier decades, and their populations have remained strong ever since. These two species are continuously the most abundant and frequent weeds in the Nordic countries as an entirety. Because *S. media* is much more abundant than *C. album* in autumn-sown crops and leys, it is on the whole the more abundant weed. It should be stressed again that a comparison of this type is no comparison of the economic importance of weeds.

To what extent agricultural measures other than chemical control have contributed (directly or indirectly) towards reducing weed populations since the 1950s is difficult to determine. In some cases, however, the dominating influences may be pointed out with a high degree of certainty.

Thus, the very pronounced decline in the tall-growing and nitrogen-favoured *Sinapis arvensis* and *Centaurea cyanus* is certainly predominantly a result of the frequent use of herbicides to which these species are extremely sensitive. The similarly pronounced decline of *Spergula arvensis* is, on the other hand, certainly largely caused by cultural measures having impaired the competitive situation of this weed as much as, or more than, the herbicides.

In broad outline, winter annuals have decreased less than summer annuals in the Nordic countries. Among those winter annuals that germinate in spring as easily as in autumn, many have increased, at least in great parts of Sweden, e.g. *Lamium purpureum*, *Myosotis arvensis*, *Stellaria media* and *Viola arvensis*. Germination both in the earlier and later parts of the growing season facilitates establishment and reproduction in very different situations in various cropping systems. The seeds of the species mentioned also mature and shed in different seasons. *Galium aparine* is favoured both by climbing shoots and by germination in the spring as well as in the autumn. According to previous descriptions, *Bilderdykia convolvulus* and *Polygonum aviculare* are summer-annual species with early germination and other traits facilitating their establishment in both autumn- and spring-sown crops (Table 6).

Examination of soil seed banks in 37 arable fields in Denmark in 1964 (Jensen, 1969) and 1989 (Jensen and Kjellsson, 1995) indicated that the number of viable weed seeds per unit area had been reduced from an average of 56,600 m^{-2} in 1964 to 26,400 in 1989. It had thus been halved in 25 years of intensified cropping measures including chemical weed control. Most of the common dicotyledonous weeds had decreased considerably. The grass species *Poa annua* had increased. This agrees with other experiences in the Nordic countries suggesting that this grass has increased since the 1950s (e.g. Bylterud, 1983; Erviö and Salonen, 1987). The mean number of species per field sample was reduced from 12.1 to 4.8, indicating a decrease in the diversity of species within small areas.

However, the species diversity has to be related to specified areas or situations. From a survey of 357 arable fields in Denmark in the late 1980s, Andreasen *et al.* (1996) reports a strong reduction of the general weed abundance in comparison with that recorded in the 1960s (Mikkelsen and Laursen, 1966; Laursen, 1967). The number of species that dominated the weed flora to a defined degree had also decreased strongly. At the same time, however, the total number of species found in the Danish fields investigated was almost unchanged, about 200. In comparison, many species that were frequent arable weeds in Sweden in the 18th and 19th centuries according to Fogelfors (1979) are nowadays rarely seen on arable land, but all of them can still be found there, although only in scattered places.

A reduction in the use of herbicides, including lowered doses, in many farms in recent years, has led to an increase of some weeds that had previously decreased. In organic farming, some weeds that are easily controlled by herbicides are becoming problematic, e.g. *Cirsium arvense*.

Certain species that are common weeds further south in Europe but rarely seen in Scandinavia up to now have become problematic in scattered fields in recent years. One of them is *Echinochloa crus-galli* (a C$_4$ plant: see pp. 51–52), appearing in scattered fields in Southern Scandinavia where sugar beet, potatoes and similar less competitive crops are grown recurrently. Changes in cropping systems, transports, climatic conditions, etc., will lead to many unforeseeable changes in weed flora.

In organic farming, which has been adopted by a minor proportion of farmers in Sweden and elsewhere in recent decades, the weed flora is changing towards situations like those before the 1950s (e.g. Rydberg and Milberg, 2000). Some of the weeds that are particularly sensitive to the herbicides used in 'conventional' agriculture are strikingly increasing in organic farming. Examples are *Sinapis arvensis*, *Centaurea cyanus* and *Cirsium arvense*. Weeds tolerant to poor soil, both annuals and perennials, seem to increase, probably due to the lower levels of nitrogen fertilizer use and, for some species, also to the discontinuation of chemical control. This applies to *Equisetum arvense* (e.g. Andersson, 1997) and *Spergula arvensis*.

In 1997–1999, Salonen *et al.* (2001a,b) surveyed weed infestations in spring cereals in Finland, in organic farming introduced in the past few decades and in conventional agriculture. When compared with densities registered in the 1980s (see Table 7), weed densities had increased considerably in organic farming. Infestations by *Chenopodium album*, *Elymus repens*, *Cirsium arvense* and *Sonchus arvensis* were considered of major concern (Salonen *et al.*, 2001a). Although comparisons with earlier surveys have to be made with caution, a certain increase in average infestations of weeds may have occurred in Finland in recent decades according to data presented by Salonen *et al.* (2001b), probably mainly resulting from a decreased use of herbicides even in conventional agriculture.

Northern and Southern Europe

The basic data from a European weed survey presented by Schroeder *et al.* (1993) were made available to me by these authors. They were used for analyses of relations between plant life forms and their occurrences as weeds in different crops (Håkansson, 1995b). The data were based on a questionnaire sent to weed scientists in all European countries, including Turkey. They comprise 46 species, selected in advance as particularly important weeds in a European perspective. Their occurrence in different crops was reported from each country using four score levels. *These weeds are now distributed into life forms on the basis of their appearance in Northern Europe.*

Four score levels were stipulated by Schroeder *et al.* (1993) for grading the 'frequency and abundance (economic importance)' of the weeds in different crops represented in a country. The scores, here given the values 0 (absence), 1, 2 and 3, have been used for calculations underlying the figures in Table 12. Figures in Table 13 describe the occurrence of groups of weeds representing different life forms.

The crops distinguished in the questionnaire are characterized in Tables 12 and 13 using the same principles as for the characterization of crops in Tables 3 and 5. Some crops in the questionnaire have been grouped into categories with similar relations to weed life forms, according to studies of the basic data. The following crop categories are distinguished:

- *As.*   Spring-sown annual crops.
- *As1.*   *Potatoes, sugar beet* and *vegetables* (separate categories in the questionnaire). Later stand closure and crops therefore weakly competitive, at least in early stages of development.
- *As2.*   *Spring cereals*: barley, wheat and oats (one group originally). Crop stands close early and are comparatively competitive.
- *As3.*   *Maize + sorghum* (originally in one group) and *sunflower + soybeans* (originally in one group). In early stages less competitive than As2.
- *Aw.*   *Winter crops*: winter cereals (original group: wheat + barley + rye) and winter rape. Crop stands close early and are comparatively competitive.
- *P.*   Perennial crops: *orchards* and *vineyards*.

Though there are some weaknesses in the basic data when used for the present purposes (see Håkansson, 1995b), Tables 12 and 13 will, on the whole, offer important qualitative information of reasonable validity. The tables strongly suggest that, in Europe as a whole, the occurrence of weeds with different traits in relation to different crops exhibits patterns similar to those in the Nordic countries. There may be exceptions regarding the occurrence of summer and winter annuals in Southern Europe (see below).

Annual plants are, as expected, important weeds in annual crops (As, Aw). They occur proportionally more in the perennial crops (P), orchards and vineyards, than in Swedish perennial crops, leys for cutting. This is not surprising. The perennial plantations constitute a more heterogeneous category of crops, which, as an entirety, can be seen as a mosaic of sites representing regularly cultivated areas and areas that are ecologically near parallels to grass leys.

No *biennial species* occur among the 46 species considered of such importance as weeds that they were selected for the questionnaire; neither are biennials mentioned in complementary comments asked for in that questionnaire.

Stationary and *stoloniferous perennials*, characterized above as sensitive to tillage, are rarely mentioned as weeds of importance in annual crops in the answers to the questionnaire. They are mostly mentioned among the weeds in the perennial plantations (P). Leys of the type used in the Nordic countries are not included in the questionnaire.

No *species with rhizomes sensitive to tillage* according to the above occur among the 46 weeds selected for the questionnaire, neither are any of them mentioned in complementary comments asked for.

Among perennial plants, only those with strong *rhizomes* or *creeping roots* are characterized as weeds of real importance in annual crops. They are described as important weeds in orchards and vineyards. Particularly plants with creeping roots seem to be, on average, more conspicuous in orchards and vineyards than they are in North-European leys. They might be particularly favoured where differently disturbed areas are mixed in the same fields. Species with fast-growing runners often have particularly good growth conditions along lines where tilled areas meet areas with less disturbed vegetation. Where, in addition, this vegetation is less competitive than uniform ley stands, plants with creeping roots may have better growth conditions than in such ley stands.

It should be stressed again that all species are grouped into life forms on the basis of their appearance in Northern Europe. The perennial species listed in Table 12 have largely similar perennating

Table 12. The 'frequency and abundance' of 46 important weed species in different kinds of crops. Species grouped here with regard to life forms, as judged from their habit in Northern Europe. Figures are calculated on the basis of data from a European weed survey (Schroeder *et al.*, 1993). Two regions are distinguished, i.e. 'Northern, Central and Western Europe' (16 countries)[a] and 'Southern Europe' (eight countries).[b] Species named according to *Flora Europaea* (Tutin *et al.*, 1964–1980). Crop groups (As, As1, As2, As3, Aw and P) as defined on p. 40 (Håkansson, 1995b).

| Life form (including lifespan) and species of weeds | Northern, Central and Western Europe | | | | | | Southern Europe | | | | | |
| | As | | | | | | As | | | | | |
	1	2	3	Aw	P	Total	1	2	3	Aw	P	Total

The sum of score values of each weed species as a percentage of the sum of their maximum scores (each of which = 3) in individual crops and as a total:

Summer annuals

Life form (including lifespan) and species of weeds	1	2	3	Aw	P	Total	1	2	3	Aw	P	Total
Abutilon theophrasti	2	2	18	2	8	5	16	6	29	6	10	15
Amaranthus retroflexus (C_4)	29	6	60	3	46	26	77	17	64	12	55	54
Ambrosia artemisiifolia	6	2	16	3	6	6	16	17	17	15	14	16
Atriplex patula	35	12	51	6	31	26	44	11	32	18	26	31
Avena fatua (1)	25	60	29	27	15	29	25	44	10	67	24	30
Bilderdykia convolvulus	48	41	53	27	42	42	16	17	13	15	10	14
Chenopodium album	82	49	80	27	42	59	83	33	67	24	64	62
Digitaria sanguinalis (C_4)	16	4	53	4	40	19	43	22	42	15	60	40
Echinochloa crus-galli (C_4)	43	16	84	4	42	35	72	22	69	18	57	56
Erysimum cheiranthoides	13	22	11	13	23	15	0	0	0	0	0	0
Fumaria officinalis	44	31	27	20	35	33	28	33	19	52	21	29
Galeopsis tetrahit	39	59	27	28	17	35	9	17	10	27	5	12
Galinsoga parviflora (2)	38	6	51	4	31	26	26	6	13	6	12	15
Polygonum aviculare	57	45	49	33	40	46	50	28	39	52	45	45
Polygonum persicaria	63	45	53	27	40	48	31	17	33	21	26	28
Setaria pumila (C_4) (3)	16	8	56	7	27	19	41	22	50	18	45	39
Solanum nigrum	51	12	67	6	29	34	63	11	42	12	31	39
Sonchus asper	40	20	36	17	29	30	19	11	15	12	31	19
Sonchus oleraceus	33	16	33	16	31	26	36	11	23	18	43	29
Spergula arvensis	31	29	29	18	25	27	17	22	13	30	21	20

Winter annuals

Life form (including lifespan) and species of weeds	1	2	3	Aw	P	Total	1	2	3	Aw	P	Total
Alopecurus myosuroides	14	25	10	50	11	23	13	11	19	48	21	21
Apera spica-venti	4	24	4	46	4	17	0	0	0	26	0	4
Capsella bursa-pastoris	41	29	33	50	44	41	39	17	25	55	40	37
Galium aparine (4)	52	55	34	71	46	54	20	33	19	61	26	28
Lamium purpureum (5)	38	47	31	59	42	44	24	33	13	45	31	27
Lapsana communis	21	22	22	21	19	21	4	6	4	15	12	8
Myosotis arvensis	24	29	18	43	15	27	12	17	8	21	17	14
Poa annua	64	51	49	59	52	58	30	17	10	24	57	29
Stellaria media	83	73	56	81	60	75	32	22	15	45	64	36
Thlaspi arvense	42	39	27	48	36	42	16	22	17	39	26	22
Viola arvensis	45	51	24	66	25	46	16	22	13	30	17	18

Stationary perennials

Life form (including lifespan) and species of weeds	1	2	3	Aw	P	Total	1	2	3	Aw	P	Total
Rumex acetosa (6)	3	6	4	3	21	6	9	6	4	12	21	10
Rumex crispus (6)	8	10	16	12	38	14	14	17	10	18	40	19
Rumex obtusifolius (6)	7	6	13	9	42	13	6	6	4	6	19	8
Taraxacum officinale coll.	16	12	18	13	65	21	14	6	8	9	36	16

continued

Table 12. *Continued.*

Life form (including lifespan) and species of weeds	Northern, Central and Western Europe						Southern Europe					
	As						As					
	1	2	3	Aw	P	Total	1	2	3	Aw	P	Total
Creeping perennials: *stoloniferous* (with above-ground prostrate shoots or stolons)												
Ranunculus repens	11	8	11	10	40	14	10	22	8	21	29	16
Creeping perennials with underground plagiotropic shoots, rhizomes												
Cynodon dactylon (C_4)[c]	4	4	20	2	23	8	26	22	25	15	81	35
Cyperus esculentus (C_4) (7)[c]	7	2	13	2	6	6	16	11	13	6	31	16
Elymus repens	50	52	58	58	60	55	12	11	15	24	43	20
Equisetum arvense	31	24	33	22	47	30	17	11	16	18	33	20
Polygonum amphibium	26	22	33	16	29	24	10	11	8	9	7	9
Sorghum halepense (C_4)	3	2	27	2	13	7	46	11	54	15	79	46
Creeping perennials with plagiotropic thickened roots												
Cirsium arvense	39	53	36	49	52	45	48	67	38	67	74	55
Convolvulus arvensis	23	22	47	20	77	32	45	44	49	45	81	53
Rumex acetocella	11	14	16	12	27	14	7	11	4	18	40	15
Sonchus arvensis	36	31	33	22	42	33	46	28	29	30	40	36

[a]Austria, Belgium + Luxemburg, the former Czechoslovakia, Denmark, Finland, France, Germany, Hungary, Ireland, The Netherlands, Norway, Poland, Sweden, Switzerland and the UK.
[b]Bulgaria, Greece, Italy, Portugal, Romania, Spain, Turkey and the former Yugoslavia.
[c]*Cynodon dactylon* forms above-ground stolons in addition to the rhizomes. *Cyperus esculentus* and *C. rotundus* have swellings (tubers) in their rhizome systems.
Comments: (C_4) means C_4 photosynthesis. (1)–(7): In some cases, answers to the questionnaire state that a given species name may include related species to some extent or that other conditions should be noted, according to the following: (1) in Southern Europe, information on *Avena fatua* includes *A. sterilis* (the main *Avena* species in Greece), and '*A. ludoviciana*' and/or '*A. macrocarpa*' (= *A. sterilis*, e.g. in Spain). (2) To some extent, the rather similar *G. ciliata* (e.g. in Belgium). (3) To some extent, *S. viridis* and *S. verticillata*. (4) In some areas, *G. spurium* to some extent. (5) In some cases, *L. amplexicaule*. (6) Many answers, particularly from Northern Europe, state that these species mainly, or exclusively, are weeds in grassland or pastures. (7) *C. rotundus* to some extent, mainly in parts of Southern Europe.

structures throughout Europe. Annual species may differ more with respect to their appearance as summer or winter annuals in different parts of Europe, but available information does not provide good grounds for reliable characterization of possible differences.

In the answers to the questionnaire underlying Tables 12 and 13, control difficulties or other problems with individual weeds may sometimes have overshadowed strict judgements of frequencies or abundances. As an example, *Avena fatua* has a much higher value in As2 (spring cereals) than in As1 (potatoes, sugar beet and vegetables) in Northern Europe, probably because it is considered less problematic in As1 than in As2. The values of some other weeds lead to analogous conclusions.

Species confusion may also result in wrong interpretations in certain cases. For example, the comparatively high value of *A. fatua* in Aw (winter crops) in Southern Europe is probably partly a result of *A. fatua* being mixed up with other *Avena* species (see comments below Table 12). At the same time, however, *A. fatua*, which appears as a true summer annual in Scandinavia, seems to be a facultative winter annual in many other areas (e.g. Thurston, 1982). To what extent this different behaviour is due to genetic diversity within the species or

Table 13. Weeds with different life forms (including lifespan) in various kinds of crops. Average values of their 'frequency and abundance' are calculations based on data of the individual weed species presented in Table 12. The distinction between summer annuals and winter annuals (facultative) is based on characteristics of the species in Northern Europe. Crop groups (As, As1, As2, As3, Aw and P) are defined on p. 40.

| | Northern, Central and Western Europe | | | | | | Southern Europe | | | | | |
| | As | | | | | | As | | | | | |
Life form of weeds	1	2	3	Aw	P	Total	1	2	3	Aw	P	Total
Sums of reported score values of all weeds of a given life form in per cent of the sum of their corresponding maximum scores (each = 3) in individual crops and as a total. *For horizontal comparisons only.*												
Summer annuals	36	24	44	15	30	29	36	18	30	22	30	30
Winter annuals (facultative)	39	40	28	54	33	41	19	18	13	37	28	22
Stationary perennials	7	8	13	9	41	13	11	8	7	11	29	13
Creeping perennials												
Stoloniferous	11	8	11	10	40	14	10	22	8	21	29	16
Rhizomatous	24	21	36	20	36	26	25	16	26	18	55	29
With plagiotropic roots	23	30	33	26	49	29	36	38	30	40	59	40

mainly to climatic differences is an example of questions that ought to be thoroughly studied.

In some cases, the geographical distribution of crops and weeds certainly give rise to biased figures in Table 12. In the northern areas, for example, *Abutilon theophrasti*, *Amaranthus retroflexus*, *Ambrosia artemisiifolia*, *Cynodon dactylon*, *Cyperus esculentus* and *Sorghum halepense* are represented by larger numbers in the As3 crops (maize, grain sorghum, sunflower and soybeans) than in the As1 crops. These species become weeds mainly in the same areas where the As3 crops are grown in the first place, i.e. in areas with more 'southern' climatic conditions, which are favourable to the weeds as well as to the crops under discussion. In Southern Europe, where the weeds mentioned are more generally abundant, there are no, or negligible, differences between the numbers representing these weeds in the As1 and As3 crops. Largely similar conclusions can be drawn comparing conditions in As2 and As3 crops. Factors such as stronger competition in early stages of the As2 crops may be reasons for considerably diverging conditions for *A. retroflexus* in the As2 crops (spring cereals).

Table 13 indicates that weeds classified as (facultative) winter annuals under North-European conditions occur, on the whole, proportionally more in spring-sown crops than do the summer annuals in autumn-sown crops. This is in accordance with conditions in the Nordic countries except Finland (Tables 6, 7, 9 and 10; see also associated discussions). In Southern Europe, plants classified as *summer annuals* seem to have a stronger position *in relation to winter annuals*, on the whole.

To be reliably explained, this requires thorough investigation. Differences in weed incidence could thus be due to genetic differences within the species between Southern and Northern Europe. However, this might be the result of climatic conditions solely. The following reasoning may provoke opposition or stimulate research.

Assume that the seeds of plants classified here as summer and winter annuals enter and leave dormancy states with an almost similar seasonality in Southern and Northern Europe. Thus, in the mild, humid winters of Mediterranean climates, winter temperatures are low enough to break dormancy but, at the same time, periodically high enough to allow germination of many species. Germination and emergence in winter and very early spring might enable many 'summer annuals' to establish and set seed more successfully in the winter crops in Southern than in Northern Europe. Although seedbed preparation for spring crops may destroy many of these early seedlings, it will still also trigger continued germination and plant establishment in the spring crops. In an average situation, this might favour the seed banks of 'summer annuals' more than those of the 'winter annuals'.

Global Perspectives

Life forms of weeds

Analyses presented here are based on information from *The World's Worst Weeds*, a book published by Holm *et al.* (1977). On the basis of answers to a questionnaire sent to weed scientists throughout the world, Holm *et al.* selected and described 76 taxa (= species, with few exceptions) representing terrestrial herbaceous plants judged to be particularly important weeds on arable land in a global perspective. Founded on these answers and literature reviews, they described the occurrence of the plants as weeds in different crops and geographical areas and presented biological characteristics of the weeds.

The information on the 76 weeds is now briefly structured to illustrate relations between plant traits and occurrence in different types of crops. Table 14 presents individual taxa. The names of the plants listed are given as in Holm *et al.* Knowledge underlying the construction of Tables 14 and 15 and the comments on the plants is based on the descriptions of the plants and their occurrence in Holm *et al.* (1977) and complementary information from many floras and various textbooks.

Table 15 describes weeds in different life-form groups seen as entireties.

Aiming at a global applicability, summer and winter annuals cannot be distinguished. The weed plants are thus grouped into annuals and perennials (no biennials occur among the 'most important weeds'). The perennials are subdivided according to previous descriptions.

The major part of the agricultural areas in the world are situated in the tropics and other warmer regions. The great majority of the 76 taxa dealt with as 'the world's worst weeds' are therefore either important mainly in such areas or are of worldwide importance. Only a few of them are restricted to areas with temperate climate (Table 15).

Crops are grouped into annual and perennial crops. In a global perspective, crops grown on areas described as arable land are, not surprisingly, strongly dominated by annual crops. Plants of the 76 taxa representing 'the world's worst weeds' therefore always occur as important weeds in annual crops; none of them solely in perennial crops (Table 15).

The main difference between annual and perennial crops, as they are distinguished here, refers to the occurrence, frequency or intensity of soil cultivation. Thus, cultivation is normally linked to annual crops, whereas perennial crops represent areas where the vegetation cover is either less completely, not annually, or never broken by soil operations.

In Table 15b, the different taxa are distributed with respect to their occurrence as weeds in different crops, i.e. to their ability to grow and reproduce in crops or land characterized by different use of soil cultivation. Their occurrence is characterized as follows:

- 'Annual crops': means occurrence mainly in annual crops.
- 'Annual–perennial crops': means occurrence particularly in annual crops, but frequently also in perennial crops or plantations, or in other areas with less disturbed vegetation.
- 'Perennial–annual crops': means occurrence predominantly in mechanically undisturbed areas, but important weeds also in annual crops.

('Perennial crops': none of the 76 taxa occurs as a weed solely in such crops and are consequently not represented in Table 14.)

Table 14. The 76 terrestrial plant taxa (of herbs, grasses and sedges) representing 'the world's worst weeds' on arable land according to Holm *et al.* (1977). Characteristics and occurrence are described by symbols and abbreviations explained below the table. Species and families are named as in Holm *et al.*

Life form, name, family and special characteristics[a] of weed species	Photosynthesis[b]	Climate[c]	Crop or field character[d]
Annuals (or short-lived perennials largely 'appearing as annual plants' because of abundant seed production, rapid seedling growth and flowering in their first year of development)			
Ageratum conyzoides, Asteraceae		1	1–2
Ageratum houstonianum, Asteraceae (per, pro)		1	2–1
Amaranthus hybridus, Amaranthaceae	C_4	1	1
Amaranthus spinosus, Amaranthaceae (per, tap)	C_4	1	1
Anagallis arvensis, Primulaceae		2	1
Argemone mexicana, Papaveraceae		1	1
Avena fatua, Poaceae		2	1
Bidens pilosa, Asteraceae		1	1–2
Capsella bursa-pastoris, Brassicaceae (also biennial)		2	1–2
Cenchrus echinatus, Poaceae	C_4	1	1–2
Chenopodium album, Chenopodiaceae		2	1
Commelina benghalensis, Commelinaceae (per, pro)		1	1–2
Commelina diffusa, Commelinaceae (per, pro)		1	1–2
Cyperus difformis, Cyperaceae (per, tuf)		1	1r-2
Cyperus iria, Cyperaceae	C_4	1	1–2
Dactyloctenium aegyptium, Poaceae (per, pro)	C_4	1	1–2
Digitaria adscendens, Poaceae (rooting at nodes)	C_4	1	1–2
Digitaria sanguinalis, Poaceae (rooting at nodes)	C_4	2–1	1–2
Echinochloa crus-galli, Poaceae	C_4	2–1	1
Echinochloa colonum, Poaceae	C_4	1	1
Eclipta prostrata, Asteraceae (per, pro)		1	1–2
Eleusine indica, Poaceae	C_4	1	1–2
Euphorbia hirta, Euphorbiaceae	C_4	1	1–2
Fimbristylis miliacea, Cyperaceae (per, tuf)	C_4	1	1r
Fimbristylis dichotoma, Cyperaceae (per, tuf)		1	1–2
Galinsoga parviflora, Asteraceae		2–1	1
Galium aparine, Rubiaceae		2–1	1
Heliotropium indicum, Boraginaceae (tap)		1	1–2
Ischaemum rugosum, Poaceae		1	1r
Leptochloa chinensis, Poaceae (per, tuf)		1	1r
Leptochloa panicea, Poaceae		1	1r
Lolium temulentum, Poaceae		2	1
Murdannia nudiflora, Commelinaceae (per, pro)		1	1–2
Pennisetum polystachyon, Poaceae (per, tuf)	C_4	1	1–2
Polygonum convolvulus (Bilderdykia convolvulus), Polygonaceae		2(–1)	1(–2)
Portulaca oleracea, Portulacaceae (rooting at nodes)	C_4	1–2	1
Rottboellia exaltata, Poaceae	C_4	1	1(–2)
Setaria verticillata, Poaceae (rooting at nodes)	C_4	1–2	1–2
Setaria viridis, Poaceae	C_4	2(–1)	1
Solanum nigrum, Solanaceae (also biennial)		1–2	1(–2)
Sonchus oleraceus, Asteraceae (also biennial)		1–2	1(–2)
Spergula arvensis, Caryophyllaceae		2(–1)	1(–2)
Sphenoclea zeylanica, Sphenocleaceae		2	1r
Stellaria media, Caryophyllaceae (per, pro)		2(–1)	1–2
Striga lutea, Scrophulariaceae (semi-parasitic)		1	1(–2)
Tribulus terrestris, Zygophyllaceae (per, tap)	C_4	1	1(–2)

continued

Table 14. *Continued.*

Life form, name, family and special characteristics[a] of weed species	Photosynthesis[b]	Climate[c]	Crop or field character[d]
Xanthium spinosum, Asteraceae		1–2	1–2
Xanthium strumarium, Asteraceae		2–1	1–2
Stationary perennials			
Taproot			
Chromolaena odorata, Asteraceae (seed)		1(–2)	2–1
Rumex crispus, Polygonaceae (seed)		2	1–2
Sida acuta, Malvaceae (seed, often shrub)		1	1–2
Tribulus cistoides, Zygophyllaceae (seed)	C_4	1	2–1
Short rootstock			
Plantago lanceolata, Plantaginaceae (seed)		2–1	1–2
Plantago major, Plantaginaceae (seed)		2–1	1–2
Tufted			
Panicum maximum, Poaceae (seed, also stol. and/or rhiz., tall)	C_4	1	2–1
Paspalum dilatatum, Poaceae (seed, rooting at nodes, tall)	C_4	1	2–1
Pennisetum pedicellatum, Poaceae	C_4	1	2–1
Creeping perennials			
Stolons or prostrate stems			
Axonopus compressus, Poaceae (seed)	C_4	1	2(–1)
Brachiaria mutica, Poaceae (stol., wet areas)		1	2(–1)
Oxalis corniculata, Oxalidaceae (seed)		2–1	1–2
Paspalum conjugatum, Poaceae	C_4	1	2(–1)
Rhizomes			
Agropyron repens (*Elymus repens*), *Poaceae*		2	1–2
Cyperus rotundus, Cyperaceae (with tubers)	C_4	1	1(–2)
Cyperus esculentus, Cyperaceae (with tubers)	C_4	1–2	1–2
Digitaria scalarum (*D. abyssinica*), *Poaceae*	C_4	1	1–2
Equisetum arvense, Equisetaceae (with tubers)		2	1–2
Equisetum palustre, Equisetaceae (with tubers)		2	2(–1)
Imperata cylindrica, Poaceae	C_4	1	2–1
Panicum repens, Poaceae	C_4	1	1–2
Pennisetum purpureum, Poaceae	C_4	1	2–1
Sorghum halepense, Poaceae (seed)	C_4	1(–2)	1–2
Rhizomes plus stolons or prostrate stems			
Cynodon dactylon, Poaceae (rarely seeds)	C_4	1(–2)	1–2
Leersia hexandra, Poaceae (seed and/or pro, swamps)		1	1r-2
Pennisetum clandestinum, Poaceae (rarely seeds)	C_4	1	2(–1)
Creeping roots (plus rhizomes)			
Cirsium arvense, Asteraceae		2(1)	1–2
Convolvulus arvensis, Convolvulaceae		2(–1)	1–2

[a]Abbreviations in parentheses: per, sometimes surviving as a short-lived perennial; pro, prostrate shoots, at least to some extent; seed, abundant seed production; tap, taproot; tuf, tufted.
[b]C_4, C_4 photosynthesis according to results of screening.
[c]Climate: 1, tropical – subtropical; 1–2, 1(–2), tropical – temperate; 2–1, 2(–1), temperate – tropical; 2, temperate.
[d]Crop: 1, annual crops; 1r, rice; 1–2, 1(–2), annual – perennial crops; 2–1, 2(–1), perennial – annual crops.

Table 15.   Life forms and occurrence of weeds representing 76 species characterized in Table 14 on the basis of descriptions in Holm *et al.* (1977). Numbers of species classified with regard to their life form, their occurrence in different crops and climatic areas and their plant family affiliation. Species known as C_4 plants are distinguished. All the others are (or are assumed to be) C_3 plants.

Description of distribution	Annual weeds	Stationary perennials	Stolons	Rhizomes	Rhiz. + stol.	Creeping roots (+ rhiz.)	Absolute number[a]	Percentage distribution[a]
(a) All the 76 weed species and the C_4 plants among them								
All weed species	48	9	4	10	3	2	76	100
C_4 plants	18	4	2	7	2	0	33	43
(b) Distribution of the weeds with respect to their occurrence in fields with annual and fields with perennial crops, the last category of fields including non-cropped areas without, or with infrequent, soil cultivation								
All weed species								
Annual crops	18	0	0	0	0	0	18	24
Ann.–per. crops	29	4	1	7	2	2	45	59
Per.–ann. crops	1	5	3	3	1	0	13	17
Perennial crops	0	0	0	0	0	0	0 76	0 100
C_4 plants								
Annual crops	7	0	0	0	0	0	7	21
Ann.–per. crops	11	0	0	5	1	0	17	52
Per.–ann. crops	0	4	2	2	1	0	9	27
Perennial crops	0	0	0	0	0	0	0 33	0 100
(c) Distribution with respect to occurrence in different climatic areas								
All weed species								
Trop.–subtrop.	27	5	3	5	2	0	42	55
Trop.–temp.	6	1	0	2	1	0	10	13
Temp.–trop.	9	2	1	0	0	1	13	17
Temperate	6	1	0	3	0	1	11 76	15 100
C_4 plants								
Trop.–subtrop.	13	4	2	5	1	0	25	76
Trop.–temp.	2	0	0	2	1	0	5	15
Temp.–trop.	3	0	0	0	0	0	3	9
Temperate	0	0	0	0	0	0	0 33	0 100
Grass weeds (*Poaceae*), all species								
Trop.–subtrop.	10	3	3	4	2	0	22	70
Trop.–temp.	1	0	0	1	1	0	3	10
Temp.–trop.	3	0	0	0	0	0	3	10
Temperate	2	0	0	1	0	0	3 31	10 100
Grass weeds (*Poaceae*), C_4 plants								
Trop.–subtrop.	7	3	2	4	1	0	17	74
Trop.–temp.	1	0	0	1	1	0	3	13
Temp.–trop.	3	0	0	0	0	0	3	13
Temperate	0	0	0	0	0	0	0 23	0 100

continued

Table 15. *Continued.*

Description of distribution	Annual weeds	Stationary perennials	Stolons	Rhizomes	Rhiz. + stol.	Creeping roots (+ rhiz.)	Absolute number[a]	Percentage distribution[a]
							All weeds	
			Perennial weeds					
			Creeping perennials					
(d) Distribution with respect to plant family affiliation (figures in parentheses: C_4 plants)								
Poaceae	16	3	3	6	3	0	31 (23)	41 (30)
Asteraceae	8	1	0	0	0	1	10 (0)	13 (0)
Cyperaceae	4	0	0	2	0	0	6 (4)	8 (5)
Others (21 families)	20	5	1	2	0	1	29 (6)	38 (8)
Total (24 families)							76 (33)	100 (43)

[a]Underlined numbers indicate totals for each category.

The relationships between life forms of plants and their occurrence as weeds in different crops in a global perspective according to Table 15b are in good agreement with those found under European conditions according to Tables 3–13.

Thus, plants listed as annuals in Table 14 occur as weeds predominantly in annual crops. Particularly in warm climates, many of them sometimes survive the unfavourable (usually dry) seasons and can flower and set seed in more than one year as short-lived perennials (see remarks in Table 14). However, they can flower and set seed in their first year of growth and will mostly die at the latest in the year following their first year of growth. As understood from the literature, all of them can grow rapidly in early stages, set seed abundantly and build up strong soil seed banks. As they can also easily establish new plants after soil cultivation, they have the characteristics typical of invasive annual plants. In broad outline, it therefore seems reasonable to characterize them as annuals.

When these plants perennate, many of them do so with the characteristics of stationary perennials: e.g. *Amaranthus hybridus*, *A. spinosus* and *Heliotropium indicum* developing branched taproots and *Eleusine indica* and *Cyperus difformis* forming tufts. However, there are several species with prostrate shoots, rooting from the nodes of their stems, e.g. *Ageratum houstonianum*, *Commelina* spp., *Setaria*

verticillata and *Stellaria media*. An important question is to what extent perennating within these species is restricted to certain biotypes or solely a result of phenotypic plasticity. (This question is discussed by Håkansson, 1968c, regarding *Andropogon abyssinicus* R. Br., a seed-producing, loosely tufted grass in areas with shifting cultivation in Ethiopia. The species increases rapidly in young grazed areas after the abandonment of a cropped field. It then strongly declines in a few years leaving types with seemingly stronger vegetative structures than those dominating in the young grassland.)

Stationary and *stoloniferous perennials* occur as weeds both in fields with annual crops and in areas with perennial crops or plant stands less disturbed by tillage (Table 15b). In Sweden and Europe, similar life forms are more seldom important weeds in annual crops. Here, it should be stressed again that only species capable of growing vigorously in annual crops will reach the status of important weeds in a global perspective. The plants represented in the present analysis are therefore probably a somewhat biased selection of species, particularly as far as stationary and stoloniferous perennials are concerned. An additional reason why such plants stand out as more important weeds in annual crops in a global perspective than in Europe may be that arable fields, for the globe as a whole, are less effectively cultivated than in Europe.

Out of 15 *perennial plants with rhizomes or creeping roots* (Table 15a), 11 are weeds in 'annual–perennial crops' (Table 15b). These are thus important weeds in annually cultivated fields, where they may even grow and reproduce better than on less disturbed ground. Only four of them are described as weeds in 'perennial–annual crops'.

C₄ plants among arable weeds

Plants can be classified into three major groups regarding their photosynthetic pathways, i.e. plants with C_3 (Calvin cycle), C_4 (C_4-dicarboxylic acid) and CAM (crassulacean acid metabolism) pathways. The C_3 and C_4 plants are of interest here. C_4 plants are, on average, better adapted than C_3 plants to hot and dry climates. The optimum temperatures are about 10°C higher for C_4 plants than for 'comparable' C_3 plants. C_4 plants further exhibit a better water use efficiency, i.e. a higher dry matter production per unit of water used.

A great number of cultivated plants and important weeds have been screened, distinguishing C_3 and C_4 plants. Results of that screening are reported and reviewed by many authors (e.g. Downton, 1975; Raghavendra and Das, 1978; Elmore and Paul, 1983; Collins and Jones, 1985; Pülschen and Koch, 1990; Andrés, 1993; Hoffmann, 1994). Among the species listed in Tables 12 and 14, those known to be C_4 plants are marked 'C_4'.

It may be of interest for the following discussion to present a study of the rhizomatous perennial grass *Elymus repens* (*Agropyron repens*) regarding its response to temperature and light. *E. repens*, a C_3 plant, is a particularly important weed in northern temperate areas with cool to moderately warm summers. It can be asked why this is so, while so many rhizomatous grasses are very important weeds in the tropics and other warmer areas according to Table 14.

It is well known that the temperature optimum for production of underground plant structures is lower than the optimum for aerial shoot growth. In a Swedish pot experiment in growth chambers, plants from rhizome fragments of *E. repens* were grown at a light intensity around its young foliage similar to ordinary levels in a wheat stand (Table 16). At temperatures raised above average mid-summer temperatures in Scandinavia, the biomass production of the *E. repens* plants was lowered. Both an increased temperature and a decreased period of light thus resulted in a lowering of the dry matter production. This lowering affected new rhizomes much more than aerial shoots. The allocation of photosynthates to underground plant structures thus decreased more than the allocation to the above-ground shoots. In Table 16, this is illustrated by the percentage dry weight ratios of new rhizomes to aerial shoots.

Thus, at temperatures of 35/25°C and a light period of 12 h, i.e. at regimes closer to conditions in cereal stands in tropical areas, no rhizome branches were formed within 70 days of planting. In growth chambers with a higher light intensity (Håkansson, 1969d), a very low rhizome production was registered even at 35/25°C.

According to flora and related literature describing the geographical distribution of plants, *E. repens* occurs in the

Table 16. Occurrence of aerial shoots and young rhizomes of *Elymus repens* plants developed during 70 days from eight rhizome fragments, 8 cm long, planted in soil in pots placed in growth cabinets at different temperatures and daily periods of light (6000 lux). (Data from Håkansson, 1969d.)

Temperature (°C)		Daily light period (h)		
Light period	Dark period	12	15	18
		Aerial shoots, dry wt (g)		
20	10	8.2	10.7	10.9
27.5	17.5	7.0	8.4	8.7
35	25	3.0	4.3	3.6
		Young rhizomes, dry wt (%)[a]		
20	10	2.2	9.9	15.7
27.5	17.5	1.6	2.4	2.3
35	25	0.0	0.2	0.3

[a]Per cent of dry weight of aerial shoots.

Mediterranean countries and the southern parts of the USA. There, however, it is rarely an arable weed of importance, whereas it becomes important further north, in America as in Europe. Moreover, according to flora, *E. repens* seems to be completely absent in the hot tropics, but has been found in scattered highlands with cooler climate even at lower latitudes (e.g. Holm *et al.*, 1977).

As rhizomes are the structures making *E. repens* a persistent and troublesome weed in temperate areas, restricted rhizome formation at high temperatures is certainly one, perhaps the only, reason why this plant is absent in hot climates. Although it occurs as an arable weed in most agricultural areas in the northern states of the USA and in Canada, its frequency and abundance are mostly more moderate than in North-western Europe.

High temperatures and drought in the summer may be the main factors restricting the occurrence of *E. repens* in large parts of Canada and the USA, even in areas where annual mean temperatures are similar to those in North-western Europe. In Scandinavia and Finland, for instance, where summers are cooler and more humid, *E. repens* is on the whole a considerably more important weed than in the continental parts of North America. Descriptions by Alex (1982) indicate that it is not as common in the most continental parts of Canada as in the eastern areas, where summers are cooler and less dry and winters less cold.

Now back to the question why so many rhizomatous perennials are widespread, important weeds in the tropics at temperatures that obviously prevent or restrict rhizome growth in *E. repens*. A plausible answer is that they are C_4 plants (Håkansson, 1969d, 1982). In C_4 plants, the maximum net photosynthesis occurs at higher temperatures than in comparable C_3 plants, and logically also the maximum translocation of photosynthates to underground structures. This will be of a special importance to those creeping perennials whose persistence as weeds in regularly tilled fields mainly depends on their ability

to produce *underground* structures with strong regenerative capacity.

In comparison, annual plants and most stationary perennials rely on proportionally smaller underground structures requiring lesser amounts of photosynthates; the annuals exclusively reproduce and spread by seeds, the stationary perennials largely reproduce and spread by seeds. Although C_4 photosynthesis may be favourable to any category of plants in hot climates, many C_3 annuals and, also, stationary C_3 perennials are thus important weeds even in the hot tropics (Table 15c). However, it seems that perennials that persist as arable weeds mainly through vigorous rhizomes – and, at the same time, represent major weeds in the tropics – are plants with C_4 photosynthesis.

Thus, among the 76 plants presented by Holm *et al.* (1977) as 'the world's worst weeds' and dealt with in Tables 14 and 15c, seven rhizomatous perennials (without stolons in addition) are widespread weeds in the tropics, all of them being C_4 plants. Three rhizomatous perennials, including *E. repens*, are C_3 plants. These are important weeds only in temperate areas. As outposts from the tropics, four rhizomatous C_4 plants, i.e. *Cynodon dactylon*, *Cyperus esculentus*, *C. rotundus* and *Sorghum halepense*, sometimes occur as important weeds in Mediterranean and subtropical regions (Table 12). With the exception of *C. rotundus*, they also occur to minor extents in temperate areas with warm summers.

In their climatic border regions, plants may become important weeds under special local conditions with regard to climate and/or in less competitive crops. Local conditions thus allow *E. repens* to occur as an arable weed at scattered sites in typical Mediterranean areas of Europe. *C. esculentus* and *S. halepense* are seen as weeds as far north as in Canada.

Table 15a shows that 33 out of the 76 'worst weeds' in the world are reported to be C_4 plants. The seven C_4 perennials that develop rhizomes but not stolons occur predominantly in the tropics, but to a certain extent also in other areas with high temperatures in the growing seasons (Table 15c).

Five of them are grasses (*Poaceae*), the other two are sedges (*Cyperus* spp.) with their rhizome structures partly formed as tubers.

There are three grass species forming both rhizomes and stolons or prostrate, rooting shoots. They also occur in areas with high temperatures in the growing season. Two of them, *Cynodon dactylon* and *Pennisetum clandestinum*, are C_4 plants. The third one, *Leersia hexandra*, mainly occurring on moist soil, has probably not been tested regarding its photosynthesis. This plant has a rich seed production. Seeds and/or rooting from prostrate shoots might allow such a plant to persist in hot climates, particularly on moist soil, even if it proved to be a C_3 plant.

C_3 plants relying on creeping roots for their persistence will have difficulties in becoming weeds in hot climates similarly to those relying on rhizomes. Two species with creeping roots, i.e. *Cirsium arvense* and *Convulvulus arvensis*, are represented in Tables 14 and 15. Both of them are C_3 plants. They mainly occur in temperate to subtropical or Mediterranean areas.

It was suggested above that annual plants and stationary perennials on the whole do not need to allocate as much of their photosynthates to underground structures as do the rhizomatous perennials. This is probably one (perhaps the main) reason why many annuals and stationary perennials that are important arable weeds in the warm tropics are C_3 plants. Out of the 33 (27 + 6) annual species predominantly distributed in tropical areas according to Table 15c, there are 15 (13 + 2) known to be C_4 plants; 11 of these are grasses (*Poaceae*) and two are sedges (*Cyperus* spp.).

Five of the nine stationary perennials represented in Table 15b, are weeds in 'perennial–annual crops', i.e. they are judged to be relatively competitive in perennial plant communities. Of the six species described as important weeds in tropical areas, four are C_4 plants and two C_3 plants (Table 15c).

Of the 46 species listed as important European weeds by Schroeder *et al.* (1993), seven are C_4 plants. They occur mainly in Southern Europe, but also further north (Table 12). There are five species of *Poaceae*

(the annuals *Digitaria sanguinalis*, *Echinochloa crus-galli* and *Setaria pumila* and the rhizomatous perennials *Cynodon dactylon* and *Sorghum halepense*), one species of *Cyperaceae* (*Cyperus esculentus*) and one of *Amaranthaceae* (*Amaranthus retroflexus*). In answers to the questionnaire of Schroeder *et al.* (1993), the C_4 plant *Cyperus rotundus* (not included in the questionnaire) is stated to be as important as *C. esculentus* in parts of Southern Europe.

C_4 photosynthesis is not an advantageous plant trait in areas with cool and moist growing seasons, such as in North-western Europe. Consequently, C_4 plants are rarely important weeds there. Yet, even in those areas north of the Alps where summers are relatively warm, they occur as weeds (Table 12) (see below).

Both in Europe and in North America, annual C_4 plants (all of them summer annuals) are important weeds further north than the rhizomatous C_4 perennials. The reason may be that the summer annuals are not influenced by winters as much as perennials relying on winter survival of their vegetative structures.

This argument has support in the results of experiments with the C_4 perennials *Cyperus rotundus* and *C. esculentus* conducted by Stoller (1973) in the USA. His results indicate that the perennating structures of *C. esculentus* survive frost much better than those of *C. rotundus*. Stoller suggested this to be the reason why *C. rotundus* is restricted to the south-eastern and south-western regions of the USA, where the soil seldom freezes, while *C. esculentus* can be found almost throughout the USA, i.e. also in areas where the soils normally freeze in the winter.

In Europe, *Cyperus rotundus* is not reported as an arable weed north of the Alps, whereas *C. esculentus*, as well as the rhizomatous perennials *Cynodon dactylon* and *Sorghum halepense*, are weeds in scattered places even there. Table 12 illustrates that the three last-named perennials can be troublesome weeds, mainly in maize and other 'As3' crops exerting a weak competition in their early period of growth. It should be noted again that these crops are largely grown in places with climates resembling 'southern conditions'. Table 12 further suggests that, in Northern Europe, the perennial C_4 plants are less noteworthy weeds in 'As1' crops (potatoes, sugar beet, vegetables) than the four C_4 annuals. The As1 crops are, on average, grown at 'more northerly' sites than the As3 crops. This is further support for the suggestion that C_4 annuals persist better than the rhizomatous C_4 perennials in areas with harder winters. None of the C_4 perennials has been reported as far north as in Sweden. On the other hand, two of the four C_4 annuals presented in Table 12, i.e. *Amaranthus retroflexus* and *Echinochloa crus-galli*, can be seen as weeds at scattered sites in Southern Sweden, in sugar beet, potatoes and vegetables (As1 crops) and in maize for silage (an As3 crop).

As stated previously, annual C_4 plants occur further north than rhizomatous C_4 perennials also in North America. Various C_4 annuals are important arable weeds in most parts of the USA. Alex (1982) mentions many of them as important weeds in annual crops in large parts of Canada. *Setaria viridis* is by far the most widespread of them, being one of the most abundant weeds in the country. According to Alex, the rhizomatous C_4 perennial *Cyperus esculentus* is a widespread but seldom important weed in Canada, where the rhizomatous C_4 grass *Sorghum halepense* also occurs, although rarely.

The two C_4 annuals *Amaranthus retroflexus* and *Echinochloa crus-galli* were mentioned above as weeds at scattered sites in Southern Sweden. In addition, *Setaria viridis* is sometimes seen in vegetable gardens and, rarely, in agricultural fields, but up to now it is a harmless weed. Hoffmann (1994) presents 48 C_4 species as spontaneously appearing in Germany and of these, 23 summer annuals also occur in Sweden. These plants appear sporadically in Southern Sweden, mainly in coastal areas, as ruderals, on cultivated ground, on seashores, etc. In addition to the above-mentioned C_4 species appearing as weeds in Sweden, there are scattered findings of another few annuals, i.e. seven *Amaranthus* species, two *Atriplex*, two *Digitaria* and four *Setaria* species and, further, *Eragrostis minor*, *Euphorbia peplus*

(fairly common), *Panicum miliaceum*, *Portulaca oleracea* and *Salsola kali*.

In their climatic outposts in Sweden, these plants obviously only persist on ground with weak competition from other plants. In arable fields, annual C_4 weeds establish mainly after tillage and in crops whose stands are weakly competitive in early stages (sugar beet, potatoes, vegetables, maize). In these open stands, they are probably also favoured by high nutritional levels. Outside arable land, C_4 annuals establish on frequently disturbed ground (wastes, seashores, etc.) with weak competition from mostly short-lived vegetation. C_4 plants, on the whole, also withstand dry conditions better than C_3 plants, and many of them therefore can also grow at high osmotic pressures in the soil solution ('salt-drought') where other vegetation is sparse. Some of the places where C_4 plants establish are not only rich in mineral nutrients; they are also 'dry' due to high salt concentrations in the soil. A seashore plant such as *Salsola kali* is competitive under such conditions.

Plants of different families as arable weeds

Table 17 names some of the more prominent plant families represented among the arable weeds.

The grass family, *Poaceae*, is always strongly represented among the important arable weeds. From the percentage figures in Table 17, it seems that the proportion of grasses as components in the weed flora increases from temperate to tropical climates. If closer studies confirm this, it is of interest to determine whether this also applies when we go from cold to warmer areas on the southern hemisphere.

According to Table 17, *Asteraceae* exceeds *Poaceae* as regards the number of species mentioned as important weeds in Sweden (which is not to say that *Asteraceae* represents the economically most important weeds). Going from Sweden over Europe to a global perspective (comprising increasingly large areas with warm climates), *Poaceae* exceeds *Asteraceae* more and more, but

Table 17. Plant families represented among important arable weeds. Sweden: figures outside parentheses represent weeds important in annual crops, those within parentheses also include weeds that are important in perennial crops only, i.e. in leys for fodder, according to Table 5.[a] Europe: 46 important species listed by Schroeder *et al.* (1993). N, Northern, Central and Western Europe; S, Southern Europe, according to Table 12. Global overview: 'world's worst weeds' as presented by Holm *et al.* (1977) comprising 76 taxa listed in Table 14.

| Family | Number of species | | | | Percentage distribution among families[b] | | | |
| | Sweden | Europe | | 'World's worst weeds' | Sweden | Europe | | 'World's worst weeds' |
		N	S[c]			N	S[c]	
Poaceae	7 (10)	10	10	31	13 (16)	20	26	41
Asteraceae	12 (14)	8	8	10	20 (17)	15	15	13
Polygonaceae	5 (6)	8	8	2	11 (11)	15	12	3
Brassicaceae	5 (7)	3	2	1	10 (11)	7	5	1
Caryophyllaceae	2 (2)	2	2	2	7 (5)	8	5	3
Chenopodiaceae	1 (1)	2	2	1	4 (3)	6	8	1
Lamiaceae	5 (6)	2	2	0	8 (8)	6	3	0
Cyperaceae	0 (0)	1	2	6	0 (0)	+	1	8
Other families	21 (28)	10	10	23	27 (29)	23	25	30
Total	58 (74)	46	46	76	100 (100)	100	100	100

[a]Taxa as presented in Table 5.
[b] Figures under Sweden are based on sums of gradings (–, 1; I, 2; II, 3; III, 4) of taxa according to Table 5. Figures under Europe are based on sums of the score values under 'Total' in Table 12. Figures under 'World's worst weeds' are based on numbers of taxa (largely species) in Table 14.
[c] *Erysimum cheiranthoides* is not reported among the weeds in S. Europe. On the other hand, *Cyperus rotundus* is added here. This means a total of 46 species both in N. and in S. Europe. (Cf. Table 12.)

Asteraceae holds its position as family number two among the 'world's worst weeds', where *Cyperaceae* is number three and is, like *Poaceae*, represented by many C_4 plants.

The suggestion that the proportion of grasses and sedges among important weeds increases from temperate to tropical areas is supported when we compare different groups of species in Table 15c. Among the 42 species in tropical–subtropical areas, 52% are grasses and 12% sedges. Among those 34 (10 + 13 + 11) species that are important weeds in areas including temperate regions, only 26% are grasses and 3% sedges.

Of the 'world's worst weeds', 31 out of 76 are grasses and six are sedges, i.e. 41 and 8%, respectively (Table 17). Holm *et al.* (1977) characterized 17 species as particularly widespread and important as arable weeds. Ten of them, i.e. 59%, are grasses.

Holm *et al.* (1997) describe 126 taxa in their book *World Weeds*. These taxa represent major weeds ranked 'in importance' next to the 76 'worst weeds' of the world. They comprise a higher proportion of weeds important in temperate climates. The great majority of them are arable weeds, 14% belong to *Poaceae* and 15% to *Asteraceae*.

Of 167 species registered by Åfors (1994) as arable weeds in Northern Zambia, *Poaceae* represents 23%, *Asteraceae* 14%, *Fabaceae* 13% and *Cyperaceae* 7%. For 33 species selected by Åfors as particularly important weeds, the corresponding values are 45, 21, 3 and 6%, respectively.

Information in Holzner and Numata (1982) supports the opinion that *Poaceae* and *Asteraceae* are dominant families among the arable weeds throughout the world. In East and South Africa, *Poaceae* and *Asteraceae* each represent from 15 to 30% of the species, varying between areas (Wells and Stirton, 1982; Popay and Ivens, 1982). In Brazil, Hashimoto (1982) characterized 17 species as aggressive weeds; five belong to *Poaceae*, four to *Asteraceae*, with other families being represented by one or two species. If we look at smaller areas or regions, very different families may dominate the weed flora. Among families not mentioned in Table 17 but represented by very important weeds, particularly in warm climates, are *Amaranthaceae*, *Euphorbiaceae*, *Leguminosae* (*Fabaceae*) and *Solanaceae*. One of the top-ranked genera with respect to arable weeds is *Amaranthus*, represented by many C$_4$ annuals, from tropical regions to warmer temperate areas. Two of them occur among the 76 weeds in Table 14.

Parasitic weeds

More than 3000 plant species are known to be parasites on other plants. A few are parasites on arable crops. The most important parasitic weeds occur in the tropics or subtropics, but some appear in temperate areas (Parker and Riches, 1993; Press *et al.*, 2001).

Parasitic plants derive part or all of the nutrition they need from their host plants. Some of them are arable weeds, appearing as parasites on crop plants and sometimes also on weeds. They reproduce and spread mainly by seeds. They take up substances through haustoria attaching them to the tissues of their host plants. *Hemiparasites* are green plants with photosynthetic ability, taking up only part of the nutrition they need from their host plants. Others are *holoparasites*, lacking chlorophyll and thus depending on their host plants for all of their nutrition.

Here, plants from four genera are briefly discussed: *Striga* (fam. *Scrophulariaceae*), *Orobanche* (fam. *Orobanchaceae*), *Cuscuta* (fam. *Convolvulaceae*) and *Rhinanthus* (fam. *Scrophulariaceae*). There are many taxonomic uncertainties within these genera. The parasitic weeds discussed in the following are therefore, with one exception, only mentioned under genus names. It

should be stressed, then, that plants in the genera discussed may represent quite different species/genotypes in different geographical areas or as parasites on different crops. For details, see Parker and Riches (1993) and Holm *et al.* (1977, 1997).

Striga spp. are hemiparasites attached to the roots of their hosts. They predominantly occur in the tropical parts of Africa and Southern Asia as hemiparasites on maize, sorghum, pearl millet, upland rice, sugar cane, cowpea and many other plants. Holm *et al.* (1977) count the *Striga* species, looked upon as a group, among 'the world's worst weeds'.

Orobanche spp. are holoparasites attached to the roots of their hosts. Plants representing the genus are found on all continents and in different climates. As weeds they are most frequent in Europe, particularly in the Mediterranean countries, in Africa and in parts of Southern Asia. A great number of dicotyledonous plants from several families are among their host crops.

Cuscuta spp. are obligate shoot parasites with yellowish stems twining around the stems of their hosts. They are not strict holoparasites because they contain some chlorophyll, but photosynthesis is negligible in most of the species. The genus is represented in many climatic areas with plants occurring as parasites on a great variety of herbs, bushes and trees. In the Nordic countries, *Cuscuta* spp. occurred earlier as shoot parasites on clover and flax, in the first place, but are nowadays very rare in these areas.

Rhinanthus spp. are annual green herbs. The plants are mostly attached as hemiparasites to the roots of other plants. They can grow, but less vigorously, without being attached to a host plant. They occur in most of Europe, in Western Asia and in North America. Their host range is wide. In Scandinavia, a variety of *R. angustifolius*, previously often named *R. major* var. *apterus* (see *Flora Europaea*, Tutin *et al.*, 1964–1980) was earlier a locally serious weed, mainly in winter rye. The plant is nowadays almost extinct in Scandinavia.

Which Crops in a Crop Sequence Favour the Population of a Specified Weed?

A crop sequence with a great proportion of crops favouring the reproduction of a specified weed species naturally allows that species to develop a stronger population than a sequence with a smaller proportion of favouring crops. However, which crops are favouring in a crop sequence?

We may first compare situations in crops where active control of growing weeds is excluded, but with other cropping measures being accomplished as usual. This corresponds to situations described in Table 5. Let us now look at a crop where the emergence, growth and reproduction of the weed under consideration are particularly favoured according to Table 5, i.e. in the absence of direct control. If the growth of the weed is effectively suppressed by direct control, this crop could be one in which the population of the weed is even more effectively suppressed than in other crops in the crop sequence. This is emphasized here and will be illustrated below, because it is often overlooked in general discussions on the long-term effects of different crops in crop sequences.

Consider the following example: in the absence of effective direct control of weeds, the population of a summer-annual weed is more favoured where spring crops dominate over winter crops in a crop sequence than where winter crops dominate. If, on the other hand, the growing plants of the weed are effectively controlled by direct measures in each crop, the opposite can be expected.

This, in fact, has been experienced in Sweden in the case of *Sinapis arvensis*, a species very effectively controlled by the herbicides regularly used from the mid-1950s in both summer and winter cereals. Before that time, the summer annual *Sinapis arvensis* was abundant mainly in spring-sown crops. As a cereal weed, it was particularly abundant in spring cereals, but could also be seen in winter cereals. After regular chemical control started in the 1950s, *S. arvensis* declined rapidly. In the beginning, *S. arvensis* was more abundant in areas where spring cereals were frequently grown in the crop sequences than where they were less frequently grown. However, during a few years of regular chemical control, it declined *relatively* more rapidly in the former case, i.e. in areas where spring cereals were grown most frequently.

Analogously, the winter-annual ('winter-cereal') weed *Centaurea cyanus*, which is also very sensitive to the herbicides used, declined most rapidly where winter cereals were frequently grown.

The reason for the situations described above is obvious. In a cropping system where the soil is tilled conventionally, more seeds of *S. arvensis* germinate in spring-sown crops than in autumn-sown crops, and more seeds of *C. cyanus* germinate in autumn-sown crops than in spring-sown crops. If reproduction is effectively prevented in the respective 'favouring' crops, the soil seed bank is more strongly reduced in these crops than in the others. Parallel situations concerning other weeds and other crops are easily found.

6

Germination, Emergence and Establishment of Crop and Weed Plants

Introduction

Germination, seedling emergence and plant establishment is the initial sequence of events in the development of a crop established from seeds. Knowledge of the fundamental processes in this sequence and how they are influenced by various factors is necessary for the guidance of rational measures in seedbed preparation, seed treatment and sowing. Both crops and weeds must be considered even when the management of weeds is in focus. Weeds and crop–weed interactions are treated in more detail in chapters dealing with plant competition and soil tillage. In principle, most of the present discussions relating to growth starting from seeds also apply to, or include, situations where plant growth starts from vegetative organs.

Seed Dormancy and Germination and Soil Seed Banks

Several phases in the germination process can be distinguished. For the sake of simplicity, seed *germination* will be characterized here as the expansion of the embryo that makes roots and shoots break the seed coverings (i.e. testa and other coverings that may occur) and become visible outside them. (Any dispersal unit containing a seed is called *seed* here.) Germination can be regarded as the first phase in the development and growth of a plant after the initial formation of the embryo in the seed on the mother plant.

For seed germination, as for plant development and growth in later stages, various basic requirements must be satisfied:

- The *water* content in the living seed tissues must be sufficiently high to allow biochemical processes and translocation of dissolved substances. Dry seeds therefore must imbibe water and swell before germination.
- Oxygen must be supplied at rates sufficient for the aerobic processes in the seed and, at the same time, the carbon dioxide produced must not accumulate to unfavourable levels. This means that a *gas exchange* between the seed tissues and their surrounding medium must be possible.
- The *temperature* conditions must satisfy specific requirements.
- Seeds of various plant species often have additional, specific requirements that must be satisfied to enable germination.

Seeds of most wild plants and arable weeds can survive for many years in various states of rest or dormancy. Germination readiness then requires distinctive temperatures and other environmental conditions. In many species, not least among weeds, light influences are crucial. Substances inhibiting germination occur in many seeds or seed coverings and must be decomposed or be leached out by the soil water before germination can take place.

Drought and unfavourable temperatures are normal obstacles to germination. The living tissues of most seeds (i.e. those called orthodox seeds) are characterized by their ability to survive extreme dehydration. This dehydration restricts biochemical activities, allowing the seeds to retain energy reserves for long periods of time until germination, followed by successful plant growth, is possible. Even after water imbibition and swelling, many types of seed can remain dormant in the soil for long times with low life activities. Then, biochemical alterations requiring certain medium conditions have to take place in these seeds before germination is possible. As mentioned above, specific influences of light are often crucial. Diurnally fluctuating temperatures often strongly favour germination. Conditions allowing sufficient supply of oxygen and removal of carbon dioxide are necessary.

Whether a seed germinates or remains dormant in a given period of time depends on its response both to the past and to the present environmental conditions. It is therefore not possible to make a strict distinction between dormancy caused by inherent (constitutional) factors and dormancy or rest enforced by the environment. However, it may be practically reasonable to characterize dormancy on the basis of the factor or factors understood as the main obstacles to germination. This has been done in various ways by different authors (e.g. Crocker, 1916; Bibbey, 1948; Harper, 1977; Nikolaeva, 1977; Mayer and Poljakoff-Mayber, 1978; Baskin and Baskin, 1998).

The following types of dormancy have been frequently distinguished. They are named with respect to their main causes, although varying terms have been used in the literature, as is seen in the following.

1. *Inherent, innate, constitutional* or *organic* dormancy. The different terms presented here may be considered synonymous, characterizing dormancy caused by biochemical or anatomical conditions in the seed or seed coverings. Depending on the main reason for the dormant state, the following terms are often used. *Physiological* (*endogenous*) dormancy is caused by biochemical or other mechanisms in the embryo or in other structures inside testa, mechanisms that obstruct radicle emergence and further embryo growth. *Chemical* dormancy is often understood as resulting from germination inhibitors occurring in seed coverings outside the testa, inhibitors that have to be decomposed or leached out before germination can take place. However, it is sometimes difficult to distinuish physiological and chemical dormancy and both types may coccur simultaneously in the same seed. *Physical* (*exogenous*) dormancy results from obstacles to the passage of water and/or gases through the dead seed coverings. *Morphological* dormancy (or 'long delay of germination') is caused by the embryo being very small and its tissues only slightly differentiated at seed maturity on the mother plant. The embryo therefore has to develop further until its expansion is visible as germination. *Mechanical* dormancy means that seed coverings, mostly woody fruit walls, prevent embryo penetration. Different types of dormancy frequently concur together (e.g. *morphophysiological* dormancy and *physical* plus *physiological* dormancy).

- *Primary* dormancy. The seed is dormant right from its separation from the mother plant. Specific environmental influences are needed to break this dormancy.

- *Secondary* or *induced* dormancy. This dormancy is induced by environmental factors in a previously non-dormant seed. It is always of the physiological dormancy type and lasts even under conditions that have become optimal for germination until specific environmental factors have brought about dormancy-breaking changes in the seed.

2. *Enforced* or *environmental* dormancy. This kind of dormancy, or state of rest, is the result of environmental conditions obstructing germination. The seed germinates as soon as not obstructed by the environment.

This makes the distinction between enforced and inherent dormancy.

When the testa or other coverings of the seed are impermeable to water and/or gases, the microenvironment of the living seed tissues is less easily influenced by the environment surrounding the seed. Some authors therefore do not regard dormancy caused by this impermeability as inherent (Category 1) dormancy but as a kind of environmental or enforced dormancy (Category 2). However, since it is caused by the constitution of the seed, this dormancy is considered by most authors to be dormancy of Category 1, termed physical dormancy.

Physical dormancy among weed seeds in the soil is broken when the seed coverings are made more permeable by microbes or other organisms, by frost, by scratching at tillage, etc. Inhibiting substances in the coverings are gradually washed out by water or decomposed. Physiological and morphological dormancy may last for a long time in combination with restricted water supply and gas exchange through the seed coverings. These kinds of dormancy are immediately, or more rapidly, broken when the seed coverings have become more permeable through some degree of destruction.

Long survival of dormant seeds has been described as a wild-plant trait enabling a plant species to build up a bank of seeds of different ages in the soil. As understood from the discussion on p. 7, the persistence of most annual weeds depends on their ability to build up a soil seed bank. Although perennial weeds can rely more or less on vegetative survival (by 'bud banks'), sexual reproduction is always of great importance for their long-term survival through genotypic plasticity. Most of them, in fact, produce seeds that can build up seed banks.

In our cultivated plants, dormancy is a positive seed trait as long as the seeds develop on the mother plant and often also in a short period after harvest. However, as we want crop seeds to germinate rapidly after sowing, long-lasting innate dormancy is unfavourable in these seeds. In plants cultivated for many centuries or millennia, selection has long since resulted in plants producing seeds without strong germination barriers. Sowing of the same plant year after year has resulted in harvests with an increasing proportion of such seeds. Plants established early following rapid germination have a competitive advantage over later plants. To a gradually increasing extent, early plants from non-dormant seeds produce seeds without long dormancy. Selection will finally eliminate plants producing seeds with long dormancy. Absence of innate seed dormancy over years is considered a cultural trait or a result of domestication. However, a short period of dormancy or germination tardiness often occurs in seed lots of, for example, cereals. This tardiness is normally overcome during storing, more or less rapidly depending on the storage conditions.

Seeds of plants taken into cultivation relatively recently are often dormant for long periods if not subjected to dormancy-breaking treatments. This applies to clover, whose seeds to some extent have coats impermeable to water ('hard seeds', physical dormancy). These coats can be made permeable by scarification, mechanically or chemically. Clover seeds are frequently scarified, reducing the number of 'hard seeds' from, for example, 50 to 10%. Substances inhibiting germination, such as water-soluble nitrogen compounds in sugar beet seed, often occur in seed coverings. They are removed by soaking or decomposed by other measures. Many ornamental plants have seeds requiring specific treatments to eliminate dormancy. Seeds of many vegetables, even among species cultivated from ancient times, have germination barriers that must be eliminated by specific treatments to make the seeds good for sowing.

The type and degree of seed dormancy vary greatly both between and within weed species and may also vary considerably between seeds produced on the same individual plant, depending on position on the plant, time of maturation and environmental conditions during maturation. Mature seeds may become ready to germinate after varying lengths of time, depending on their constitution or environment. This secures a spread of the germination over more than 1 year, which makes the weed population persistent.

The seeds of many weeds have become dormant during the process of development on the mother plant, thus separating from the plant in the state of primary dormancy. Seeds of other weeds can germinate immediately or shortly after their separation from the mother plant if temperature and moisture conditions suit germination. However, environmental factors often obstruct immediate germination and secondary dormancy may then be induced. A seed in innate, primary or secondary dormancy, cannot germinate until specific environmental influences have broken its dormancy. If the environment does not then allow a follow-up through germination, innate dormancy is often reinduced, again lasting until broken through environmental influences.

Some of the seeds in a soil seed bank remain uninterruptedly dormant for many years. Others become ready to germinate after shorter periods and then germinate if the environment is suitable or, if not, again become dormant or die, etc. A great many seeds, usually the majority, do not produce established plants owing to death through predation, decay by fungi, bacteria, etc. and, in different stages of germination or seedling growth, often also because of drought.

The seeds in the soil seed bank of most weeds in temperate areas exhibit a more or less distinct seasonal variation in their readiness to germinate. This causes the seasonality in the germination and seedling emergence of the more typical summer and winter annual weeds; Fig. 1 shows average patterns of this seasonality in the Nordic countries. In areas with strong differences between summer and winter temperatures, degrees of innate dormancy or of readiness to germinate are strongly governed by the seasonal temperature variation. Temperatures facilitating the break of dormancy often differ from those promoting germination. The low winter temperatures, which break or weaken the dormancy in many seeds in temperate areas, largely do not allow (rapid) germination. This starts

when temperatures have increased in the spring.

Dormancy and non-dormancy are not absolute, contrasting states. Dormancy is instead differently strong or deep, depending on the environment (Vleeshouwers et al., 1995; Baskin and Baskin, 1998). Measuring the degree of dormancy and/or the readiness to germinate in specified environments may be done by the determination of the germination rate, i.e. the accumulated germination percentage over time. According to a recent hypothesis (Vleeshouwers and Bouwmeester, 2001), dormancy is related to the amount of a phytochrome receptor, which fluctuates in an annual pattern. A higher or lower degree of dormancy corresponds to a wider or narrower range of temperatures, respectively, in which germination can occur.

An extensive literature deals with the complex reactions underlying dormancy, germination and life length of different seeds measured as responses to the environment. A few examples may be given. Nikolaeva (1977), Thompson et al. (1997) and Baskin and Baskin (1998) have carried out extensive experimental work and also reviewed important literature on soil seed banks, dormancy and germination. Among authors presenting experiments with special relevance to arable field conditions in humid, temperate climates, another few may be mentioned separately. They are: from Germany, Koch (1969); from the UK, Roberts and Feast (1970), Roberts and Neilson (1980), Roberts and Potter (1980) and Wilson and Lawson (1992); from the Netherlands, Karssen and co-workers (e.g. Karssen, 1982); from Sweden, Håkansson (1983a, 1992), Milberg and Andersson (1997) and Andersson et al. (2002); from France; Barralis et al. (1988); and from the USA, Burnside et al. (1996). Important information on weed seedbanks from many scientists has been collected by Champion et al. (1998).

Aspects of the influence of seedbed preparation on weed seed germination and seedling development in different seasons is discussed in greater detail later in this book (pp. 75–80).

From Seed Swelling to Plant Establishment

Figures 3 and 4 illustrate the sequence of events in early plant development, starting with the swelling of a seed by water imbibition. The swelling is followed by germination, defined here as the expansion of the embryo leading to roots and shoots appearing outside the seed coverings. Figure 3 describes the further expansion of the seedling leading to shoot emergence from the soil, up to the stage when the plant is considered to be established; when total leaf areas (or green shoot areas) are large enough to enable photosynthesis to compensate for the consumption of substances by respiration. Before this stage, which is the *compensation point* of the entire plant, consumption exceeds assimilation. Thus, the dry weight decreases to a minimum at the compensation point, and then increases (Fig. 4); the fresh weight of a plant, on the other hand, increases continuously from the very beginning of embryo growth. After the dry-weight minimum has been passed, the rate of positive net photosynthesis increases over a long period of undisturbed growth. The minimum dry weight is passed earlier in the entire plant than in the underground plant parts. This is analogous to conditions in early plant development from vegetative structures of perennial weeds (Håkansson, 1963a, 1967, 1969d, 1982) summed up on p. 168. It may be a general pattern in early plant growth.

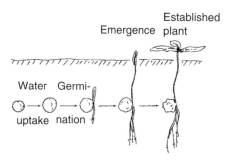

Fig. 3. From a dry seed to an established plant.

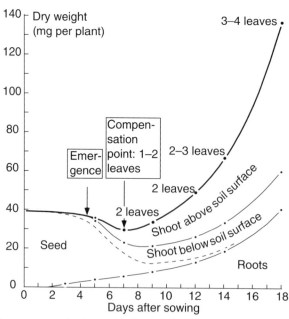

Fig. 4. Dry weight changes at germination and early plant development. Measurements in a growth-chamber experiment with two-row barley at Uppsala (Håkansson, 1981). Barley seeds sown in pots at a depth of 4 cm in fine sand repeatedly watered with a dilute nutrient solution. Light from halogen lamps providing 15–18% of maximum mid-summer light at Uppsala for 16 h day^{-1}. Temperature 20–22°C in the light period, 15°C in the dark.

A plant should not be considered *estab-lished* until it has passed its compensation point after a period following emergence, i.e. when net photosynthesis has become positive (cf. Håkansson, 1979).

If light in combination with other environmental factors does not allow a plant to pass the compensation point, this plant fades away. How long it takes and how large the green leaf (or shoot) areas must be until net photosynthesis becomes positive depend on many factors. Adequate supplies of water and nutrients, light and tempera-ture are the main determining factors.

When light and temperature fluctuate, as is normal in the field, the compensation point may be reached and then be followed by a variation between states of negative and positive net photosynthesis. This should be kept in mind when experimental results illustrated in the following are evaluated, as these results represent stable environmental regimes.

Experiments Exemplifying Influences on Plant Emergence and Establishment

Experimental conditions

In a number of pot experiments, seeds were placed at different depths in homogeneous loam soil with a high content of fine sand and moderately rich in humus and mineral nutrients. (The experiments were conduc-ted for the courses in Crop Production and Weed Science at Uppsala; some of them are described in detail by Håkansson, 1979, others in an internal report available according to Håkansson, 1993, in the reference list.) The soil was compacted to a suitable firmness as uniformly as possible. It was watered daily such that water contents suitable for germination and plant growth were kept as constant and as homo-geneous as possible in the soil profile. Repeated weighing of the pots made it possible to check the moisture contents. To prevent excessive watering, all pots had drainage holes. With the same intention, water application was not repeated until

the soil surface had become dry in an individual pot.

Unless otherwise stated, the experi-ments were carried out either in growth chambers, at temperatures of 17°C in 19-h periods of light and 12°C in the intermediate dark periods, or in greenhouses (in winters) at approximately similar mean temperatures and at day lengths of 19 h regulated by fluorescent tubes. *Greenhouse experiments were conducted in winters, i.e. in periods of years when temperature could be regulated in greenhouses.* Unless otherwise stated, the light intensity was about 10% of maxi-mum daylight at midsummer in Southern Sweden. This provided a daily light supply to the plants near averages in the field in the growing season.

By keeping the water content relatively stable and homogeneous throughout the soil profile, the effect on plant emergence and establishment caused by the depth location of the seed in the soil could be studied with-out water gradient influences. Only when seeds were placed at the very surface of the soil was germination and plant establish-ment disturbed by moisture variation (see the following).

Dry matter losses before plant emergence

Table 18 concerns plants from seeds placed at different depths in the soil. It presents times of emergence and dry weights at emergence expressed as percentages of seed dry weights at sowing.

Seeds of a uniform size were sorted out from high-quality seed lots of peas (*Pisum sativum*) and wheat (*Triticum aestivum*). The pea seeds sorted out had a mean weight of 242 mg and a dry matter content of 93%. In the wheat seed lot, two fractions of kernels of different sizes were selected. The fractions had equal and 'normal' exteriors and mean weights of 28.3 and 48.3 mg, respectively. Both fractions showed the same dry matter content of 92%. The seeds were sown in 20 cm deep 3-l pots at depths presented with the results in Table 18. The pots were placed in growth

Table 18. Emergence of seedlings from seeds at different depths in soil[a] – number of days from sowing to 50% emergence and dry weight of entire plants (including seeds) at emergence as a percentage of the dry weight of seeds at sowing.

| Depth of sowing | Emergence | |
	Days from sowing	Percentage dry weight
Peas		
3 cm	5	92
12 cm	7	82
Wheat		
2 cm		
large seeds	4.7	94
small seeds	4.7	93
8 cm		
large seeds	6.8	84
small seeds	6.8	76

[a]Sandy loam soil, homogeneous, with water contents sufficient for germination and plant growth at all depths. From Håkansson (1979).

chambers under conditions described above.

Plants were collected and weighed when shoots from 50% of the seeds had emerged, that time being regarded as the time of emergence. Practically all seeds had germinated and produced normal shoots.

Table 18 shows that both the time from sowing to emergence and the dry matter consumption before emergence increased with increased depth of sowing. The different seed size in wheat did not affect the time of emergence at the depths compared. (Judging from other experiments, all shoots from both the smaller and the larger wheat kernels were able to emerge both from the 2-cm and the 8-cm depths. At greater depths, where shoots from many smaller kernels displayed greater difficulties in emerging than shoots from larger seeds, the former shoots also emerged later than the latter.)

A greater percentage consumption of dry matter was recorded in the emerging plants from the smaller kernels than in those from the larger ones at the 8-cm depth. This agrees with the logical expectation that *the percentage consumption of dry matter before emergence increases with increasing*

depth and decreasing seed size in a homogeneous soil profile. It should be stressed that this only applies to depths below the very superficial soil layer where various disturbances are inevitable.

In the experiment with wheat, dry weights were also determined as percentages of the fresh weights of the emerging plants. In plants representing the 2-cm depth, the percentages were 23 and 22% for large and small kernels, respectively. Corresponding values for the 8-cm depth were only 15 and 13%, respectively. This indicates that *the dry weights of the plant tissues differ more than the fresh weights and the volumes.* From other observations of plants from seeds at different depths, it is obvious that leaf areas in the early period following emergence differ less than the plant dry weights.

Plastic plant adaptations of the type now being discussed promote photosynthesis relative to respiration and therefore partly compensate for increasing losses of substances caused by an increasing depth of the seed in the soil and, probably, also for a decreasing seed size. Such adaptations are decisive for the competitive ability and continued growth and survival of plant individuals.

Plant establishment

Examples representing cultivated plants

Seeds of six species were placed at different depths in the soil in 10-l pots placed in the greenhouse (Fig. 5). They had been sorted out from high-quality seed lots and represented seeds with very uniform sizes and exteriors. Clover seeds were scarified, enabling the great majority of them to imbibe water and germinate. Plant emergence and establishment were recorded. Observations continued until all emerged plants capable of continued growth were judged to have established.

Figure 5 illustrates the influence of the depth of seed location on the relative number of established plants. Figure 6

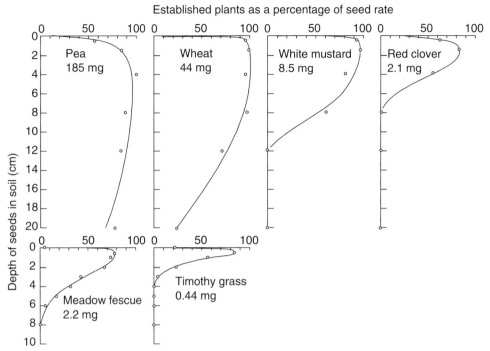

Fig. 5. Plants established from seeds at different depths in soil in a growth-chamber pot experiment at Uppsala (Håkansson, 1981). Soil homogeneously moist (except in the very superficial layer). Mean weights of seeds are given below the species names.

shows that the deeper the seeds had been placed in the soil, the later the plants emerged (cf. Table 18).

Below the superficial soil layers where various disturbances occur, there was a range of depths from which plants established maximally. This range was wider for species with larger seeds than for those with smaller seeds. The same type of relationships are illustrated in other studies, including experiments with vegetative organs at different depths, as exemplified in Fig. 55.

Additional studies show that curves representing seemingly equal material of one and the same species may present very different positions. Such curves can unmask differences in seed quality, soil conditions, etc. Numbers of established plants described as functions of seed size (dry weight) and depth location can be expected to exhibit a strictly stable pattern only for genotypically similar seeds in the same uniform soil.

No seeds of the cultivated plants represented in Fig. 5 were dormant, with

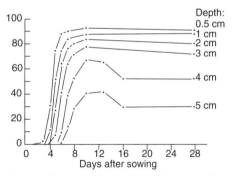

Fig. 6. Changes in the number of living aerial shoots of red clover within 28 days of placing seeds at different depths in moist soil. Curves illustrate means from the experiment, other results of which are described in Table 21. They represent emergence minus death after emergence. In the first period of 10 days following sowing, only a few shoots died. The numbers on the 28th day represent plants judged to be established. The soil profile was homogeneously moist except in the very superficial layer. (From Håkansson, 1979.)

the exception of a few clover seeds that remained 'hard'. The great majority of the seeds thus germinated at all depths and then died. With depths increasing below the optimum depths, the seedlings increasingly failed to emerge and establish.

Seedlings from seeds placed on, or close to, the soil surface largely died of drought in early stages. Having failed to penetrate the soil, the roots often dried out above the soil surface. Seeds placed only slightly below the soil surface were often heaved up by the expanding roots, which then became more or less exposed to the air and killed or damaged by drought. This means that death caused by drought can occur even among young seedlings from seeds placed in fairly moist layers close to the soil surface.

The main reason for the restricted establishment of plants from non-dormant seeds placed at greater depths is, no doubt, that food reserves are used up before emergence or to some extent after emergence, but before the compensation point is reached. This latter case is further discussed on the basis of additional studies presented in a following section (Tables 20 and 21, Fig. 6).

A decreasing plant emergence with increasing depth of the seed placement may also have anatomical reasons. Shoot elongation may be restricted or the shape of the shoot may change with elongation in such ways that it has difficulties in penetrating the soil. For instance, the coleoptile of cereal and other grass seedlings cannot elongate above certain lengths. In cereals, these lengths are 5–8 cm, depending on species, variety and seed size (according to frequent observations in my own experiments). After sowing of cereals at depths exceeding maximum coleoptile lengths, the soil layer above the coleoptile is penetrated by foliage leaves, which are less well suited for soil penetration than coleoptiles. None the less, they are able to emerge from considerable depths if the soil is not too compact. According to Fig. 5, wheat established to a minor extent even from 20 cm. It was seen, however, that some plants that had emerged from greater depths died before being counted as established plants.

The ability of a shoot to penetrate the soil and emerge from critically great depths is strongly influenced by its morphology and by the soil compactness. Observations in our greenhouses suggest that the cotyledons of dicot plants have greater difficulties than grasses in penetrating a compact soil or emerging from critical depths. Further studies on the ability of a seedling to emerge and establish from seeds of various plants ought to involve systematic investigations considering the influence of morphological/anatomical properties and the related energy consumption. The ability of a seedling to penetrate the soil and the time it takes for penetration and possible emergence are strongly influenced by the degree of soil compaction. At slow penetration and late emergence, the seedlings are also more often weakened or killed by fungi or other agents than at rapid emergence. It is of similar interest to study these topics in weeds as it is in crop plants.

An example representing annual weeds

Seeds of three species were placed at different depths in the soil under conditions previously described. Table 19 presents materials and results.

Maximum emergence and establishment were recorded at the 0.5-cm depth.

Table 19. Plant establishment from seeds of three weed species placed at different depths in soil.[a]

Depth (cm)	Established plants as a percentage of the number of seeds		
	Sinapis arvensis 2.3 mg[b]	*Chenopodium album* 0.7 mg[b]	*Matricaria perforata* 0.4 mg[b]
0.5	84	6	29
1.5	38	0	3
4	38	0	0
8	8	0	0
12	0	0	0
20	0	0	0

[a]Sandy loam soil, homogeneous, with water contents sufficient for germination and plant growth at all depths. (From Håkansson, 1979.)
[b]Mean weight of seeds.

Seedlings of *Matricaria perforata* and *Chenopodium album* only established from rather shallow depths. Seedlings of *Sinapis arvensis* with considerably larger seeds established from greater depths. At the 0.5-cm depth, light might have been influencing, although very weakly in a soil of the type used.

Unlike seeds of the cultivated plants dealt with above, those of the weeds under discussion remained dormant to a large extent at the end of the experiment. The number of dormant and dead seeds could not be counted exactly, but approximate counts indicated that *the proportion of seeds remaining dormant became increasingly large when shoot emergence decreased with increasing depth* (from shallow soil layers). In the weed species, a decreased emergence thus resulted both from increased proportions of seeds remaining dormant and from an increased mortality due to failure in shoot emergence. In *Matricaria* and *Chenopodium*, most seeds remained dormant even at the 0.5-cm depth. The shoot emergence from weed seeds at different depths in the soil has been dealt with in numerous publications, recently in papers by Mohler and Galford (1997) and Benvenuti *et al.* (2001) considering shoot emergence, dormancy and mortality of seeds in response to depth location and other factors.

Plant emergence and early mortality among emerged plants

It can be understood from the above that an emerged seedling dies if shortage of food reserves does not allow it to pass the compensation point. Seedlings with only small amounts of reserves remaining after emergence run greater risks of dying early than plants with larger amounts of reserves. Furthermore, any factor that increases photosynthesis relative to respiration in an emerged seedling increases the chances of the seedling establishing and continuing its growth. The risk of starving to death before establishment logically depends on factors such as the depth location of the

seed in the soil, the seed size and, last but not least, the light and temperature conditions after emergence. Experiments to test the veracity of this were carried out at Uppsala and some results are presented below.

From a seed lot of red clover (diploid), seeds of two sizes and with similar exteriors were selected and placed at three depths in the soil in plastic pots which were placed under three light regimes (Table 20). Light intensities lower than 'full light' were achieved by placing layers of loose white weave above the pots. The lower light intensities represented the light conditions encountered by low-growing weeds in ordinary cereal stands. Emergence and mortality after emergence were recorded up to the time when the surviving plants were judged to have established in the meaning defined above. Each combination of seed size, depth of seed location in the soil and light regime

Table 20. Plant emergence and subsequent death preventing establishment. Red clover (cv. Disa) from seeds of different sizes (large, 2.43 mg; small, 1.32 mg) at different sowing depths[a] and light intensity (day length 19 h). Growth chamber experiment. (From Håkansson, 1979.)

Light intensity and sowing depth	Emergence (percentage of seeds sown)		Mortality (percentage of emergence)	
	Large seeds	Small seeds	Large seeds	Small seeds
Full light[b]				
1 cm	78	73	0.8	0.9
3 cm	74	72	0.4	1.8
5 cm	66	57	1.0	4.5
35% of full light				
1 cm	71	77	0.5	0.4
3 cm	76	68	0.4	2.8
5 cm	59	46	0.0	5.4
10% of full light				
1 cm	73	72	2.2	0.5
3 cm	65	63	1.0	7.4
5 cm	68	49	0.5	20.7

[a]In sandy loam soil, homogeneous, with water contents sufficient for germination and plant growth at all depths.
[b]Full light: about 10% of maximum daylight at midsummer in Southern Sweden.

was represented by four 1-l pots with 25 seeds per pot.

Although the seeds were scarified, 20–25% of them remained hard at each of the three depths. Table 20 presents a maximum emergence of only 78%. It shows, on the whole, a decreased emergence with an increased depth, more pronounced for smaller than for larger seeds. The emergence was not, or only slightly, influenced by the light conditions.

The mortality figures in Table 20 show that most of the emerged plants established, even those from the smaller seeds at the greater depths. Only very few plants from the larger seeds died at any combination of depth and light. Among the plants emerging from the smaller seeds, however, the mortality increased with an increased depth and a decreased light intensity.

The results support the suggestion that the risk of plant death after emergence, due to shortage of energy reserves, increases with a decreasing seed size, an increasing depth location of the seed in the soil and a light intensity decreasing to critically low levels.

This suggestion is further supported by results shown in Table 21 and Fig. 6, which present an experiment with red clover (tetraploid). Clover seeds of two sizes were placed at different depths in soil in 2-l pots. Each combination of depth and seed size was represented in six pots with 50 seeds in each. In three of these pots, only clover seeds were sown; in the other three pots, 16 seeds of barley were added, in all cases at a depth of 2.5 cm. In this experiment, only a few clover seeds were hard and almost all barley seeds gave rise to established plants.

Table 21 illustrates strong influences of depth and seed size on the emergence and subsequent plant mortality. The influences were stronger than those shown in Table 20, but were of the same character. Thus, the emergence decreased and the mortality after emergence increased with a decreased seed size and an increased depth below certain levels. Clover seeds that had been 'half-way' pressed down into the surface soil (depth '0 cm') 'emerged' (= germinated) to a great

Table 21. Plant emergence and subsequent death preventing establishment. Red clover (cv. Rea) from seeds of different sizes (large, 3.18 mg; small, 1.72 mg) sown at different depths.[a] Greenhouse pot experiment. (From Håkansson, 1979.)

Sowing depth (cm)	Emergence (percentage of seeds sown)		Mortality (percentage of emergence)	
	Large seeds	Small seeds	Large seeds	Small seeds
Pure stands				
0	≈ 90	≈ 90	≈ 99	≈ 99
0.5	94	87	2	1
1	88	89	0	2
2	86	89	2	8
3	82	79	2	5
4	85	60	4	32
5	61	21	26	48
With barley				
0	≈ 95	≈ 95	≈ 85	≈ 85
0.5	96	91	1	3
1	93	88	0	6
2	86	74	1	12
3	82	66	0	23
4	77	51	34	27
5	65	18	11	61

[a]Sandy loam soil, homogeneous, with water contents sufficient for germination and plant growth at all depths, except in the superficial layer.

extent, but most of the seedlings died rapidly.

The presence of barley plants obviously favoured the establishment of clover plants from seeds placed at the soil surface (lower mortality). The probable reason is that the barley plants raised and loosened the surface soil. They thereby improved the conditions for clover germination and root penetration into the soil. The barley plants otherwise influenced clover emergence and mortality rather moderately, probably because repeated watering prevented water stress and because the barley plants in the pots shaded the clover seedlings only moderately. However, growth and dry matter production of the clover plants were strongly reduced by competition from the barley plants (see further Table 33).

Figure 6 (p. 63) represents the same experiment as Table 21. The curves illustrate the number of living aerial shoots of clover from seeds placed at different depths, averaged over the seed size and the presence and absence of barley. The first parts of the curves, up to their maximum levels, approximately represent the numbers of emerged plants. They therefore clearly illustrate the gradual delay in plant emergence with increasing depth of the seed location in the soil. The downward slope of some curves after their maximum levels thus also satisfactorily describes the times and magnitudes of plant death.

The relationships discussed above concern factors that are not only decisive for the start of plant growth but also strongly influential on the competitive position and further growth of the individuals in a plant stand. This is dealt with from different aspects in subsequent parts of this book.

Overview of Factors Influencing Plant Emergence and Establishment

Seed size and depth location in soil, light intensity and temperature

The following discussion focuses on photosynthesis and respiration, the processes of major influence on the amount of organic matter and energy reserves in the plant. Influences of seed depth location in the soil, seed size, light intensity and temperature are discussed. Graphs in Figs 7–10 are idealized descriptions based on experiments and general physiological knowledge (Håkansson, 1979).

It is assumed that the soil is homogeneous with respect to texture, structure and moisture content (except in its very superficial layer, where moisture inevitably varies) *and that all seeds are non-dormant.*

Figure 7 describes the dry weight changes in the tissues of an entire plant (including the seed) in the early period of development from seed germination (G) on. The changes at four depth locations of the seed in the soil are illustrated by curves 1–4. The curves describe different abilities of the seedlings to emerge and establish (pass the

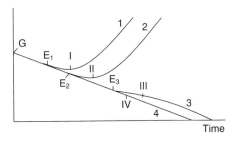

Dry weight relative to that at the start of germination

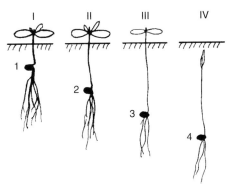

Fig. 7. Plant development following germination in situations I, II, III and IV, as defined below. The diagram describes changes in the total dry weight of plant tissues (including seed tissues) following the start of germination of seeds placed at different depths (1, 2, 3, 4) in soil assumed to contain water sufficient for germination and plant growth at all depths. (From Håkansson, 1979.)

G. Start of germination.
E1, E2, E3. Emergence from depths 1, 2, 3.
I. Depth 1: Plant at compensation point.
II. Depth 2: Plant at compensation point.
III. Depth 3: Food reserves emptied; compensation point not reached; tissues are dying.
IV. Depth 4: Food reserves emptied before emergence; tissues are dying.

compensation point) when the seeds are placed at different depths.

On the basis of general physiological knowledge and minor experiments of my own, some illustrations for the students in our courses at Uppsala have been outlined (Håkansson, 1979, 1993). A few of them are presented below (Figs 8–10).

Dry weight relative to that at the start of germination

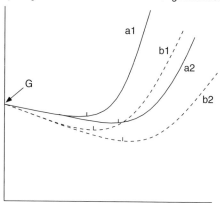

Time

Fig. 8. Changes in the total dry weight of plant tissues (including those of the seed) following the start of germination (G) of seeds of different sizes (a > b) placed at two depths (2 > 1) in soil assumed to contain water sufficient for germination and plant growth at all depths. (From Håkansson, 1979.)

Living emerged shoots relative to seed rate

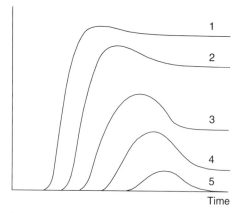

Time

Fig. 9. Relative numbers of living aerial shoots of plants from seeds placed at different depths in soil – following the start of germination. Depths: 1 is considered to be the optimum depth for plant establishment, 2–5 are gradually increasing depths. The soil is assumed to contain water sufficient for germination and plant growth at all depths. After emergence, death is recorded mainly in a period immediately following maximum shoot number; after that, plant numbers stabilize. (From Håkansson, 1979.)

The diagram in Fig. 8 is of the same type as that in Fig. 7. It describes different dry-weight changes in plants during early

Living emerged shoots relative to seed rate

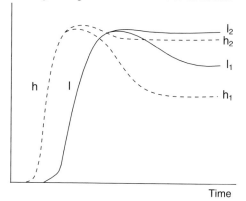

Time

Fig. 10. Relative numbers of living aerial shoots of plants emerged from seeds placed at a depth considered to be the optimum depth for plant establishment – following the start of germination. Two temperatures (l, lower, h, higher) and two light intensities (1, lower, 2, higher). After emergence, death is recorded mainly in a period immediately following maximum shoot number; after that, plant numbers stabilize. (From Håkansson, 1979.)

growth following germination (G) of seeds representing two sizes placed at two depths in the soil. All cases represent situations where the plants are able to establish, i.e. to pass the compensation point. The curves illustrate different degrees of dry-weight changes and different positions of the compensation point on the time axis.

Figure 9 illustrates the number of living above-ground shoots emerged from seeds at different depths. With increased depth (curves 2–5) below the optimum (1), the number of emerged shoots decreases and the shoot emergence is increasingly delayed. With increasing depth of the seeds in the soil, the curves exhibit more and more pronounced maxima. The reason for this is an increasing death of emerged plants with increasing depth. It is assumed that the final death of emerged shoots is largely delayed relative to emergence such that the curves, up to their maxima, approximately describe the course of emergence.

Figure 10 illustrates conceivable plant responses to a lower and a higher temperature (say 12 and 24°C) and to two suboptimally low levels of light intensity.

Most life processes run faster at the higher than at the lower temperature. However, the respiration rate then increases more than the rate of photosynthesis. Therefore, at critically low light intensities, a higher temperature will result in a higher mortality after emergence.

In temperate climates, temperatures in the seedbeds are mostly between 5 and 15°C at the times of spring sowing (N. Rodskjer, personal communication). Rodskjer and Tuvesson (1976) measured temperatures at a depth of 5 cm in bare clay soil on arable land at Uppsala, Sweden. Five-year mean temperatures in periods from sowing to emergence of crops were 9.5°C for spring cereals and 9.2°C for winter wheat.

Even among plants adapted to temperate climates, the most rapid emergence after sowing largely occurs at temperatures exceeding 20°C. (These should not be confused with reactions in the balance of dormancy/non-dormancy among seeds being physiologically dormant for long periods.) Temperatures that are optimal with regard to the establishment of vigorous plants are, however, lower than those leading to the most rapid emergence.

A temperature greatly exceeding this optimum may lead to an increased mortality both before and after emergence. The established plants usually become stronger at moderately low temperatures than at higher temperatures, although their life activities proceed less rapidly.

Lower temperatures favour root growth relative to the growth of aerial shoots. This facilitates the tolerance of the plant to drought in later stages. In most temperate regions, it is favourable to carry out spring sowing as early as possible with regard to weather and crop character, because an early start also means a better use of the growing season. However, very low temperatures for long periods following sowing – even temperatures above zero – may be harmful due to injuries by certain fungi or other organisms active at low temperatures. The relative time of germination and emergence of crops and different weeds after (early) seedbed preparation and sowing followed by long periods of low temperatures ought to be studied systematically.

Water conditions in the soil

A decrease of the soil water content in ranges suboptimal for germination leads to a decreased velocity of germination and seedling growth. Moreover, at contents near the wilting point, mortality increases among seedlings before and after emergence. When moisture contents vary rapidly and strongly around the wilting point, the living tissues are particularly subjected to injuries, directly by drought, but also by parasites and predators in the soil (e.g. Gulliver and Heydecker, 1973; Champion et al., 1998).

Even at soil water tensions of 1–1.5 MPa, i.e. slightly above the wilting point (1.5–1.6 MPa), germination can proceed, although very slowly (Manuhar and Heydecker, 1964; Matsubara and Sugiyama, 1965; Kaufman, 1969; Currie, 1973; Hallgren, 1974b; Aura, 1975; Baskin and Baskin, 1998). The minimum water content allowing germination is, theoretically, the lowest content enabling the seed to take up enough water for germination. At that content, however, water uptake and resulting germination proceed very slowly. The minimum water content allowing acceptably rapid processes in the field is therefore considerably higher. The level of this water content depends on seed properties and factors such as the texture and structure of the soil.

In the field, the soil water content usually varies strongly between depths. This results in complex and varying paths of water movements and supply to the seeds. In a soil that is not so wet that movements of free water occur, water can be translocated both by capillarity and as vapour in an interplay of evaporation and condensation.

Where the water content in the soil surrounding a seed is low, movement of water in the form of vapour from (usually deeper) soil layers with higher water contents is very important (Currie, 1973; Aura, 1975). This may be the main reason why germinating seeds and young seedlings survive when the

soil water content around the seed fluctuates
between levels favouring germination and
further seedling growth and levels too low
for the survival of the growing seedling. The
lowest water content enabling early germination processes is lower than the minimum
content allowing further development and
growth of seedlings before and after emergence (e.g. Matsubara and Sugiyama, 1965;
Aura, 1975).

The anatomy and size of the seed, as
well as the properties and water contents
of the soil, influence the rate of water
translocation into the seed. As larger seeds
must imbibe more water for germination
than comparable smaller ones, they require
movement of water from wider distances
than do smaller seeds. Moreover, their surfaces are smaller relative to their volumes,
which also makes the imbibition more difficult. Therefore, when there is shortage of
water in the surrounding soil and when, at
the same time, long-distance water movements cannot proceed rapidly, larger seeds
have greater difficulties than smaller ones in
taking up the amounts of water needed for
germination (Fig. 11).

In experiments with peas on clay soils
at Uppsala, A. Bengtsson (personal communication) recorded different times of
emergence of larger and smaller seeds,
which he interpreted as a result of their
different ability to take up water. The
pea cultivar Flavanda with large seeds
(270–370 mg) always emerged a few days
later and at considerably lower percentages
than did the cultivar Torsdags III, the
seeds of which are smaller (180–200 mg).
The average emergence of Flavanda and
Torsdags III relative to the number of sown
seeds was 72 and 95%, respectively.

Thus, according to Fig. 11, small seeds
run lower risks of suffering from drought
under hazardous moisture conditions than
do larger seeds. Large seeds do not only tolerate deeper sowing than small seeds; they
also often require deeper sowing to avoid
disturbances from drought. Seeds with
small amounts of food reserves are subjected
to special difficulties when their living
tissues are surrounded by thick water-
absorbing dead coverings. Because of the

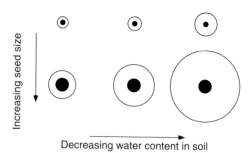

Increasing seed size

Decreasing water content in soil

Fig. 11. To be able to germinate, larger seeds
have to imbibe more water than smaller seeds with
otherwise similar properties. Larger seeds therefore
require transfer of water from larger soil volumes
than smaller seeds. These soil volumes also increase
with decreasing water contents in the soil. This is
illustrated in a simplified form by circles of different
sizes surrounding smaller and larger seeds at higher
and lower contents of water in the soil. (From
Håkansson, 1981.) At low water contents, large
seeds have greater difficulties germinating than
small seeds, or they germinate more slowly. (The
possibilities and pathways of water movements in
the form of liquid and/or as vapour in the soil
strongly vary with soil texture, structure and
compaction.)

water absorption by these layers, the seeds
have greater difficulties imbibing enough
water for germination than comparable
seeds with thin coverings. Untreated beet
'seed', for instance, presents this problem,
which can be reduced by grinding off parts
of the dead coverings.

The optimum water content in the soil
with regard to germination varies both with
seed and with soil properties. In extreme
situations, this optimum is close to the
maximum water-holding capacity of the
soil. However, the optimum is usually much
lower (Clark, 1954; Gulliver and Heydecker,
1973). Both soil texture and structure
influence the methods and rates of water
movement, as liquid and/or as gas. Because
the content of water influences the diffusion
rate of oxygen and carbon dioxide in the
soil, it interacts with any factor affecting gas
movement in the soil. Its optimum level
thus varies with variation in these factors.
Decreasing water contents make osmotic
and colloidal conditions increasingly
influential.

In coarse soils, the amount of plant-available water per unit of soil volume is lower than in soils with a finer texture. In clay soil layers with a coarse and cloddy structure, the supply of water to the seed may be hazardous, even when there are high contents of bound water.

Germination in the field is influenced by complex and fluctuating interactive processes in the environment (e.g. Hillel, 1972; Gummerson, 1986). The soil structure as well as the water content may vary strongly and differently from layer to layer in the soil profile. This is a problem that has been studied empirically as a basis for rational seedbed preparation (e.g. Kritz, 1983; I. Håkansson and von Polgár, 1984).

Pot experiments were carried out in growth chambers at 20°C (at Uppsala, M. Tuvesson, unpublished) to study the plant emergence from seeds of red clover and timothy grass at different water contents in soil. This was fine sand moderately rich in humus. The maximum rate of shoot emergence from seeds placed 1.5 cm below the soil surface was recorded at the next highest water content tested (water tension 0.025 MPa). In both species, a somewhat lower number of plants emerged at the highest content (water tension 0.0015 MPa, which was near the maximum water-holding capacity of the soil). When the seeds were placed only a few millimetres below the soil surface, maximum emergence occurred at the highest water content, probably because aeration is less restricted there. At a water tension of 1.0 MPa, representing the lowest water content tested, the final numbers of emerged shoots were about 50% of the numbers emerging at higher water contents. The average time from sowing to emergence was doubled, or more, relative to the time at 0.025 MPa. It increased by 4 days for timothy grass and by 9 days for clover at the experiment temperature, 20°C. Experiments with seeds of 18 cultivated plants and 12 weed species illustrate effects following the same principles (Håkansson, 1979). With weed seeds, effects of/on dormancy were included in the measured results.

Conditions dealt with above are discussed below regarding influences on the relative time of emergence and the importance of this time in plant competition.

Soil Moisture Content – Seed Germination – Plant Growth

Under field conditions, water contents normally vary with the depth in the soil and also horizontally from spot to spot. Moreover, the variations differ from field to field and from time to time depending on soil and weather. In addition to varying temperatures, this causes a considerable variation in the rates of germination and seedling growth of both weeds and cultivated plants. Where the water content decreases to low levels during germination, or in the period immediately following germination, the germinating seeds or seedlings may die or be injured.

There are often great variations in the microenvironments surrounding neighbouring seeds and small seedlings. These may, for instance, have different access to water, leading to variation in the rate of their early growth. This, in turn, may cause differences in the future competitive status and growth of neighbouring individuals. In crops with rapidly closing stands, such as cereals, competitive effects appear from early periods of plant growth.

Effects of the relative time of emergence of neighbouring plants in a cereal stand

A number of experiments illustrate that differences of only a few days in the time of emergence of neighbouring plants cause great differences in their growth and production (Håkansson and Larsson, 1966; Håkansson, 1972, 1979, 1986; Dahlsson, 1973). As stressed above, the properties of the microsites may vary considerably, leading to a variation in the rate of germination and time of emergence of the plants. This is almost inevitable even in a well prepared seedbed, considered to be uniform. Without the effect of competition, this early variation would only lead to a slight variation in the final size and production of the individual plants. However, in ordinary cereal stands, competition maintains, or increases, the variation in their size in early growth. This does not necessarily reduce the quantitative yield per unit area, but does sometimes lead to an uneven quality of the harvested product (see Chapter 7).

In a small-plot field experiment with spring wheat and two-row barley, the production of neighbouring plants emerging at different times was studied (Håkansson and Larsson, 1966). Differences in time of emergence were arranged by a special technique of putting seeds into the soil at different times. As can be understood from Table 22, plant emergence at different times was set to occur either in separate stands or alternating in the same rows in mixed stands.

Table 22 indicates that, when the two categories grew in mixed stands, the seed production of plants delayed by 8 or 15 days was much lower than that of earlier plants; drastically more so at a delay of 15 days than

at an 8-day delay. The seed production reflected the size of the plants. When plants emerging at different times were grown in separate stands, their production was of the same magnitude. The total seed production per unit area was of the same magnitude in the different stands. However, where early and late plants grew in mixed stands, their seed size differed more, and the quality of the yield was therefore in certain respects lower than where early and late plants grew in separate stands.

The relative time of emergence of crop and weed plants is of the greatest importance for the outcome of their competition. This is illustrated in the discussions in Chapter 7 (e.g. Table 28 and Fig. 30). As shown there, a delay of barley by only 3 days in relation to the weeds may result in a doubling of the weed biomass and a somewhat reduced barley production.

The above illustrates the importance of carrying out seedbed preparation and sowing in such a way that the crop plants emerge both as simultaneously as possible and, in relation to the weeds, as early as possible.

Table 22. Seed production in spring-sown cereal plants with delayed emergence as a percentage of the production in corresponding earlier plants.[a] Early and delayed plants grew either in mixed stands, alternating in rows, or in separate stands. Small-plot field experiment.

Plant stand arrangement and seed rate	Seed production in delayed plants (%)	
	Wheat	Barley
Delayed plants mixed with early plants		
Delay 8 days		
200 seeds m^{-2}	28.3	
400 seeds m^{-2}	20.1	13.4
800 seeds m^{-2}	12.7	
Delay 15 days		
200 seeds m^{-2}	9.5	
400 seeds m^{-2}	3.9	2.0
800 seeds m^{-2}	2.6	
Delayed and early plants in separate stands		
Delay 8 days		
400 seeds m^{-2}	102	105
Delay 15 days		
400 seeds m^{-2}	86	95

[a]Production in stands with entirely early plants was 410 g m^{-2} in wheat and 544 g m^{-2} in barley. (From Håkansson and Larsson, 1966.)

Sowing depth and growth of crop and weed plants

It was illustrated previously in this chapter that the emergence is delayed and the vigour of the emerged plants reduced when the depth of the seed location is increased in a soil that is uniform throughout the profile. In such a soil, an increased sowing depth of a crop could then be expected to reduce the competitive power of the crop relative to weeds, as the weed emergence is unaffected or less influenced by the sowing depth of the crop. Results shown in Table 23 agree with this expectation.

Table 23 presents results of a greenhouse pot experiment with wheat sown at three depths in a soil where weed seeds had been uniformly mixed in. An increase of the sowing depth from 2 to 8 cm resulted in a delay of the wheat plant emergence by more than 2 days. Although the number of emerged wheat plants was almost similar, the number

Table 23. Effects of sowing depth: emergence and growth of spring wheat and annual weeds after sowing wheat at different depths in a sandy loam soil kept moist throughout the profile. Weed plants, strongly dominated by *Chenopodium album*, developed from seeds uniformly distributed in the soil. Experiment in 12-l pots. Plant numbers and weights, determined 49 days after sowing, are sums of numbers from four pots. (From Håkansson, 1979.)

Sowing depth of wheat (cm)	Time of 50% emergence (days from sowing)	Wheat plants				Weight reduction by weeds (%)	Weed plants	
		Number		Dry weight (g)				
		Without weeds	With weeds	Without weeds	With weeds		Number	Dry weight (g)
2	4.9	95	96	57	47	18	209	12
5	6.2	91	89	45	29	36	206	21
8	7.2	85	85	30	18	40	206	31

of established plants decreased moderately with increased sowing depth. The joint effect of a reduced plant density, a delayed emergence and an increased consumption of energy reserves before establishment, resulted in a considerable decrease of the dry matter production in wheat, particularly so in wheat competing with weeds.

Competition from weeds caused a proportionally stronger biomass reduction in wheat plants developed after deeper than after shallower sowing. At the same time, the weeds grew more vigorously and produced a greater amount of dry matter in the former than in the latter case.

Sowing depth – principles regarding crop and soil in the field

Seedbed preparation and sowing should be looked upon as one complex. Some principles relating to the need for adaptations to soil characteristics are discussed.

Figure 12 illustrates, in principle, the establishment of seedlings as affected by seed size and depth location in the soil. It is thus presupposed that the seeds only differ in amounts of food reserves. It is also still assumed that the moisture content in the soil is uniform (except in the very superficial layer).

The question may now be raised whether it is possible to determine the optimum depth of sowing for a specified seed in a specified soil on a specified occasion from

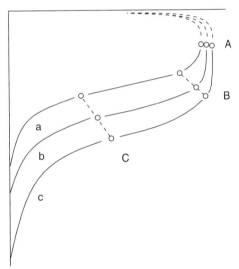

Established plants relative to seed rate

Depth of seeds in soil

Fig. 12. Number of plants established from seeds of different sizes (a < b < c) placed at different depths in soil assumed to contain water sufficient for germination and plant growth at all depths, except in the very superficial layer. Circles at A denote reasonable sowing depths when the risk of drought permeating from the soil surface is very low. Circles at B represent depths that may be suitable when this risk is greater. Circles at C mark depths advisable when there is a great risk of drought permeating to deeper soil layers. At C, an increased seed rate may also be advisable. (From Håkansson, 1979.)

curves – such as those in Fig. 12 – worked out for various typical situations. In the field, more or less unforeseeable weather conditions influence moisture contents and gradients in the soil. However, in combination with prognoses of the moisture conditions and risks of drought in a given climate and soil, curves of these types certainly help in increasing the frequency of successful choices of sowing depth.

In the field, surface drought is a frequently disturbing factor reaching to different depths in the soil. Circles A, B and C in Fig. 12 mark, in principle, differences in sowing depths that could be recommended at different risks of drought permeating the soil from the surface to different depths.

The optimum sowing depth is the depth from which a maximum number of sown seeds can produce established plants in a minimum period of time. The optimum depth in the field is then the shallowest depth where germination is not obstructed by drought and where, at the same time, the seedlings can root easily and become sufficiently anchored in the soil. (See comments on Fig. 5 (p. 64) concerning rooting problems when the seeds are placed at a very shallow depth.)

Due to risks of drying from the soil surface, the preferable sowing depth is often greater than the optimal depth in a homogeneous soil profile (Fig. 12). This is particularly true when small seeds are sown. When, for instance, a clay soil has a coarse and cloddy surface structure, shallow sowing is risky even when the clods are moist. Deeper sowing results in a reduced emergence, but this can be to some extent compensated for by a higher seed rate. Light permeating through the openings in a coarse soil may also reduce the negative effects of deep sowing. When the surface layer of a clay soil is broken into crumbs by winter frost, this layer must dry out considerably before sowing by traditional means is possible. Small seeds often have to be sown in a shallow and 'too dry' soil layer 'in the hope of rain'. However, supply of water in the form of vapour translocated from deeper layers is more important than is usually taken into account (e.g. Aura, 1975). This may be particularly true where small seeds are concerned (cf. Fig. 11 and associated discussion on p. 70).

The sowing depth on different soil types is briefly reviewed below. Discussions apply to areas with humid temperate climate and regular winter frosts, such as Northern Europe.

In non-aggregated soil rich in fine sand and silt, capillary water movements usually supply moisture sufficient for rapid germination in shallow soil layers even after comparatively long periods of dry weather.

In bare clay soils, freezing and thawing create stable fine crumb structure in the surface layer to depths of 3–6 cm. Practically no movements of capillary water occur in this layer, which therefore rapidly dries out in dry weather. At the same time, fields with such clay soils cannot be sown using traditional equipment until this layer has become relatively dry. To ensure sufficient water supply, crops with large seeds, such as cereals, are therefore preferably sown at depths below this layer, 'on a firm, moist seedbed bottom' (Kritz, 1983; Stenberg, 2000).

Coarse mineral soils, such as those with large proportions of coarse sand, easily dry out to great depths in dry weather. Both on these soils and on clays, deeper sowing is therefore usually recommended than on fine-textured non-aggregated soils under the climatic conditions considered here. In a clay soil, the depth of harrowing and sowing could be 4–5 cm if crumbs smaller than 4 mm could be achieved above this depth and if the underlying layer is moist, but both harrowing and sowing depths often have to be greater (I. Håkansson and von Polgár, 1984). In Sweden, Kritz (1983) recommended sowing of cereals at depths about 6 cm in clays and coarse mineral soils and 4 cm in fine-textured non-aggregated mineral soils. This particularly applies to spring sowing in areas where the late spring and early summer are often dry, but it largely applies to autumn sowing as well.

Fine-textured soils high in silt and low in organic matter easily harden in their surface layer. In such soils, shallower sowing than in clays with stable crumb structure is not only possible, but is also often necessary in order to get the crop plants to emerge before hardening of the soil makes plant emergence difficult (I. Håkansson and von

Polgár, 1979; Stenberg *et al.*, 1994). In clay soils too, heavy rain followed by dry weather after sowing may cause difficult surface crusts. Under these circumstances, seedlings of dicotyledonous plants have greater difficulties in emerging than have cereals and other grasses, the coleoptiles of which are particularly well suited for penetrating a compact soil (Heinonen, 1982).

Risks of long periods of dry weather after sowing are lower in autumn than in spring. Although the soil is sometimes dry to greater depths in late summer and early autumn than in spring, the chances of rain after sowing facilitating germination are greater in the autumn. It is often easier to avoid coarse and cloddy structures in clay soil seedbeds in the spring, following winter frost, than in late summer and autumn. Shallow sowing is risky in fields with such cloddy soil, but the seedbed structure can be considerably improved by rolling, which is therefore frequently practised both before and after sowing in the autumn.

Sowing on organic soils often has to be done at greater depths than would be necessary considering only the water supply. Because organic soils are comparatively loose, shallow sowing easily results in unsatisfactory anchoring of the plants and problems with lodging.

Seedbed Preparation and Sowing Considering Relative Times of Crop and Weed Emergence

The following discussions, like those in the above, mainly apply to *humid temperate areas with regular winter frosts*. They are focused on seedbed preparation and sowing depth aimed at favouring rapid emergence of crop plants and minimizing early and extensive emergence of weeds. We cannot fully control the seedbed conditions and the resulting responses of seeds and seedlings of crop and weeds, but by conscious adaptations to soil, climate and weather expectations, we can get a high frequency of desired plant responses.

Ordinary spring sowing – examples regarding different soils

Operations on *clay soils with stable crumb structure* developed in the upper soil layer in winter are dealt with first. As understood from the foregoing, seedbed preparation and sowing normally have to be carried out in dry weather after the upper crumb soil layer has become dry. Crops such as cereals usually germinate and emerge rapidly when sown below this layer. In continued dry weather, weeds from seeds in this layer emerge slowly, or not at all, and weeds from deeper soil layers emerge later than those from shallow layers emerging in cases of rain shortly after sowing. On the clay soils now discussed, e.g. on those in the eastern parts of Sweden, weed emergence in spring-sown crops varies strongly between years due to variation in the distribution and amounts of rain.

In *non-aggregated soils rich in fine sand and silt and with low clay contents*, the moisture in the surface layers varies less strongly from year to year than in aggregated clay soils. In fine-textured but non-aggregated soils, capillary water can rapidly move to the soil surface. The superficial soil can therefore be supplied with plant-available water during long periods of dry weather. This, of course, applies to areas where the content of plant-available water remains comparatively high throughout the spring, at least up to the lower top-soil layers. Because capillary water rapidly moves up to the soil surface, weeds can usually emerge rapidly from seeds in the shallow layers even in years with long periods of dry weather. Providing that equipment suited for shallow sowing is available, sowing should preferably be shallower than on clay soils with aggregated surface layers.

Where the soil surface has been sufficiently levelled in the preceding autumn, seedbed preparation in spring can be light and shallow. If suitable equipment is used, soil operations even then can destroy early weed seedlings efficiently and, at the same time, stimulate further weed seed germination to a minimum extent. Soils sufficiently levelled in the autumn can be sown in the spring without preceding spring cultivation. Then, however, many of the previously emerged weed plants will become more competitive than plants emerging after spring cultivation. They may therefore require stronger control in the crop than those emerged after such cultivation.

Effects of spring tillage on the vegetative structures of perennial weeds are dealt with in Chapter 10. In short, it can be said here that plants from strong rhizomes and creeping roots are less affected by shallow seedbed preparation in spring than are seedlings of annuals or perennials. Plants of perennials such as *Elymus repens* and *Cirsium arvense* developing from strong vegetative structures are usually also less influenced by weather variations. They are mostly in better contact with moist soil layers than seedlings.

'Delayed' spring sowing

A delay of the spring sowing in relation to 'normal' times of sowing can seldom be recommended. None the less, late spring sowing has attracted interest under special conditions. Cultivation on two occasions, about 3 weeks apart, followed by sowing is a way of reducing the emergence of weed plants in the crop. Effects have therefore been studied in field experiments.

The first cultivation, which should be made as early as possible, kills early weed seedlings and stimulates further seed germination of annual weeds, not only summer annuals (Fig. 1). On certain soils, it can also be deep tillage, sometimes ploughing, particularly with regard to perennials. The second soil operation about 3 weeks later should kill seedlings and kill or injure seedlings developed after the first cultivation. At the same time, it should be shallow so as to minimize further germination of weed seeds. On the whole, however, this late cultivation stimulates weed seed germination less strongly than the early cultivation can do (Fig. 1). At the time of the second soil operation, many of the seeds ready to germinate in the current spring have already

germinated or have reverted to dormancy. Sowing should be done immediately following the second cultivation to favour an early crop emergence relative to weeds.

Effects of delayed sowing could be exemplified by the reduced number of *Avena fatua* plants in spring cereals. Thus, in various experiments reported by Granström (1972) and Gummesson (1982), the plant emergence of this weed has been reduced by 50% in relation to that after ordinary, early sowing. The effect on different weeds differs, but the annuals largely behave like *A. fatua*. Perennial weeds such as *E. repens* and *C. arvense*, which are, on average, less affected by ordinary, shallow seedbed preparation, can be considerably weakened by two soil operations at intervals before sowing.

Autumn sowing

Winter annuals are usually strongly dominant among the annual weeds emerging in autumn-sown crops, and they emerge more abundantly and can grow more vigorously after early sowing (mid-August to early September) than after late autumn sowing (late September and October) under Scandinavian conditions (Fig. 1).

Shoots from vegetative structures of perennial weeds, such as *E. repens* and other perennial grasses, which are not dormant in late summer and autumn, may grow vigorously in autumn-sown crops, particularly after early sowing. Many dicotyledonous perennials enter a state of physiological dormancy in the later part of the growing season, e.g. *Sonchus arvensis* and *Tussilago farfara* and maybe to some extent also *Cirsium arvense* (see Chapter 10). Such perennials rarely develop new shoots in the autumn, but in the following spring they may develop vigorous plants from their vegetative organs, competitive not only in spring-sown crops, but also in autumn-sown crops.

In those areas of Southern Scandinavia where winter rye is grown, this cereal is often sown around 3 weeks earlier than winter wheat: rye from 25 August to 15

September, wheat from mid-September to mid-October. As understood from Fig. 1, more weeds therefore often emerge in rye than in wheat. In spite of this, the weeds have a greater ability to compete and grow vigorously in winter wheat than in winter rye, providing well wintered stands are compared. Winter rape is normally sown considerably earlier than rye, i.e. in the first half of August. In this period of the year, the soil moisture content varies strongly between years. The times of germination and emergence therefore vary considerably. This applies not least to the winter-annual weeds, which, in cases of sufficient moisture, germinate rapidly and establish vigorously. The degree of winter survival also differs more strongly from year to year in winter rape than in the winter cereals; in rye, wintering varies more than in wheat. All in all, this leads to considerable variations in the opportunity for weeds to grow strongly in winter crops in the following growing season.

If autumn sowing is preceded by early harvest, there may be a long period between the first tillage following harvest and the final seedbed operation before sowing. If, moreover, the soil is not too dry in the meantime, repeated tillage brings more weed seeds to germinate and more seedlings to be destroyed before sowing than where all tillage has to be done within a very short period after late harvest. Where tillage can be repeated at long intervals, its increased effect in many respects corresponds to that achieved at delayed spring sowing. Intense mechanical measures accomplished or started a relatively long time before autumn sowing can also considerably weaken perennial weeds such as *Elymus repens*. Perennials such as *Sonchus arvensis*, with regenerative structures becoming dormant in the later part of the growing season, are less weakened through immediate consumption of food reserves. Increased food consumption caused by breakage through tillage influences growth in the following spring. An immediate effect of tillage is the check it exerts on photosynthesis and translocation of substances to the roots, which otherwise continue even when dormancy prevents the development of new shoots (e.g. Håkansson, 1969e; Håkansson and Wallgren, 1972a; Fykse, 1974a). To what extent differences in the effect of tillage are of interest depends for example on how and to what extent chemical weed control is used in the cropping system and on how competitive the crops are.

Effects of rolling

In the ideal situation, the seeds sown should be surrounded by fine-textured and fine-structured soil, not very compact but firm, facilitating the supply of water. Furthermore, the deeper soil layers should be moist enough for the root activities and not so compact that root penetration is obstructed. The soil texture, structure and moisture contents should also allow satisfactory movements of oxygen and carbon dioxide.

Rolling before sowing is frequently justified on loose soils in order to achieve an adequate firmness of the upper soil layers. It is sometimes an important measure before sowing. Rolling after sowing is justified in many cases and is a particularly important seedbed improvement after shallow sowing of small seeds. It facilitates the supply of water to the seeds, and levels out the soil surface and presses down stones in stony soils. The compaction of the surface soil sometimes also improves the anchoring of the plants.

In Swedish experiments (von Polgár, 1984), rolling immediately following sowing resulted, on average, in a yield increase of 3% in spring cereals. However, the effect varied greatly with soil type and weather conditions. To exert full effect, rolling should be done immediately following sowing. If delayed by only a few days, it injures the young seedlings and reduces emergence, often causing a lowered yield.

Many qualitative observations, including those in the experiments mentioned, indicate that rolling often increases the emergence of weed seedlings. In field experiments in the USA (Larson *et al.*, 1958) and England (Roberts and Hewson, 1971), rolling increased the emergence of weed seedlings by about 60% and 30–40%, respectively. Different species are expected to respond differently. In the US experiment, grass weeds increased by 88%, broadleaved weeds by 34%.

It seems that in dry weather, weed emergence after spring sowing is sometimes more than doubled by rolling on the aggregated clay soils at Uppsala; in other situations, the effect is often indistinct. Rolling could be expected to stimulate the germination of weed seeds in the superficial soil layer in particular; observations support this. Rolling may therefore facilitate weed seed germination as it facilitates the germination of shallowly sown seeds of ley plants (cf. Känkänen *et al.*, 2001). The effects of rolling on weeds require more systematic investigations.

Late Emergence of Weeds – Consequences in Differently Competitive Crops

In *crops such as cereals whose stands close rapidly* and therefore exert a strong competition against weeds from early stages, weed plants emerging late largely exhibit a very weak growth and sometimes a high mortality. Late weed plants, in turn, exert a weak effect on the earlier crop plants. However, many weeds can grow and reproduce vigorously when light improves in the maturing crops.

Crops with stands closing late, such as sugar beet, potatoes and vegetables, exert a weak competition in a long period of early growth. Many of them are weakly competitive all the time. Late emerging weed plants can therefore also often grow vigorously and create problems in these crops. Even at low plant densities, early emerged weeds must be actively controlled. This also largely applies to late weed plants, which may otherwise produce a vigorous growth and also add great amounts of seed to the soil seed bank; perennials such as *Elymus repens* are able to strengthen their regenerative

structures considerably. Even late emerged weeds may sometimes also influence crop production, quantitatively and/or qualitatively. In some cases, they create problems mainly by obstructing the harvesting process. Late emerged weed plants may also become problematic as they are less conveniently controlled than early plants.

Certain species prone to late germination, e.g. *Solanum nigrum*, are troublesome in weakly competitive crops in many areas, whereas they are of minor importance as weeds in cereals due to the competitive conditions.

Fertilizer Placement

Experiments in many countries have shown that placing fertilizers in strips below the sowing depth usually results in a better crop growth and production than an even placement in the superficial soil. In a somewhat deeper soil layer, the fertilizer is more easily available to the crop. Its availability is also less weather-dependent. Nitrogen fertilizers placed on, or close to, the soil surface sometimes have a good effect, but may have slight, or unfavourably delayed, effects in dry weather. Huhtapalo (1982) presented very positive average effects of N-fertilizers placed below the sowing depth by combi-drilling of spring cereals. At the same time, he discussed measures to avoid negative tillage effects of the fertilizer coulters. In these Scandinavian field experiments, responses of the weeds were not studied systematically because weed problems were solved by chemical control.

Effects of different fertilizer placement on crop and weeds were studied in some pot and box experiments (Håkansson, 1979; Espeby, 1989). A sandy loam, moderately rich in humus, was used. In a pilot experiment in 12-l pots (Håkansson, 1979), a complete fertilizer (N, P, K) was mixed in soil layers, 2 cm thick, which were placed either as a surface layer reaching a depth of 2 cm or as a layer between 3 and 5 cm below the surface (Table 24). Two-row barley was sown in some of the pots, 25 seeds per pot at a depth

Table 24. Number of plants and dry matter of aerial shoots of two-row barley and annual weeds[a] per 0.90 m² (the total area of 15 pots) in a greenhouse pot experiment within 50 days of sowing. At two concentrations, a complete fertilizer was mixed in 2-cm soil layers situated at two depths in a loam soil, rich in fine sand, poor in humus and available plant nutrients. Experiment in winter, when temperature could be regulated around 18°C in the daily periods of light (18 h, 2000–9000 lux) and 10°C in the dark periods. (From Håkansson, 1979.)

Complete fertilizer, NPK 20–5–9, mixed in a 2-cm soil layer		Number of plants			Dry weight (g)		
Amount of N applied in soil layer (g l⁻¹)	Depth of fertilized layer (cm)	Barley	Chenopodium album	All weeds	Barley	Chenopodium album	All weeds
0	–	0[b]	313 a[c]	999 a[c]	0[b]	182 a[c]	283 a[c]
0.5	0–2	0	269 a	700 b	0	203 a	313 a
0.5	3–5	0	303 a	827 ab	0	218 a	318 a
0	–	337 a[c]	285 a	899 a	205 a[c]	24 c	55 bc
0.5	0–2	343 a	269 a	614 b	293 b	43 b	71 b
0.5	3–5	340 a	263 a	767 b	365 c	24 c	51 c
0.25	0–2	336 a	310 a	784 b	242 d	35 bc	71 bc
0.25	3–5	341 a	279 a	836 ab	288 b	29 c	65 bc

[a]Species with the highest densities: *Chenopodium album* (dominating), *Matricaria perforata*, *Bilderdykia convolvulus* and *Thlaspi arvense*.
[b]Barley not sown.
[c]Different letters in a column indicate significant differences between values within the column ($P \le 0.05$).

of 2.5 cm, which meant either immediately below or immediately above a fertilized layer. The fertilizer was applied on the same day as barley was sown, in amounts corresponding to 50 or 100 kg N ha⁻¹ (0.25 or 0.5 g l⁻¹ soil in the fertilized layers). Seeds of common annual weeds were mixed in the soil to a depth of 5 cm, equally in layers with and without fertilizer and in pots with and without barley. Water was applied in small amounts in pots where the soil surface appeared dry at daily inspection. In this way, watering could be done with moderate amounts of water each time, thus minimizing vertical water movements. The plant responses to the treatments, determined 50 days after sowing, are summed up in Table 24. They may be commented on as follows.

The number of *barley* plants was not markedly influenced by fertilizer application or placement. On the other hand, the dry weight of aerial shoots of barley was significantly increased by the addition of fertilizer, more at the higher fertilizer level than at the lower. When the fertilizer was placed below the sowing depth, it resulted in a greater dry weight increase in barley than when it was placed in the surface layer.

Different *weed species* responded differently to the fertilization as such, to the fertilizer placement and to the competition from barley. However, their responses to this complex of influences were mostly unclear.

Chenopodium album dominated the weed flora, particularly by weight. The weight of this weed differed strongly between pots with and without competition from barley. Its responses strongly coloured the responses of the total weed flora. Due to its dominance, its competitive effects probably levelled out the responses of other weeds to the presence or absence of barley and to the amount and placement of the fertilizer.

The number of *C. album* plants was not significantly, although perhaps slightly, influenced by the treatments. In stands without barley, the dry weight of its aerial shoots increased slightly, but not significantly, through the fertilization. The reason for its seemingly slight response to the depth location of the fertilizer in stands free of barley is certainly that the great majority of the plants originated from seeds near the soil surface. The roots of these plants may have had as good access to the nutrients in the shallow soil as in a deeper layer.

The germination and early growth of *Chenopodium album* in the superficial soil were not, or only slightly, influenced by osmosis or other factors linked to a high concentration of the fertilizer in the shallow soil layer. (Note that *C. album* among many other *Chenopodiaceae* plants is known to tolerate high salt concentrations in the soil solution.) In competition with barley, its dry weight was 11–21% of that in stands free of barley. It produced a lower amount of dry matter when the fertilizer was placed in the deeper layer than when it was placed in the shallower. This is suggested to be a result of stronger competition from barley for nutrients where the fertilizer was placed in the deeper layer. Table 24 thus shows that barley produced a stronger growth when the fertilizer was placed deeper than when it was placed near the soil surface.

Different weed species responded differently to fertilizer application, both by number and by weight. *Matricaria perforata* in particular, but also some other species, responded by a reduced number of plants to a high amount of fertilizer in the shallow soil layer. Weeds as a whole ('all weeds') reflected this response. By dry weight, however, weeds as a whole responded similarly to *Chenopodium album*, although to a less pronounced degree because of contrary responses of different species.

Espeby (1989) continued and extended the studies in comprehensive box experiments. Studying barley and annual weeds, he obtained, in principle, similar results. He varied both the sowing depth and the depth placement of an NPK fertilizer. He found that the fertilizer, when placed slightly below the crop seed, largely favoured barley above the weeds. Although responses differed, annual weeds were, on average, more strongly suppressed in competition with barley when the fertilizer was placed below the sowing depth than when placed above this depth. Many of the observed differences between species can certainly be ascribed to the complex interactions between more than two species mixed in the crop–weed stands.

7

Competition in Plant Stands of Short Duration

The Concept of Competition

General

Plant individuals growing near one another – in communities, in stands – mutually influence one another's growth conditions and growth. The more narrowly they are spaced, the more strongly they interfere. The spacing is usually expressed by the number of individuals per unit area, the *plant density*. When, in the following, *density* is used without specification, it means plant density and, at the same time, plant density in an initial stage of stand development. The density-determining units have to be defined from case to case. They could be seeds sown, plants emerged, etc.

'Stand density' is used here as a term for 'stand compactness'. The stand density is, of course, not only determined by the density and spatial distribution of plant individuals, but also by the environment. An optimum supply of vital resources could bring about a denser stand at a low plant density than a suboptimum supply at a higher plant density. The shape and extension of plant shoots, and thereby the stand density, are not least influenced by the supply of water and nitrogen.

Under otherwise equal conditions, the density of a monospecific stand increases faster at a higher plant density than at a lower (biomass increase, see Fig. 37). This does not always apply to a mixed stand. Thus, even a great increase in the total plant density may not lead to a more rapid increase of the density of the stand as a whole. This could happen, for instance, when the plant density is lowered in an intermixed fast-growing species at the same time as the total plant density is increased.

An increased plant density, i.e. an increased number of plants on the area (above certain 'low' levels), leads to an earlier and stronger limitation of the growth of the individual plants – due to limited supply of vital resources. With the progress of time, the growth becomes more and more restricted in all plants of a stand, or in increasingly many of them. The plants now *compete* for resources, and do so more strongly with an increased density and with the progress of time in their early period of growth.

In a narrow sense, 'plant competition' can be defined as competition for vital resources. 'Plant competition' generally refers to competition between plants, *interplant competition*. The forces driving the translocation of substances within a plant are sometimes looked upon as competition for these substances between plant organs; they are sometimes called *intraplant competition*.

Plants may influence one another – negatively, or even positively – in other ways than through competition for resources. The word *'competition' is very often used in a broader sense*, including other interplant influences, which may be both chemical and physical.

Chemical influences may be caused by 'allelochemicals' in 'allelopathic actions' – in other words, by *allelopathy*. Only exudates from living plant tissues were earlier considered allelochemicals. Influences of biologically active compounds released from decomposing dead plant tissues are nowadays usually also considered allelochemicals causing allelopathy (Putnam, 1988).

Plants may influence one another's environment by *modification of their physical conditions* in the soil (soil structure: compactness, etc.) and/or in the plant stand above ground (e.g. wind conditions and air humidity). As light is a vital resource, shading is considered competition for resources.

Competition for oxygen may sometimes occur between plant roots because of shortage of oxygen. This shortage may arise as a result of the use of oxygen by the roots themselves, by bacteria and fungi, etc., when rapid gas movements in the soil are simultaneously obstructed. More details ought to be known about this. Furthermore, little is known about the extent to which the supply of carbon dioxide within a plant stand is reduced to levels where competition for this gas really occurs.

The bacterial nitrogen fixation in the nodules of legumes in a mixed plant stand affects the biomass production in all categories of plants in the stand. This fixation often causes a higher total production in leguminous and non-leguminous plants, e.g. grasses, when they grow in mixed stands than when they grow separately.

The individual plants in a stand may be differently capable of capturing resources in competition with each other. This makes them differently *competitive*. The reason may be genotypic diversity (between or within species), but also phenotypic differences between organs in which plant growth starts. Seeds as well as vegetative propagules of genetically equal plants mostly exhibit a strong phenotypic variation, resulting in a different competitive status among the plants arising from them. Variation in the microsites of the plants in their initial stages of growth influences their competitive status in their following growing period.

The relative time of emergence affects the competitive ability of plants very strongly (Table 22).

Plants with diverse traits may respond very differently to environmental changes. Diverse plants may therefore respond very differently when they compete in the same stand and when they grow in separate stands or as solitary individuals. Plants with a low demand for a given resource may be more competitive than plants with a higher demand when they grow in habitats that are poor in this resource, but less competitive in habitats that are richer. Many plants that were competitive arable weeds under poorer nutritional conditions in earlier times are nowadays rare, or less competitive, weeds.

Under field conditions, it is hard to distinguish between competition for resources and other plant interference. All the reciprocal influences between the growing plants in a stand are, in this book, therefore termed *plant competition* (cf. Holzner, 1982). Some authors prefer to describe this package of reciprocal influences as *plant interaction* or *plant interference*. The terminology is discussed in many publications, e.g. in Grace and Tilman (1990).

In the following, *the word 'competition' is used in its broader sense* according to the above. *'Competition' thus includes any interaction between plants in a stand that causes a weaker growth of all or some of the individuals in this stand in relation to their growth as solitary individuals under comparable external conditions.* Using 'interaction' and 'interference' as terms besides 'competition' is not conflicting.

Production–density relationships described in numerous publications support the suggestion that, in plant stands of short duration on arable land, the dominant interference between the plants is mostly their competition for resources. At the same time, however, the outcome of this competition may be more or less modified by factors such as exudates from the plants, decomposition products from plant tissues and/or changes in the soil conditions caused by the plants. As will be deduced from the following, influences of allelochemicals represent a huge complex, the characteristics and importance of which have been poorly understood up to now.

Impacts of allelochemicals – allelopathy

Chemicals exuded from plants, and their impacts in a plant stand, have been increasingly studied in the recent decades. Osvald (1950), Grümmer (1961) and Müller (1953, 1969) may be mentioned among early authors dealing with this issue. The term *allelopathy* was proposed by Rice (1984) to imply any *negative or positive* response from a plant to chemicals produced by another plant. In literature reviews, Putnam (1988) and Williamson (1990) describe allelopathy caused by substances from a number of cultivated plants and weeds.

The question has been repeatedly raised as to the possibilities of suppressing weeds by selectively acting substances produced in cultivated plants or other organisms, mainly soil microbes. In rice, for instance, breeding for lines producing substances with suppressive effects on certain weeds have shown interesting results (Olofsdotter *et al.*, 1999; Olofsdotter, 2001). These results and other advances in work on allelopathy, in natural as well as in managed ecosystems, were debated in a recent symposium. The reports from this are presented in *Agronomy Journal* (Vol. 93, pp. 1–98) and summed up by Olofsdotter and Mallik (2001).

Competition and Phenotypic Plasticity

The following discussions mainly concern plant stands of short duration, or young stands of perennial plants. They are thus relevant to crop stands or crop–weed stands on arable land. The discussions are only partly relevant to conditions in stands or communities of perennial plants in more advanced stages of development.

Unless otherwise stated, 'competition in plant stands' is understood as competition between plant individuals – *interplant competition*. The different parts of a plant are not equally affected when growth and production of an individual are restrained by competition. Thus, biomass and size proportions of different parts within the plants are changed; so also are the anatomy and morphology of their various structures. Anatomical changes naturally mean changed proportions between various chemical compounds within the organs, in their tissues and in cells. They are the changes sometimes described as results of 'intraplant competition', i.e. competition between various plant structures for substances that determine their size, shape and function. These phenotypic modifications, enabled by the *phenotypic plasticity* of a plant, are *adaptations* facing different environmental conditions. It is such changes that cause the different properties of harvested products we then call changes in the product *quality*.

From general physiology and ecology we know a lot about phenotypic responses of plants to light, water, nutrients, etc. Knowing how these factors change individual plants when their densities are changed, we can explain many effects of competition on the individuals in a plant stand.

Within a narrow range of low densities, which decreases with advancing plant growth, competitive effects do not appear. Beyond this density range, competitive effects can be indicated as a decrease of the average biomass production per plant with an increased density. At the same time as the biomass production per plant decreases with increasing density, branching also decreases. Beyond certain densities, the plants only develop a primary shoot (Figs 13 and 14). Some plants, e.g. biennials, such as beets, do not develop branched shoots in their first year. The size of the beet shoot is then regulated merely by the number and size of leaves and the size of the stem part of the taproot in parallel with the root part. In most plants, the biomass ratio of stems to leaves changes when the plant density is changed. With an increased density, the ratio of below-ground to above-ground biomasses, 'the root:shoot ratio', decreases. A density increase above certain levels reduces the production of seeds more than the production of vegetative parts of a plant.

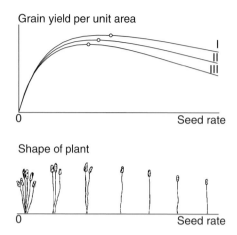

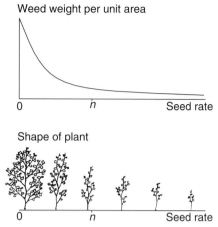

Fig. 13. Grain yield and plant shape in cereal stands at different seed rates (plant densities): general principles. The lower sketch illustrates the aerial shoots of an average plant (leaves excluded) at different seed rates. Curve I describes the gross yield of grain. Curves II and III show net yields, i.e. the gross yield minus the seed rate multiplied by a coefficient representing the price ratio of the seed sown to the grain harvested. Curves II and III describe net yields at a lower and a higher seed price, respectively. Positions of curve maxima (o) illustrate that net yield maxima fall at lower seed rates than does the gross yield maximum.

Fig. 14. Weeds in stands of crops such as small grain cereals: biomass per unit area and plant size as varying with the seed rate (plant density) of the cereals, general principles (from Håkansson, 1979). It is assumed that the initial weed plant density and the species composition of *emerged* weeds are constant. At an ordinary seed rate, here marked n, the weed biomass per unit area usually varies from (2–)5 to 15(–25)% of that in comparable unsown field plots (as measured in Swedish experiments by Håkansson, 1975b, 1979, 1984a; and Andersson, 1987). Plant sketches illustrate the size (considering branching and size of branches) of an average weed plant at different crop seed rates. The curve may represent a definite weed species or the entire population of diverse weeds in a field. The mortality of weed plants may increase more or less with the crop seed rate. The decrease in plant number due to increased mortality varies depending on species and environment, but is usually lower than the decrease in plant size (Fig. 21).

The reduced seed production is caused by a lowered seed number more than by a reduced seed size, but the mean size of viable seeds mostly decreases to some extent. At high density levels, different for different plant species, seed production may fail completely. In perennial plants, the production of underground regenerative structures decreases in relation to the aerial plant parts. This is of interest in managing perennial weeds.

The death of entire plants in response to crowding, called *self-thinning*, increases with plant density above certain levels and varies with species and environmental conditions. Because most plants are highly adaptable to density, death of entire plants in the form of self-thinning is seldom the most salient reaction to increased densities in annual plant stands (cf. Harper, 1977;

Cousens and Mortimer, 1995; see also Fig. 21, p. 96). Instead of total death of a plant, death of certain parts of it is a common response to crowding. This death of certain parts may often facilitate the surviving individual. For instance, the lower leaves of an erect plant die earlier at higher densities than at lower. This is an advantageous response, because it reduces of the number of leaves that otherwise to an increasing extent become net consumers of assimilates with the increasing shade among the crowding plants.

Intraspecific and Interspecific Competition

The plant competition in a monoculture, a *pure stand*, of individuals of a uniform category is simply called *intraspecific competition*. Analogously, competition between plants distinguished as different categories is termed *interspecific*. In a *mixed stand*, i.e. a mixture of different plant categories, *both intraspecific and interspecific competition* interact in a complex interplay. The categories distinguished are often species, but may represent other taxonomic levels. They may be discernible biotypes of a species or differently treated plants, etc. Some authors will understand 'specific' as solely related to plant species. However, when the categories are species and this is to be stressed, their interference could be described in terms of *intraspecies* and *interspecies* competition.

Individuals of the same species do not necessarily compete on the same conditions. They may be represented by populations of biotypes with different traits. The plants may originate from seeds or other propagules varying in size (amount of food reserves) or in other properties, in relative time of emergence, etc., which cause differences in their competitive position. Among the individuals of a plant category, such as a species, there are usually hierarchies with respect to their ability to assert themselves competitively.

However, even in a stand of plants from genotypically and phenotypically very uniform propagules, the growth and size of the individuals may vary strongly owing to differences in their competitive position caused by variation in their microsites. This variation particularly causes a variation in the initial growth of the plant individuals (e.g. at germination and subsequent early growth), when their roots only penetrate small soil volumes. This influences the competitive status of the plants in their continuing growth. Thus, even in a 'uniform' pure stand of plants originating from 'very uniform' seeds, there is usually a considerable size variation among the plants caused by a variation in their competitive situation.

Relative Competitiveness

We now know that even genetically equal plants seldom hold fully equal competitive positions. When the relative competitiveness of genetically differing plants is to be characterized, this is normally based on the average appearance of their individuals in mixed stands. Plants with different competitiveness are expected to perform differently in mixed and separate stands. Methods and pitfalls in measuring relative competitiveness are presented in Chapter 9.

Differences in the relative competitiveness of plants resulting from genetic differences vary with the external environment, e.g. with temperature, soil moisture and nutritional conditions. A plant that is a strong competitor at higher temperatures may be weaker at a lower temperature (cf. Table 36). Although many plants occurring as arable weeds may have a great potential for both genotypic and phenotypic adaptation to different environments, etc., all of them have a more or less restricted distribution as weeds.

We can look at the competition between different plant categories in a mixed stand in two ways: (i) the effect of competition from one plant category on the other(s) and (ii) the response of the first-mentioned category to competition from the other(s). The outcome of competition measured in the two ways is sometimes distinguished as the '*competitive effect*' of a topical component on the other(s) and as the '*competitive response*' of this component to the other(s) (cf. Goldberg and Landa, 1991). Various terminology occurs. Competitiveness in the meaning of competitive effect is sometimes called *aggressiveness*.

It is of interest to describe the behaviour of a plant both with respect to its effect on others and to its response to others in their

interference in mixed stands. A plant that is
a strong competitor in the former sense is
usually, but not necessarily, similarly strong
in the latter. Methods and problems in
measuring competitiveness are discussed in
Chapter 9.

Competition as Influenced by Plant Spacing and Relative Time of Emergence

As understood from the above, the outcome
of competition in a plant stand may change
greatly with changes in the external conditions, i.e. in plant-available resources, in
temperature, etc. *In stands with specified
plant categories under given environmental
conditions*, the outcome of the plant competition is determined by the number of
individual plants per unit area, the spatial
distribution of the plants and their relative
time of emergence (or of other early stages
easily defined). The following three factors
are thus decisive.

- *Plant density.* The plant density, i.e.
  number of plants per unit area, mostly
  implies the density of individuals in
  an early growth stage, which has to be
  defined. The density sometimes refers
  to 'plant units', i.e. defined groups of
  individuals, where the number of
  separate individuals is not easily
  determined. The *initial density* may
  be described by the number of seeds,
  emerging plants, etc. per unit area.
  All plant activities in the stand after
  the density determination, such as
  biomass production, morphological
  changes and *self-thinning*, can then be
  described as functions of the density
  determined in an early stage. When
  germination and emergence are events
  drawn out over long periods of time,
  as in many weed populations, this
  must be considered in density determination. Because plants emerging late
  have a weak competitive position
  (see below), density determination in
  earlier stages of the stand development
  will often suffice. In autumn-sown
  crops, where weed plants emerge in

both autumn and spring, density effects
are less easily evaluated.

- *Spatial plant distribution.* Like plant
  density, the plant distribution over an
  area may be related to conditions in
  some early stage of the stand's development, e.g. describe the distribution of
  the seeds sown or the plants emerged.
  The plant distribution may be looked
  upon as the variation in 'plant densities' on an area, or, better, the variation
  in plant spacing (Table 25, p. 102).
  The most uniform spacing is obtained
  at 'hexagonal distribution', where all
  distances between neighbouring individuals are equal. 'Square spacing' also
  represents a fairly even distribution. In
  drilling, the most even seed distribution is obtained when inter- and
  (average) intra-row spacings are equal,
  which is seldom practised and, often,
  not practicable. An increased row spacing relative to the spacing within the
  row means a more uneven distribution
  of the seeds and usually a similarly
  increased unevenness of the growing
  plants.
- *Relative time of emergence.* The relative time of growth initiation of the
  plants in a stand is strongly influential
  on their competitive position in relation to each other. It is most conveniently determined through their relative time of emergence. Among plants
  with equal traits, early ones always
  have a competitive advantage over
  later ones. Changes in the relative time
  of emergence of plants with different
  traits change their relative competitive
  position. Plants that are competitively
  superior when emerging at a given time
  in relation to other plants become less
  dominant or inferior when emerging
  later. A change of only a few days in
  the relative time of emergence of crops
  and weeds may drastically change
  the outcome of their competition (e.g.
  Fig. 30). As was emphasized in Chapter
  4, it is hard or impossible for seedlings to establish in a closed stand
  of previously established plants (e.g.
  Fig. 2, Table 2).

Effects of Plant Density

Effects of the three factors presented above are illustrated below. It is easily understood that these factors interact. This is illustrated when the effect of a specified factor is changed because of changes in one of the other factors. Environmental influences are also illustrated.

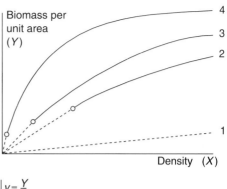

General characteristics of production–density relationships

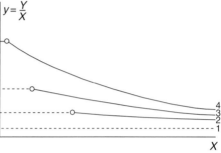

Effects of density in stands of different ages were described by Japanese scientists in the 1950s (Kira *et al.*, 1953; Shinosaki and Kira, 1956). Competitive effects measured over time have been analysed in recent years by Håkansson (1988a, 1991, 1997b). Density effects on the biomass growth of plants can be measured in young plant stands (Fig. 15).

Production–density (yield–density) *relations representing late periods of growth* in stands of genetically and otherwise similar crop plants have been presented in numerous publications. Figure 16 sums up experiences based on numerous experiments (e.g. Lang *et al.*, 1956; Bleasdale and Nelder, 1960; Holliday, 1960a,b; de Wit, 1960; Donald, 1963; Bengtsson and Ohlsson, 1966; Willey and Heath, 1969; Bengtsson, 1972; Pohjonen and Paatela, 1974; Håkansson, 1975b, 1979, 1983b, 1988b, 1997a; Erviö, 1983).

At low densities, up to certain levels, which are difficult to determine exactly (e.g. Holliday, 1960a), no competitive effects are apparent. The production–density relationship is linear in this case (Figs 15, 16). If the plants of the category studied (e.g. a crop) are mixed with other plants (e.g. weeds) at densities sufficiently high to cause competition, the relationship becomes non-linear right from 'density zero' (Håkansson, 1975b, 1983b, 1988b).

Beyond a certain (usually low) density, competition occurs and can be measured. The production–density curve (Fig. 16) gradually levels off, approaching a maximum. The density where maximum production takes place varies with plant type,

Fig. 15. Characteristics of production–density (biomass–density) relationships in a single-species plant stand. (From Håkansson, 1997a.) A very early stage of development (1) and three increasingly advanced stages (2, 3, 4) are compared. *Y*, biomass production per unit area; *X*, initial plant density, i.e. the initial number of plants or plant units (seeds, recently emerged plants, or the like) per unit area; $y = Y \cdot X^{-1}$ = the biomass per initial plant, or plant unit. *Note that vertical scales in the two diagrams are different.* Straight lines (dotted) represent density ranges without evident effects of competition. The upper curves in the two diagrams represent relationships that can usually be satisfactorily described by equations $Y = X \cdot (a + bX)^{-1}$ (Eqn 1) and $y = Y \cdot X^{-1} = (a + bX)^{-1}$ (Eqn 1a), respectively.

with the plant part considered and with the environment. Donald (1963) states that maximum production of seeds per unit area is obtained at a lower density than the maximum, or 'ceiling', production of vegetative biomass, or that of the total plant biomass (see also Håkansson, 1975b). Owing to unfavourable influences of higher densities, the production decreases sooner or later with increased density. The reason is that the decrease per plant and/or an increased mortality are no longer fully balanced by the increased initial plant density.

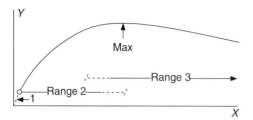

Fig. 16. General characteristics of a production–density relationship at a late stage of development of an annual plant stand. (From Håkansson, 1997a.) *Y*, the production (yield) by weight per unit area; *X*, the initial plant density. Three density ranges (1, 2, 3) are distinguished. Their margins are indistinct (1–2) or overlapping (2–3). The form of the production–density curve varies with species, plant part considered, environment, etc. Curves describing the biomass of entire plants per unit area are sometimes rather flat in a wide range of densities. Biomass maxima of generative organs and also other organs produced in advanced stages of plant development (e.g. potato tubers) fall at lower densities than do those of the entire plants or those of the vegetative plant parts as totals or as entireties above ground.

In many experiments, the production has not reached or passed its maximum within the range of densities observed. Production–density relationships are then often satisfactorily described by curves approaching a horizontal 'ceiling' line (see curve 4 in the upper diagram of Fig. 15). In stands of annual plants, however, a more or less pronounced maximum could be expected to be passed at some higher density. For stands of annual plants, the 'law of constant yield' (a 'ceiling yield') beyond a certain density will not be quite true, even considering the total biomass production. However, descriptions of the relationships using models based on the assumption of constant ceiling levels will often suffice within limited density ranges.

The production and plant composition in an old stand of perennial plants mostly depends only slightly on the densities of different plants in the young stand. The composition of an old plant community is strongly governed by the environment. However, discussions below are focused on plant stands of short duration.

Figure 14 illustrates the decrease in weed biomass per unit area caused by an increased plant density of the crop (crop seed rate, etc.). This decrease is analogous to the decrease in the biomass weight per plant shown by the curve representing curve 4 in the lower diagram of Fig. 15. When the crop plant density is increased, the biomass of an average weed individual thus decreases similarly to that per average crop plant individual, although the decrease is usually differently strong. The relative decrease of the weed biomass per average individual and per unit area is described by curves with the same slope. The curves of various species in a mixed weed population will, on the other hand, diverge strongly owing to different competitiveness, different disposition to mortality, etc.

Production–density relationships have often been described by models assuming a 'production ceiling' at higher densities. Equations of the same character have been used frequently, although they may have been originally written differently, with parameters of different biological meaning (e.g. Shinozaki and Kira, 1956; Holliday, 1960a; de Wit, 1960). All of these equations can be written in the following simple form (Håkansson, 1975b, 1983b, 1988a):

$$Y = X \cdot (a + bX)^{-1} \qquad (1)$$

where X denotes the initial density and Y the production (weight of biomass) per unit area. The course of curves representing this equation is illustrated by the upper Y-curve in Fig. 15. The mean production per individual plant (or some other plant unit underlying density values) can be written as follows:

$$y = Y \cdot X^{-1} = (a + bX)^{-1} \qquad (1a)$$

Curves of this equation decrease with increased crop density (X) approaching zero (such as the upper curve of $y = Y/X$ in Fig. 15).

In production–density equations described by maximum curves, as in Fig. 16, Holliday (1960b) added a quadratic term as follows:

$$Y = X \cdot (a + bX + cX^2)^{-1} \qquad (2)$$

The mean production per individual plant is now

$$y = Y \cdot X^{-1} = (a + bX + cX^2)^{-1} \qquad (2a)$$

Equation 2a is represented by decreasing curves similar to those of Eqn 1a but allows a greater flexibility in their course.

Willey and Heath (1969) discussed equations that are rather similar to Eqn 2 regarding their applicability, for instance the following:

$$Y = X \cdot (a + bX^c)^{-1} \qquad (3)$$

The mean production per individual plant is then as follows:

$$y = Y \cdot X^{-1} = (a + bX^c)^{-1} \qquad (3a)$$

If parameter c in Eqn 2 exceeds zero and the exponent c in Eqn 3 exceeds unity (one), both equations are represented by maximum curves.

As understood from the above, the equations do not strictly describe relationships at those low densities where competition has not set in. Furthermore, when they are fitted to experimental data, parameter a sometimes becomes negative (Håkansson, 1975b), which results in abnormal courses of the curves at low densities.

When Eqns 2 or 3 are fitted to data within density ranges where the maximum has not been reached or passed, parameter c often becomes negative in Eqn 2 and lower than unity in Eqn 3 (Håkansson, 1975b, 1988b). Such c values could be attributed to experimental errors. However, they often appear in cases where Eqn 1 provides an acceptable fit and are so often significantly lower than zero and unity, respectively, that the phenomenon should not be neglected (Håkansson, 1975b).

When production–density data represent a wide density range, it is difficult to obtain equally good fits by one and the same equation over the entire range. This is not surprising. The manner of morphological and anatomical adaptation to density differs strongly from one density range to another and seems to differ more or less between species. It is therefore difficult to find simple production–density equations enabling sufficiently good descriptions of the production–density relationship of different plant species in wide density ranges.

Three density ranges can be distinguished with regard to the production–density relationship (Fig. 16).

The low-density range represents densities lower than those where competitive effects occur. The production–density relationship is linear in this case (Fig. 16).

The mid-density range comprises densities from the lowest level, where competition appears to the level where the production approaches or reaches its maximum. In this range, Eqn 1 usually provides a satisfactory fit. The parameters added in Eqns 2 or 3 may often significantly improve the fit, but in cases of a small number of measured values, there is a risk of overparameterization. Nevertheless, a positive or negative quadratic term (Eqn 2) or an exponent differing from unity (Eqn 3) often results in reasonable adjustments and may be used when justified by the experimental data. The equations now discussed should only be considered as regression equations to be used for interpolations, never for extrapolations far outside the range of measured values.

The high-density range comprises densities from a level somewhat lower than the density of maximum production up to higher densities. If there is a distinct production maximum, Eqn 2 or Eqn 3 gives a suitable fit here (cf. Håkansson, 1975b). Sometimes, particularly when vegetative production is measured, experimental data may not show any distinct production maximum. In such cases, none of the simple equations dealt with can give a good fit.

Changes in the biomass weight of weeds – per individual or per unit area – in response to crop plant densities can sometimes be described by graphs based on equations of type 1a, 2a or 3a.

However, these equations often do not possess the flexibility needed to fit over a wide density range (Håkansson, 1997a: p. 65). Particularly in young crop stands, or in crops exerting a weak competition during a long period of growth, relationships represented by illusorily sigmoidal curves may be found (cf. Fig. 20a). The weed biomass decrease with an increased crop density is mostly reasonably well described by curves asymptotically approaching zero. Curves approaching a somewhat higher level might be reasonable under special conditions (see below).

Crop yields decrease with increasing weed densities analogously to the weed biomass decrease with increased crop plant density. The decrease in the crop yield is mostly sufficiently well described by an equation of type 1a, 2a or 3a. These equations are more often sufficiently flexible for describing the response of crop yields to weed densities than for describing the weed biomass response to crop plant density. Their curves asymptotically approach zero. Cousens (1985a,b) has discussed related models approaching levels above zero (or, which is the same, resulting in the percentage yield loss approaching a level lower than 100). Wilson and Wright (1990) presented results justifying such models.

Results of field experiments concerning effects of wild oats in barley (Bell and Nalewaja, 1967) suggested that the yield loss increased with increasing wild oat density up to a certain level and then decreased. In a greenhouse experiment at Uppsala (unpublished) conducted in barley mixed with *Sinapis arvensis* at different densities, both the weed biomass and the biomass reduction in barley passed a maximum at a certain weed density. These observations deserve further investigation to determine to what extent such phenomena occur under field conditions.

Production in crop stands

Yield in different crops

A great number of field experiments from various countries illustrate relationships between yield and plant density or seed rate. Most experiments concern crop stands where weeds have been effectively controlled.

Figure 17 describes results from such experiments. They exemplify principles discussed in association with Figs 13 and 16 and represent different density ranges in relation to the density of maximum production. This density falls either in the density range of observations or beyond this.

Effects of a given density are conditional. Both the plant genotype and the environment determine where the maximum plant production falls on a density scale. Both of these factors, of course, determine the production level in general and the response to densities. There are great differences not only between crops of different species, but also between crop varieties (Fig. 17f).

The effects of densities are strongly modified by the spatial distribution of the plants, e.g. by the row spacing (Fig. 17a, c, d). At constant densities (seed rates), the plant (seed) distribution becomes increasingly uneven when the row spacing is increased beyond the level where the intra- and inter-row spacings are equal. The yield therefore decreases with an increased row spacing beyond this level, or from a somewhat larger row spacing (Table 25 and associated discussion). At the same time, yield maximum falls at a decreasing plant density owing to adverse effects of the increasing crowding within the row (cf. Fig. 29).

Figure 17b, describing the grain yield in maize, shows that the amount of nitrogen supply not only affects the general yield level, but also the position of the maximum yield on the density scale. With increasing nitrogen supply, the maximum yield may not only become higher, but often, as in this case, also fall at an increasingly high density.

It appears to be a general rule that maximum yield is not only higher but also appears at a higher density when an environmental factor changes towards its optimum. It should be emphasized here that the optimum level of a given factor always depends on how other factors are represented. The optimum level of nitrogen fertilizer use thus depends on the supply or presence of other mineral nutrients, on water content, on temperature and light, etc., and, last but not least, on the plant density. The optimum is passed when disturbances override positive effects. In cereals, for instance, lodging often lowers the optimum level of nitrogen application. At the same time, the optimum plant

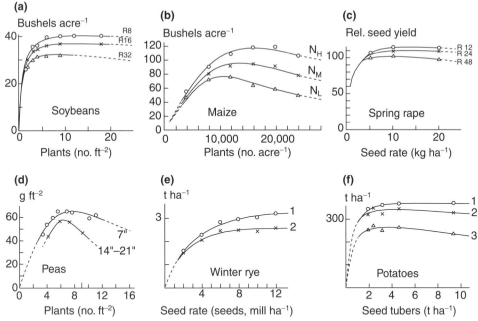

Fig. 17. Examples of yield–density relationships. Diagrams describe seed or grain yields, or, for potatoes, tuber yields; based on field experiment data. (a) Soybeans at three row spacings, 8, 16 and 32 inches; from the USA (Wiggans, 1939). (b) Maize at three levels of nitrogen supply, high (N_H), medium (N_M) and low (N_L); from the USA (Lang *et al.*, 1956). (c) Spring-sown oilseed rape at three row spacings, 12, 24 and 48 cm; from Sweden (Ohlsson, 1976). (d) Peas for canning at different row spacings; from England (Bleasdale, 1963). (e) Winter rye: 1, gross yield; 2, net yield, in this case the gross yield minus twice the seed rate (cf. Fig. 13); from Sweden (Larsson, 1974). (f) Potatoes of three varieties: 1, Bintje; 2, King Edward VII; 3, Magnum Bonum; from Sweden (Svensson and Carlsson, 1974).

density (seed rate) is lower in cultivars or species of cereals with weak straw than in those with stiff straw.

Both an increased nitrogen supply and an increased density raise the disposition to lodging and therefore interact in increasing the frequency of lodging. Results of experiments in Finland (Fig. 18) illustrate a divergence in the average response to seed rate in spring cereals with and without nitrogen fertilizer. Explanations may be that nitrogen fertilizer led to an increase in lodging (and, possibly, attacks by fungi), which is a common effect of a highly increased seed rate.

It is often asserted that increased plant densities favour *infections by fungi* in crops such as cereals. It is true that the plants are more easily infected by fungi *in dense stands* than in thin stands, because the air humidity is higher in the former stands. Now, the *stand density* increases faster at a high *plant density* than at a low one. Infections by fungi attacking young plants may therefore increase with increasing plant density. However, an increased *plant density* (seed rate, etc.) mostly means a less marked increase in *the stand density in later stages* of plant development. Large amounts of water and nitrogen, on the other hand, result in a stronger and more lasting increase in the stand density, which is due to the development of lusher plants. The anatomy of the tissues of these lusher plants makes them also more disposed to infections. Judging from my own experience, infections by fungi are more favoured by a high stand density caused by luxurious supply of water and nitrogen than by a high plant density.

The consequences of an increased 'total density' of the crop–weed stand caused by an increased weed abundance vary strongly regarding effects on lodging and attacks by

Grain yield (kg ha⁻¹)

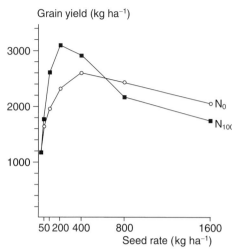

Fig. 18. Yield–density relationships in spring cereals. Means from seed rate experiments with spring-sown barley and wheat in Finland presented by Erviö (1972). N_0, without nitrogen fertilizer; N_{100}, fertilized with 100 kg ha⁻¹ of nitrogen. The combination of increased nitrogen supply and high seed rates (plant densities) resulted in extended lodging and other adverse conditions. The yield obtained at higher seed rates therefore decreased with an increased nitrogen supply (N_{100}). Maximum yield was, however, higher at N_{100} and was, at the same time, obtained at a lower seed rate than at N_0.

various organisms. These effects depend on species and varieties of crops and weeds and on the prevailing environmental conditions. However, our knowledge of these important contexts is very limited and in-depth studies on them should be welcomed.

Plasticity of crop plants and quality of harvested products – examples

The phenotypic plasticity of plants facilitates their adaptation to different environments, including the environments created by the crowding in a plant stand. Plastic responses to density are of great interest for many reasons. Responses of weeds with regard to their production of seeds or other *propagules* relative to other plant parts influence their *population dynamics*, i.e. the increase or decrease in their populations in varying situations. Knowledge of the plastic responses of crop plants is

crucial for judgments of how, and to what extent, the *quality of harvested products* can be influenced by cropping measures in various environmental situations. A few examples of responses are given below.

Changes in the proportions between different plant parts caused by changed plant densities were early illustrated in maize by Morrow and Hunt (1891). They showed that the grain yield reached its maximum level at a much lower plant density than did the total above-ground biomass. Donald (1963), reviewing his own findings and other findings in several crops, stated that the maximum production of generative organs is obtained at a lower density than the maximum of vegetative plant parts or total biomass. Vegetative and total production seemed to be fairly constant over wide density ranges. In consequence, the proportions of generative and vegetative organs change with the plant density. Experiments with different species confirm this conclusion (Spitters, 1983b). Changes with density were illustrated in a Swedish experiment (Håkansson, 1975b) with two-row barley covering a wide range of densities (Fig. 19).

The fact that seeds develop later than leaves and stems may disfavour their formation in relation to the formation of these vegetative structures at a high competitive pressure. It seems reasonable, therefore, that the grain production in cereals decreases in relation to the production of vegetative plant parts when the competitive pressure increases with increasing plant density beyond certain levels. These density levels could be lower or higher than densities in common practice today.

In cereals, there may be a maximum ratio of grain to straw at some density near, or below, that of maximum grain yield. Figure 19b supports the suggestion of such a maximum, although the data at the lower densities are uncertain. Low values of the grain-to-straw ratios at lower densities may be expected as a consequence of luxurious tillering (branching) with a long time span between the formation of early and late tillers (branches). Late tillers then produce comparatively small amounts of grain, some of them no grain at all. The maximum ratio of

(a)

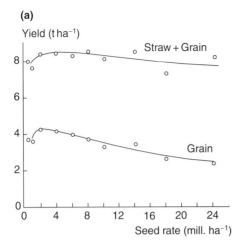

(b)

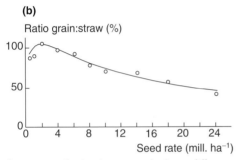

Fig. 19. Production in two-row barley at different seed rates (plant densities) in a small-plot field experiment at Uppsala. Curves represent mature plants cut at soil surface and air-dried. (Redrawn from Håkansson, 1975b.)

grain to straw is found at very different plant densities depending on the plant trait and the environment. Differences recorded are therefore sometimes indistinct, or appear conflicting (Henke, 1929; Rösiö, 1964). In experiments where plant densities vary within narrow ranges (near densities 'of interest in practice'), the density ranges may represent different positions relative to the position of the possible maximum of the grain:straw ratio.

The properties (quality) of harvested seed or grain also vary more or less markedly with plant density. At very low densities, strong branching of the plants results in a considerable variation in age among the shoots and seeds, causing an uneven degree of ripeness when harvested. At higher

densities, and a more restricted branching, the age variation is smaller and the ripening more synchronous. In years with cool and moist weather at the end of the growing period in particular, uneven ripeness due to strong branching may lie behind important quality problems. When the later growing period is warmer and dryer, ripening proceeds faster and the variation in ripeness decreases. In cereals, there is a greater variation in the size and chemical composition of grains from plants with many tillers than from those with one or two shoots only. In rape or turnip rape, high chlorophyll contents in harvested seed lots particularly often occur in the harvests from stands with low plant densities or many gaps due to high frequencies of unripe seeds.

The mean weight per seed or kernel (measured as the 'thousand-seed weight' or 'thousand-kernel weight', etc.) mostly varies moderately with the plant density. In harvested lots of cereal seeds varing greatly in size and weight, problems are mainly caused by the smallest seeds. At great differences in age, the size variation between seeds largely results from different feeding with assimilates. This leads to a variation in the proportions of different tissues and chemical compounds. In rice, the variation in seed size is comparatively small (e.g. Yoshida, 1972). In potatoes, the number of stolons and tubers – i.e. the number of 'underground branches and their apical swellings' – per individual decreases with increasing plant density. The age and the degree of physiological maturity of harvested tubers therefore usually vary less at high plant densities than at low (Carlsson and Svensson, 1975).

Crop–weed interactions

Intraspecific and interspecific competition interact. When we describe the outcome of competition between crops and weeds, we should remember that this outcome is a result of interactions between all plant categories in the stand. An increased density of crop plants means a strengthening of the general competitive pressure in the stand. The biomass proportions change, not only between crops and weeds as entire groups, but also between the individuals within each group. Individuals may differ strongly in competitive power owing to genetic diversity (different species, varieties, strains) and to differences in their relative time of emergence, etc.

The following concerns responses to densities in plant stands and is illustrated in Figs 20–25. The graphs in these figures are based on regression equations of types Eqn 1 or Eqn 2 for crops and types Eqn 1a or Eqn 2a, or modifications, for weeds.

The graphs in Fig. 20 are based on an equation (type Eqn 1) describing the production–density relation for pure stands of barley grown at different plant densities, where certain plants are looked upon as category A ('crop') and others as category B ('weeds'). The A and B plants are considered uniformly mixed. The curves are based on 'reference model calculations', treated in more detail under the heading of *Reference models* in Chapter 9. They are derived from an equation of type Eqn 1, i.e. $Y = X \cdot (a + bX)^{-1}$; in this case, $a = 0.611$ and $b = 0.00293$, where X is the initial density in number of barley seeds sown per m² and Y the biomass (dry weight) of aerial shoots of barley produced per m². The equation is represented by a curve similar to curve 4 in the upper diagram of Fig. 15. This equation, and calculations illustrated in the following and with the curves in Fig. 20, are from Håkansson (1975b; see also 1983b, 1997a).

Y_B(rel) in Fig. 20a, denotes the biomass (dry weight) of B plants (weeds) per unit area as a function of the A plant (crop) density (X_A) – expressed as a percentage of the biomass in the absence of A (i.e. at $X_A = 0$) –

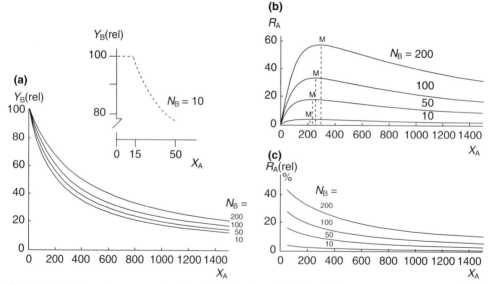

Fig. 20. Graphs based on an equation describing the dry weight of aerial shoots per unit area (Y, g m⁻²) in a uniform stand of two-row barley as a function of the barley seed rate (X_A seeds m⁻², being the 'initial density' of plant units). Plants representing certain numbers of initial units are considered crop plants (A), the others weed plants (B). Categories A and B are thought of as plants grown in mixed stands at *varying* densities of A, named X_A, and different *constant* densities of B, named N_B. Both X_A and N_B are thus numbers of barley seeds m⁻². In this *reference model* (see Chapter 9): Y_B(rel), the weight of B as a percentage of that when $X_A = 0$, i.e. when B grows in pure stands. R_A, the absolute dry-weight reduction in A (g m⁻²) caused by competition from B. R_A(rel), this reduction as a percentage of the weight of A when $X_B = 0$, i.e. when A grows in pure stands. (From Håkansson, 1975b: introducing the 'reference model' method of describing plausible characteristics of production–density relationships, biomass ratios, etc., in stands where A and B are plants with different traits.)

at different constant B densities (N_B, here representing 10, 50, 100 and 200 seeds sown per m^2). R_A in Fig. 20b means the corresponding biomass reduction in A ('crop') caused by B; R_A(rel) in Fig. 20c is that reduction as a percentage of the biomass of A in the absence of B (i.e. at $N_B = 0$). The curves are strict mathematical derivations from the production–density relationship $Y = X \cdot (a + bX)^{-1}$, as follows:

$$Y_B(\text{rel}) = 100 \cdot (a + bN_B) \cdot (a + bN_B + bX_A)^{-1}$$

$$R_A = bN_B X_A \cdot [(a + bX_A) \cdot (a + bN_B + bX_A)]^{-1}$$

$$R_A(\text{rel}) = 100 \cdot bN_B \cdot (a + bN_B + bX_A)^{-1}$$

The Y_B(rel) curves in Fig. 20a, are hyperbolic curves resembling the weed curve in Fig. 14. The lowest Y_B(rel) curve represents B at the very low density of $N_B = 10$. This curve has a different course at the lowest values of X_A, i.e. near $X_A = 0$. These low total densities ($X_A + N_B$ = densities of 'crop' plus 'weeds') are assumed to be too low to cause competition even in a stand with plants in advanced stages of growth (cf. Fig. 15, lower diagram, and discussions on reference models, pp. 138–144). Experimental data in such a situation may be data representing a sigmoidal curve. Now we may leave this special situation.

Figure 20a, shows that the decrease in Y_B(rel) caused by increasing X_A becomes stronger at lower N_B levels than at higher. This can be simply explained as follows. We have a stand where two plant categories, A and B, represent equal plants in a uniform mixture. Here, a given increase in the A density, X_A, always results in a stronger relative increase in the total density ($X_A + N_B$) at a lower level of N_B than at a higher. This, at the same time, means a stronger relative increase of the competitive pressure at the lower level of N_B.

Influences of N_B analogous to those in Fig. 20a, are found in experiments where the A and B plants in mixed stands are different species (Håkansson, 1975b, 1983b, 1997a). This can be explained in the same way as when A and B are equal plants.

This dependence of Y_B(rel) on N_B must be considered in experiments aimed at ranking weeds with respect to their relative competitiveness by comparing their response to the crop plant density. (See Chapter 9.)

Figures 20b and c show that, as expected, both the absolute and the relative yield reductions in A caused by B – i.e. R_A and R_A(rel), respectively – increase with increasing B density (N_B). With increasing A density (X_A), the absolute yield reduction (R_A) first increases and, after passing a maximum, decreases (Fig. 20b). The relative reduction, R_A(rel), on the other hand, decreases right from the lowest A density (X_A). The low values of the absolute yield reduction (R_A) at low A densities is simply explained by the fact that the production of A is low there even without competition from B. The maximum yield reduction in A (maximum R_A) falls at an increasingly high A density (X_A) with an increasing B density (N_B).

Both on logical grounds and on the basis of field and box experiments, it can be stated that the types of relationships illustrated here have a general validity. They are thus valid when plants A and B are different species, e.g. crops and weeds (Håkansson, 1988b, 1997a; cf. Fig. 24).

Besides reduced growth, increased mortality (self-thinning) is a more or less apparent response of plant individuals to an increased density (cf. Harper, 1977; Cousens and Mortimer, 1995). The regulation of the stand composition and structure by mortality differs with plant species and environmental conditions. Strong differences in traits and competitive positions (e.g. due to different times of emergence) between competing plants may result in a high degree of mortality among the weaker ones. Strongly delayed plants may thus die to a great extent. This was shown in Table 2, stressing the difficulties of seedling establishment on an area already occupied by a dense plant stand.

Figure 21 presents the result of calculations (Håkansson, 1975b) based on data from 20 field experiments on seed rates in spring cereals reported by Granström (1962). Weeds were predominantly annual weeds, typical in spring crops in Southern and Central Sweden. Regarding the weed response to the cereal seed rate, there were considerable differences between species and fields. However, the tendency was always the same. Thus, the curves in Fig. 21 illustrate the typical situation very well; the weeds respond to an increased crop plant density more frequently by a decreased plant size (growth) than by death of entire plants. Among the cereal plants, the mortality (self-thinning) usually increases much less than among annual weeds in response to an increased cereal plant density (seed rate).

Figure 22 illustrates results from a field experiment in spring barley sown with different seed rates, with and without chemical weed control. Weeds were mainly annuals common on Swedish arable land, *Chenopodium album* being dominant. Curves represent regression equations based on the measured data (Håkansson, 1979). Figure 22a illustrates a situation where maximum grain yield is reached at a commonly used, relatively low, seed rate (cf. Fig. 29, showing high yields of barley even at extremely high seed rates, probably due to weather conditions preventing lodging). In the experiment dealt with in Fig. 22, barley grew together with a rather dense weed population. Therefore, the curve in Fig. 22a represents stands where crop–weed competition was present right from the lowest crop seed rate.

Figure 22b shows that the weed biomasses per unit area decreased with an increased crop seed rate. It should be stressed that this was the case even at the high seed rates where the crop yield decreased. This is the typical weed response in cereals, observed also in other Swedish experiments, in winter cereals as well as in spring cereals (cf. Andersson, 1984, 1986, 1987). It may be the most common response even when lodging in certain cases might complicate the picture.

The curves in Fig. 22c and d describe absolute and relative yield increases resulting from effective chemical weed control. These curves are analogous to the R_A and $R_A(\text{rel})$ curves in Fig. 20. The broken line indicates an extrapolation.

Figure 22b describes the response of the entire weed population to the crop plant density (seed rate). However, a weed population usually includes species with

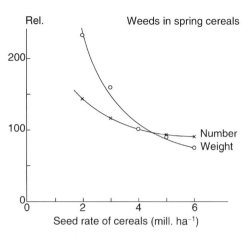

Fig. 21. Relative changes in the number and the above-ground biomass of annual weed plants per unit area with changes in the seed rate of spring cereals. Curves are regressions calculated by Håkansson (1975b) on the basis of mean values from 20 field experiments in Southern Sweden reported by Granström (1962). Determinations in advanced stages of plant development, but before the cereal crops were ripe. The spring cereals were two-row barley, six-row barley, wheat and oats in six, five, six and three experiments, respectively.

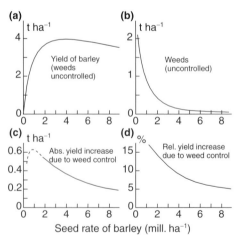

Fig. 22. Effects of seed rate in two-row barley in a field experiment at Uppsala (Håkansson, 1975b). The weed flora was strongly dominated by annuals, largely *Chenopodium album*. (a) Yield of barley is grain harvested in late August in plots without weed control. (b) The curve of weeds describes the dry weight of aerial shoots in early August. (c) Absolute yield increase and (d) relative yield increase in barley were achieved by chemical weed control resulting in effects of 95–99% by weight.

different competitive ability. The proportions of their biomasses may therefore change considerably when the competitive pressure increases with an increased density. Thus, when the crop plant density is increased, the biomasses of competitively weaker weeds decrease more strongly than the biomass of the stronger ones, the proportion of which therefore increases in the total weed biomass. This has been illustrated in many experiments (e.g. Håkansson, 1983b; Andersson, 1987). Due, for instance, to varying times of emergence, stronger and weaker plants occur within one and the same species. These plants exhibit a differently strong response to competition (see also pp. 106–111).

Figure 23a shows results of a field experiment where *Sonchus asper* and *Chenopodium album* dominated the weed flora in unsown plots. In barley plots, however, *C. album* became increasingly dominant relative to *S. asper* with increasing seed rate of the crop. The relative competitiveness of two plants largely depends on the environment, but the competitive difference between the two species now illustrated seems to persist under very varying conditions. A consequence of this is that, whereas *C. album* is a frequently dominant weed in spring cereals in Sweden, *S. asper* is often as important in less competitive crops, such as many vegetables (cf. Table 5).

Figure 23b indicates differences between three spring-sown cereals and spring-sown oilseed rape with respect to their competitive effect on weeds, as found in a field experiment at Uppsala. Differences of this type may vary considerably between fields, years and varieties. However, the differences now presented seem to have shown a rather typical appearance of the four crops in Sweden under conditions prevailing in the 1970s. When spring varieties of oats and two-row barley are sown at ordinary seed and fertilizer rates, the former mostly exerts a weaker competition against weeds than the latter. The period of effective shading is usually longer in oats than in wheat. Oats therefore sometimes suppress weeds better than does spring wheat, particularly those weeds that can grow over an extended

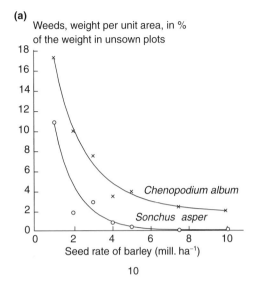

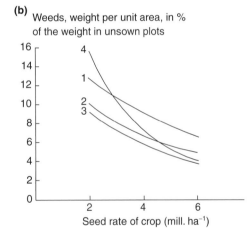

Fig. 23. Different competitive abilities of weeds and crops exemplified in field experiments at Uppsala (Håkansson, 1995d). (a) Among several species occurring in the weed flora, *Sonchus asper* dominated the weed biomass even more than *Chenopodium album* in plots where no crop was sown. The diagram shows that in plots with two-row barley, *S. asper* decreased much more strongly than *C. album* with increasing crop seed rate. This agrees with general observations that, whereas *C. album* is a predominant weed in spring cereals in the Nordic countries, *S. asper* grows vigorously mainly in cereal stand gaps and in weakly competitive crops such as sugar beet, potatoes, peas and vegetables. (b) Four spring-sown crops exhibiting different competitive effects on weeds: (1) wheat, (2) oats, (3) two-row barley and (4) oilseed rape.

period. As stressed repeatedly, however, the relative competitiveness of plants may change greatly with changes in the environment, e.g. with changed fertilizer rates. At ordinary fertilizer levels, barley exerts a stronger competition against weeds than spring wheat, but the opposite has been shown at low supply of nitrogen (example: Table 31, Fig. 34).

Figures 24 and 25 illustrate results from a greenhouse box experiment with barley and white mustard in mixed and pure stands. Barley may be considered the crop (A) and white mustard a weed (B). The graphs are based on bivariate regression equations, describing biomasses Y_A and Y_B as functions of densities X_A and X_B (Håkansson, 1983b).

The curves in Fig. 24 describe the biomass reduction in A (R_A) caused by B. They are given as functions of the A density (X_A) at different constant B densities (N_B) and are maximum curves illustrating the

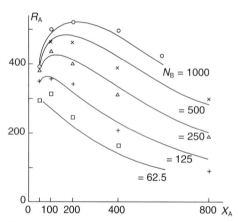

Fig. 24. Biomass reductions (R_A: g m^{-2}) in plant category A (two-row barley) caused by plant category B (white mustard) presented as functions of X_A (initial density of A, number of barley seeds sown per m^2) at different N_B (constant initial density of B, number of white mustard seeds sown per m^2). Sown seeds uniformly distributed. Greenhouse experiment in square boxes with soil (area 0.64 m^2). Curves are based on dry weights per m^2 of above-ground plant parts cut when straw elongation in barley had ceased and mustard had been flowering for a period. (The position of the lower curve certainly reveals an imperfection in the bivariate regression equation underlying the curves.) From Håkansson (1983b: pp. 60–67).

same type of responses to densities as the 'reference-model' curves of R_A in Fig. 20b.

Figure 25a illustrates relationships between the relative yield reduction, R_A(rel), in barley grown at constant densities (N_A) caused by white mustard intermixed at varying densities (X_B). The R_A(rel) curves are of the same type as the Y_B curves presented in Fig. 25b. The R_A(rel) curves increase with a gradually decreasing slope when X_B increases. This indicates that the yield reduction per weed individual decreases with increasing weed plant density (cf. Håkansson, 1983b, 1988b). Various equations for describing the relationship between yield reduction and densities are reviewed and discussed by Cousens (1985a). The curves in Figs 24 and 25 are based on production–density equations containing different $X_A X_B$ terms, indicating interactions between X_A and X_B (Håkansson, 1983b).

Figure 25c indicates an almost linear relationship between R_A(rel) and Y_B. The linear relationship is due to the fact that the R_A(rel) and Y_B curves have similar forms, as illustrated in Fig. 25a and b. In 'reference model' situations where A and B are equal plants, the relationships between R_A(rel) and Y_B are strictly linear (Håkansson, 1983b). The relationships may, however, differ strongly from linearity when plants A and B differ more than the plants represented in Fig. 25 (Håkansson, 1988b, 1997a). However, even the relationship between the grain yield of spring barley, or the seed yield of peas and rape, and the aerial shoot biomass of the perennial *Elymus repens* have proven to be linear, or nearly linear in experiments by Melander (1995). For autumn-sown wheat and rye, the relationships were more uncertain in these experiments.

Production–density relationships of weeds probably pass maxima at certain density levels analogously to those of crops discussed previously. Results indicating such a maximum seem to be rare or lacking (cf. Cousens and Mortimer, 1995). As mentioned previously, however, preliminary results from an experiment of my own with *Sinapis arvensis* (B) in two-row barley (A) suggest that both the weed biomass, Y_B, and the barley yield reduction, R_A(rel), passed a maximum at a certain B density (X_B). It would not be surprising to find such

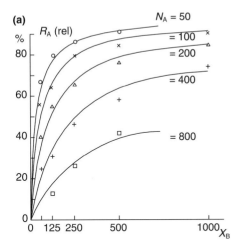

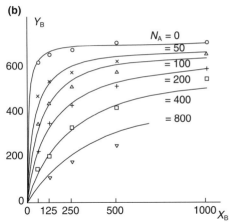

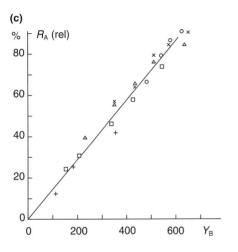

a maximum in certain weed species even within ranges of densities commonly practised. In an experiment with mixtures of barley (H), wheat (T) and white mustard (S) presented in Fig. 11 of Håkansson (1988b), the production of S presented as a function of the *total density* of two or three species, i.e. the density of H + S, T + S or H + T + S, exhibited a distinct maximum at moderate density levels.

The long-term dynamics of a plant population is, of course, determined by the formation and properties of reproductive and perennating organs of the plant. It is therefore of great interest to know how biomasses and properties of various plant organs are interrelated. Information exists on how the ratio of seed set to vegetative production changes in response to competition and other influences, but the results are not very clear. From studies of annual weed plants, Thompson *et al.* (1991) and Schnieders (1999) report that the relationships between seed and vegetative production are fairly well described by straight lines. There is, indeed, a positive correlation between the number or weight of seeds, on the one hand, and the total biomass or vegetative production, on the other, but the real relationship is not necessarily linear. In any case, it is probably not linear over wide density ranges (cf. Wilson *et al.*, 1995; Van Acker *et al.*, 1997a).

The conditions certainly vary between species (e.g. Schnieders, 1999). From my own field observations, two examples may be given. Thus, whereas very small plants of *Chenopodium album* in dense cereal stands frequently set one or a few seeds, even comparatively big plants of *Centaurea cyanus* in dense cereal stands often do not flower.

Fogelfors (1973) studied the response of weeds to light intensity. *Chenopodium*

Fig. 25. Relationships based on the same experiment as that presented in Fig. 24 with two-row barley (A) and white mustard (B) grown in mixed and pure stands. R_A(rel), relative biomass reduction in A caused by B; Y_B, biomass of B (g m^{-2}); N_A, constant density of A, and X_B, varying density of B, both denoting number of seeds sown per m^2.

album, Galeopsis tetrahit and *Stellaria media* were among species flowering at very low light intensities, whereas species such as *Capsella bursa-pastoris, C. cyanus* and *Avena fatua* exhibited a more restrained flowering. Granström (1962) reports on plant and seed weights of *C. album, Sinapis arvensis* and various crop species from a great number of plant density experiments in fields and pots. In calculating ratios of the seed to the total plant weight at different crop plant densities, I have found seemingly conflicting influences of the density on these ratios. With increasing density, these ratios have either decreased or increased, or even exhibited a more or less pronounced maximum (calculated values often uncertain). This issue requires thorough study.

From my collective experiences, I suspect that the ratios of the seed weight to the total plant weight in reality follow maximum curves when described as functions of the crop plant density (cf. Fig. 19 and associated discussion). If there are maxima, these might fall at different densities and be differently pronounced in different plants and environments. Note that relationships represented by maximum curves may easily show seemingly conflicting results when studied within a narrow density range, because they may then represent density ranges falling either below or above the maximum, or they may include the maximum.

The production of underground regenerative organs of perennial weeds normally decreases more than the production of aerial shoots at a lower than at a higher light intensity and decreases more in competition with a crop than without such competition (Håkansson, 1968a, 1969d; Håkansson and Wallgren, 1972b; Skuterud, 1984). Although it could be stated that an increased shading from a crop lowers the growth of underground regenerative organs proportionally more than the growth of aerial shoots, it will be of interest to study in more detail how these proportions change.

The plastic adaptation of plants to environmental conditions, including competition, deserves thorough studies in both crops and weeds. In crops, this adaptation largely determines the quality of harvested products. The ability of weeds to adapt by plastic responses determines their population dynamics and is indicative of their long-term survival.

Effects of the Spatial Distribution of Plants

Theory

We may intuitively believe that plants uniformly distributed over an area utilize the area more efficiently than similar plants growing in scattered patches. We can understand this as a logical consequence of the non-linear form of production–density relationships (solid curves in Fig. 26). Suppose that a field is composed of alternating squares or strips, equal in size or width, with uniformly distributed plants at different densities. Suppose, further, that *the squares or strips are so large or wide that their mutual influences on each other's production are negligible*. Then, their mean production per unit area is the mean of the production at different densities according to the solid curve (the three upper graphs in Fig. 26; cf. Hudson, 1941). This mean (*B, C* or *D*) is lower than the corresponding production (*A*) in the uniform plant stand.

Based on an analogous discussion, it is understood that weeds respond to competition from an uneven crop plant distribution by an increased mean weight per unit area (bottom graph in Fig. 26). *If the alternating areas with different crop plant densities are not very small*, production *A* becomes higher for the crop and lower for the weeds than levels *B, C* or *D*. With decreasing size or width of the areas with different crop densities, their mean production approaches *A*. This is illustrated by experimental results in Fig. 27.

Now, if we interchange crop and weeds in Fig. 26, we understand that uniformly distributed weeds produce more biomass per unit area than weeds with a more patchy distribution (cf. upper graphs). They then also cause a stronger yield reduction in the crop than when growing in patches (cf.

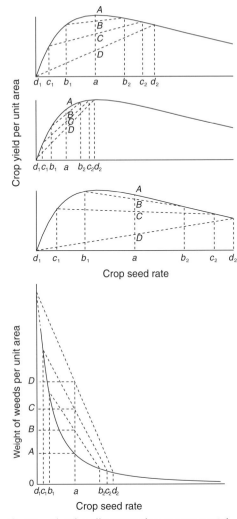

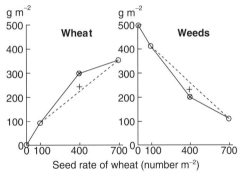

Fig. 27. Illustration of influences of the spatial distribution of crop plants (or seeds sown) on the biomass production of crops and weeds per unit area. Dry weights of the aerial shoots of spring wheat and annual weeds are presented as results of a greenhouse experiment in square boxes with an area of 0.64 m² (Håkansson, 1975b). In certain boxes, wheat was not sown, in others wheat was sown, either with 100, 400 or 700 seeds m⁻² evenly distributed (square distribution), or with 100 and 700 seeds (the mean of which being 400) m⁻² in alternating equal-sized squares. The areas of these were either 4 or 16 dm². In the squares, distances between margin seeds and square edges were half of those between the seeds within the squares. Weeds emerged spontaneously and similarly in each box.
○   Production in even stands of wheat.
×   Production in uneven stands of wheat; squares 4 dm².
+   Production in uneven stands of wheat; squares 16 dm².

Fig. 26. Sketches illustrating that an even spatial distribution of crop plants results in a higher crop production (upper diagrams) and a lower weed weight (lower diagram) per unit area than does an uneven distribution. (From Håkansson, 1975b.) Solid lines describe production–density relationships at an even plant distribution. For both crop and weeds, *A* marks the production at density (seed rate) *a*, whereas *B*, *C* and *D* denote mean productions in a field with alternating squares or strips that are equal in size or width but differ in plant densities as described by paired density levels: b_1–b_2, c_1–c_2 and d_1–d_2. The mean of each pair of densities is always *a*. The squares or strips are assumed to be so large or so wide that they do not mutually influence each other's production markedly (thereby not reducing differences in their production).

bottom graph). *Influences of the spatial plant distribution should not be forgotten when responses and effects of weed plants at different (mean) densities are measured and evaluated.*

Row spacing and plant distribution in the row

Suppose that the optimum distribution of seeds sown in rows is a uniform distribution in the row and a row spacing that equals the spacing in the row and, further, that in a crop–weed stand, not only is the highest yield then achieved, but also the lowest weed weight.

A number of experiments in cereals at Uppsala verify this supposition with some modifications according to Table 25 (Håkansson, 1984b). Seeds were sown by hand at different spacings between the rows and different distributions within the row. Seedbeds were treated equally and 400 seeds m^{-2} were sown throughout. Cereal and weed plants were harvested by cutting at the soil surface level when the cereals had reached the stage of early milk ripeness. Both living and dead plant parts with recognizable structures were collected, dried and weighed. Weeds were annual species typical of cereal fields in the area. Further details on the experiments are presented with Table 25.

As shown in Table 25, seeds sown one by one with a 5-cm spacing both between and within the rows represented the most

Table 25. Weight of grain (air-dried) and weeds (above-ground dry matter) per unit area in cereals as influenced by row spacing and seed distribution in the row. Cereal seeds were sown in rows, one by one or in dense groups at distances depending on the number of seeds per group. Relative values presented are means of 27 experiments with barley, oats, spring wheat, winter wheat and rye sown in small field plots (24 experiments) or in large aluminium boxes (three experiments) at Uppsala. Seed rate 400 seeds m^{-2}. (From Håkansson, 1984b.)

Seed distribution		Dry matter produced per unit area (rel. values)[a]	
Number of seeds per position/group in the row	Distance between positions in the row (cm)	Grain	Weeds
Row spacing 5 cm			
1	5	100	95
Row spacing 10 cm			
1	2.5	100	100
2	5	98	105
4	10	97	118
8	20	93	145
Row spacing 20 cm			
1	1.25	94	148

[a]Absolute mean weights representing the 27 experiments at the row spacing 10 cm: grain yield 397 g m^{-2}, weed weight 53 g m^{-2}.

even seed distribution. Where the seeds were sown one by one, there was no significant difference between row spacings of 5 and 10 cm, either in grain yield or in weed weight. In a few individual experiments, there were significant differences in grain yield and/or weed weight in favour of either the 5-cm or the 10-cm row spacing. There was a tendency for higher yield and lower weed weight at the 5-cm row spacing at higher weed amounts than at lower (in response to intra-row spacing, see Fig. 28). At the 20-cm row spacing, the grain yield was significantly lower and the weed weight significantly much higher.

An increased seed clumping in the rows, from single seeds to eight seeds per position, resulted in a gradually decreased grain yield and a steadily increased weed weight.

On the whole, the results indicate that moderate divergences from the most even spatial distribution only lead to minor changes in plant production in the stands (cf. discussion on sowing technique on p. 104).

Figure 28 shows that an increased unevenness of the sowing in the row caused a stronger decrease in the yield of grain and straw at a stronger weed population than at a weaker. The calculations shown in Fig. 28 are based on seven pairs of experiments in spring cereals with different amounts of weeds. These pairs were placed close to each other in fields with even soils and comparatively even weed floras. The ways of achieving different degrees of weed abundance in the experiments are presented in the legend to Fig. 28.

It should be noted that weed biomasses changed proportionally more than cereal yields with changes in the distribution of the crop seed sown. This is seen both in Table 25 and Fig. 28 (note that scales differ between graphs). Figure 28 also illustrates, as a tendency, a stronger *relative* increase in the weed weight in the weaker than in the stronger weed population. This tendency might reflect a difference in the response that could be expected as a consequence of different competitive power of a weed population. Weaker weed stands – due to

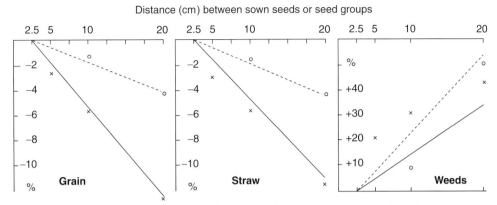

Fig. 28. Changes in the grain and straw yield (air-dried) of cereals and the weight of weeds (above-ground dry matter) per unit area with changed distribution of the cereal seeds sown. Calculations are based on seven pairs of small-plot field experiments arranged close to each other in seven fields. Different degrees of weed abundance were achieved in the experiment pairs, either by treating weeds with a moderate herbicide dose in one of the pairs (in five fields) or by sowing weed seeds in one of them (in two fields with otherwise small amounts of weeds) as described by Håkansson (1984b). Plotted mean values and regression lines represent the seven pairs of experiments with different degrees of weed abundance. Absolute values of grain and straw yields and weed weights were as follows (g m^{-2}, averaging experiments and treatments).
o   Lower abundance of weeds: grain 394; straw 524; weeds 21.
×   Higher abundance of weeds: grain 364; straw 485; weeds 84.
Seed rate 400 seeds m^{-2}, row spacing 10 cm. Seeds sown either one by one at a spacing of 2.5 cm in the rows, or in groups of two by two, four by four or eight by eight at spacings of 5, 10 and 20 cm, respectively, as seen on the horizontal axes of the diagrams. (*Note that vertical scales differ.*)

competitively weaker plants or lower plant densities – thus respond more strongly to changes in the crop plant density or distribution than do weed stands that are competitively stronger (cf. Håkansson, 1975b, 1979). At the same time, changed competition from a crop has a selective effect on differently competitive weed plants (cf. Fig. 23a) and weed individuals (cf. Fig. 75 in Håkansson, 1988b).

Strand (1968) reviewed results from row-spacing experiments in barley, oats, rye or wheat, in a total of 42 experiments from many countries. He calculated the percentage yield change caused by a given increase of the row spacing. The changes varied greatly between experiments. Possible differences between cereal species could therefore not be clearly shown. However, the average yield decreased in all cereals when the row spacing was increased beyond certain low levels, consistently so beyond 10 cm. In the range 7.5–40 cm, a row spacing increase of 1 cm led to a mean

yield decrease of 0.64%. Within the most common ranges of row spacing in cereals practised in humid areas, i.e. in the range 10–20 cm, the corresponding decrease was 0.72%. In this range, Bengtsson (1972) calculated an average decrease of 0.65% in Swedish experiments with spring wheat and barley.

In the great majority of experiments reviewed by Strand, weeds had been controlled or their occurrence was not considered. Among early experiments studying row-spacing effects on cereals and uncontrolled weeds, those of Granström (1963) may be mentioned. They were carried out in spring-sown barley, oats and wheat at ordinary seed rates. Calculations based on data from these experiments can be summarized as follows: a row spacing increase from 12.5 to 20 cm resulted in an average grain yield decrease of 5% (0.7% per cm spacing increase) and an increase in the weight of annual weeds per unit area of about 67% (9% per cm increase of row spacing). It can be seen again that the relative weed increase was much greater than the relative grain yield decrease.

Andersson (1986) reports results from a series of 50 multifactorial field experiments in spring-sown wheat and barley carried out in Sweden over 5 years. The experiments included five seed rates (from about 50 to 150% of 'ordinary' rates) and three row spacings (6, 12 and 18 cm). The rows were either drilled with shoe coulters or sown in bands with wing coulters. These arrangements were carried out *with* and *without* chemical weed control. The weight of weeds decreased proportionally much more than the grain yield increased with increased seed rate. An increased distance between rows or band centres from 6 to 12 cm caused varying changes in yield and weed weight, on average a very uncertain decrease in grain yield and a moderate increase in weed weight. There were similar, but greater, changes with an increase of the spacing from 12 to 18 cm. Band sowing with wing coulters resulted in slightly higher grain yields and lower weed weights than drilling with shoe coulters. The different coulters caused significant differences only at the wider row spacing, 18 cm. At this spacing, the spread of the seed in bands improved the plant distribution relatively more than at smaller spacing.

The improved horizontal distribution of the seed when sown with wing coulters was probably counteracted by deteriorated seedbeds and an increased vertical spread of the seed. Slightly higher yield increases than in Andersson's experiments were obtained by Heege (1977) and also by Pehkonen and Sipilä (1984), who compared sowing with shoe coulters and improved wing coulters.

However, in competitive crops, such as cereals, improvements in the horizontal seed or plant distribution near the theoretical optimum mostly result in small and uncertain effects on crop yields and only moderate effects on weeds.

Optimum plant distribution is obtained with a smaller row spacing at higher plant density than at lower. The higher the plant density, the stronger the crowding in the row increases when the row spacing is increased above its theoretical optimum. It is therefore not surprising that an increased row spacing above this optimum leads to a greater yield reduction at a higher plant density (seed rate) than at a lower. This is seen in Fig. 17a, c and d, for soybeans (Wiggans, 1939), spring-sown oilseed rape (Ohlsson, 1976) and peas for canning (Bleasdale, 1963). Experiments in winter wheat (Whybrew, 1958) and spring cereals (Bengtsson, 1972) are other examples illustrating this. Effective weed control was performed in all these experiments.

A field experiment in barley *with* and *without* weed control was conducted at Uppsala to study the joint effects of density and row spacing in broader lines (Håkansson, 1975b). The experiment comprised densities (seed rates) and row spacings over wide ranges. The design is apparent from Fig. 29, where the main results are also summarized. Annual weeds typical of the area developed in the absence of weed control. In comparable plots, weeds were effectively controlled by a mixture of herbicides adapted to the weed flora. Special effects of seed rate, row spacing and weed control are presented in Tables 26 and 27.

In Fig. 29, grain yields and weed dry weights (above-ground plant parts) are described by curves based on non-linear, bivariate regression equations (Håkansson, 1975b). (Because of dry weather in the latter part of the growing period, there was no lodging even at the highest seed rates. In barley, maximum yield is otherwise usually obtained at lower seed rates than in this experiment; cf. Fig. 22.) The degree of self-thinning among the barley plants was low even at the highest seed rate.

Figure 29 shows that an increase in the row spacing (from 10 cm, which probably exceeded the optimum throughout) resulted in more harmful effects at higher seed rates (crop plant densities) than at lower. This is most easily seen in Fig. 29b and d, which illustrates grain yields and weed weights as functions of the row spacing. The graphs show that the relationships are non-linear, but, at the same time, that linear regressions may be acceptable for describing row spacing effects within certain limited ranges.

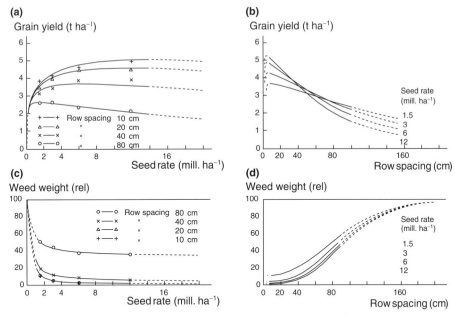

Fig. 29. (a, b) Grain yield of two-row barley per unit area. (c, d) Dry weight of the aerial shoots of annual weeds per unit area expressed as a percentage of the weight in unsown plots. Field experiment at Uppsala with different seed rates and row spacings of two-row barley (Håkansson, 1975b). Results from plots where weeds were not controlled. Weed dry weight, determined in early August, included dead parts of the shoots. It was 316 g m^{-2} in unsown plots, i.e. in the absence of competition from barley. Curves represent bivariate non-linear regression equations fitted to weight data. The weed flora was almost entirely composed of annual species dominated by *Stellaria media, Fumaria officinalis, Galium aparine, Lamium purpureum* and *Polygonum aviculare*.

Table 26. Percentage decrease in grain yield of spring barley at a specified row spacing increase, as influenced by the seed rate and the range of row spacings in stands without or with weed control (3 l ha^{-1} of Oxytril 4). Values calculated from regression equations based on the same field experiment as Fig. 29.

Range of row spacings (cm)	Percentage grain yield decrease by a 1-cm increase of the row spacing at the following seed rates (in million seeds ha^{-1}):			
	2	4	6	8
No weed control				
10–12.5	0.26	0.47	0.62	0.73
15–17.5	0.37	0.59	0.75	0.86
20–22.5	0.44	0.67	0.83	0.95
Weed control				
10–12.5	0.23	0.44	0.61	0.73
15–17.5	0.32	0.52	0.66	0.80
20–22.5	0.36	0.56	0.72	0.84

Table 26 illustrates the percentage decrease in grain yield caused by a row spacing increase of 1.0 cm, as calculated from non-linear, bivariate regression equations. The decrease differed with the seed rate and the row spacing range. Furthermore, the yield decrease was also greater where no weed control was performed than where weeds had been effectively controlled. This is analogous to the stronger yield decrease caused by an uneven seed distribution at greater amounts of weeds than at smaller (Fig. 28).

Table 27a and b illustrates grain yields and weed weights per unit area in stands of barley in the absence of weed control. The percentage weight of weeds occurring after 'effective chemical control' is presented in part c of the table. The values in part c illustrate interactions between herbicide effects and effects of competition from the

Table 27. Grain yield and above-ground bio-mass of annual weeds in stands of spring barley sown at different row spacings and seed rates, not followed by weed control. In addition, relative weed biomasses in stands with weed control (3 l ha^{-1} of Oxytril 4). Results from the same field experiment as those presented in Fig. 29 and Table 26.

Row spacing of barley (cm)	Seed rate of barley (million seeds ha^{-1})				
	0 (unsown)	1.5	3	6	12

(a) Grain yield of barley (g m^{-2}) in stands without weed control

10		384	411	443	490
20		357	394	442	442
40		312	344	385	381
80		263	265	232	209

(b) Dry weight of weeds[a] (g m^{-2}) in stands without weed control

Unsown	316				
10		38	17	10	6
20		44	23	12	7
40		61	36	29	21
80		159	140	117	113

(c) Dry weight of weeds in stands with chemical weed control as a percentage of this weight in untreated stands

Unsown	60.9				
10		4.5	2.1	2.1	1.5
20		3.4	3.5	3.2	1.7
40		6.1	5.1	4.7	5.0
80		27.1	23.1	21.5	17.3

[a]Weed flora almost entirely composed of annual species presented in legend to Fig. 29.

crop. The herbicide application followed recommendations concerning weeds and crop.

In stands with more normal row spacings (between 10 and 15 cm) and seed rates (between 3 and 6 million barley seeds ha^{-1}), the joint effect of herbicides and competition on weeds was 97–98% by weight (Table 27c). It was possibly slightly weaker at the lowest seed rate of 1.5 million seeds ha^{-1}. This joint effect decreased considerably when the row spacing was raised to the extreme of 80 cm. In unsown plots, without competition from barley, the weed weight was only lowered by about 40%. This may be commented on as follows.

Regarding doses, herbicides are usually tried out in field situations similar to those where they are intended to be used. 'Normal doses' of a herbicide tried out for use in competitive crops are therefore sublethal to many of the weed plants in the absence of strong plant competition. Although they are weakened by the herbicide, these weeds largely recover in the absence of such competition.

Several authors have recorded increased effects of herbicides with increasing plant density in cereal stands (e.g. Pfeiffer and Holmes, 1961; Suomela and Paatela, 1962; Hammerton, 1970; Hagsand, 1984). Intended effects on weeds can thus be achieved by smaller herbicide doses at higher seed rates than at lower.

The figures in Table 27c, indicate that both an increased seed rate (crop plant density) and an improved spatial distribution of the crop plants interact positively with the herbicide in reducing weed growth. However, the positive effect of increased competition at application of a foliage herbicide may be more or less counteracted by an increased crop leaf cover reducing the herbicide amount reaching the weeds. Despite this, however, suitable combinations of densities and spatial distribution of the crop plants and various measures favouring a rapid and simultaneous crop plant emergence result in strong interactions that suppress the weeds. Such interactions should be used to reduce the need for large herbicide doses, intense soil tillage, etc. The great importance of measures leading to early emergence of the crop relative to the weeds is illustrated in the following.

Effects of the Relative Time of Emergence

General

Figure 2 and Table 2 illustrate that new plants have difficulties in establishing in a closed stand of older plants. In an ordinary seedbed, weeds may emerge at very different times in relation to the crop. In a good seedbed, the great majority of the plants of cereals and many other crops emerge within 1–2 days following the first emergence, but the period is often longer.

As shown in Table 22, a difference of one week in the time of emergence of the plants in a cereal stand can influence their growth conditions very strongly. Table 23 shows that the growth and competitive power of wheat plants were weakened at an increased depth of sowing (in soil sufficiently moist for plant germination and growth at all depths). At the same time, the existing weeds grew more vigorously. This was interpreted as a joint effect of delayed emergence of the wheat plants and weakening of them due to increased consumption of food reserves before their emergence.

Effects of the relative time of plant emergence have been studied in many experiments, regarding the outcome of competition between plants of both the same and different species (e.g. Mann and Barnes, 1947; Håkansson and Larsson, 1966; Dahlsson, 1973; Håkansson, 1979, 1986, 1991, 1997b; Peters, 1984; O'Donovan et al., 1985; Cousens et al., 1987; O'Donovan, 1996). At the same time as the importance of these effects are discussed in the following, the importance of adapting seedbed preparation and depth of sowing to soil and climatic conditions are discussed.

Experiments with cereals and weeds

A number of experiments were carried out at Uppsala in order to study effects of the relative time of emergence of cereal plants and annual weeds (Håkansson, 1979). Some of the main results are presented in Tables 28–30 and Figs 30–32.

In a small-plot field experiment (Table 28, Fig. 30) two-row barley was sown by pressing seeds, one by one, to a depth of 3 cm in the soil, using a thin rod. This was done on four occasions, leading to different times of barley emergence. Repeated moderate irrigation ensured moisture conditions suitable for the germination and early growth of both crop and weeds. Barley seeds

Table 28. Effects of shifting the time of emergence of two-row barley in relation to the emergence of annual weeds (*Chenopodium album* 80–90% by biomass) on the same area. Barley sown at 'low normal' and 'high normal' seed rates. A small-plot field experiment at Uppsala. Dry weights of above-ground plant parts (alive + dead) of barley and weeds determined at the time of early milk ripeness of barley. Barley, when 'not delayed', and weeds were sown on 25 May, 'Day 0'. (From Håkansson, 1979.)

Seed rate of barley (seeds ha⁻¹)	Delay of barley (days)			
	0	3	8	12
Barley (g m⁻²)				
250	573	433	231	109
500	725	629	322	187
Weeds (g m⁻²)				
0	551	551	551	551
250	110	216	384	486
500	61	154	334	443
Barley (rel. weight)				
250	100	76	40	19
500	100	87	44	26
250	100	100	100	100
500	127	145	139	171
Weeds (rel. weight)				
250	100	196	349	441
500	100	251	545	724
0	100	100	100	100
250	20	39	70	88
500	11	28	61	80
Barley (as a percentage of barley + weeds, by weight)				
250	84	67	38	18
500	92	80	49	30

were sown with 250 and 500 seeds m⁻² at a row spacing of 10 cm. Seeds of six annual weeds, common in the area, were mixed into the superficial soil layer on the day of the earliest sowing of barley ('Day 0'). In this way, weeds emerged similarly throughout the experimental area. *Chenopodium album* became the dominant weed in all plots, both by number of plants (about 400 m⁻²) and by biomass (80–90% of the weed biomass).

With the sowing technique used, seeds and plants of weeds were negligibly disturbed when barley was sown late. The soil

Table 29. Effects of the time of emergence of *Chenopodium album* in relation to the emergence of two-row barley in the same 8-l pots in a greenhouse experiment. Calculations based on dry weights per unit area of above-ground plant parts (living + dead) of barley and weeds at termination of flowering in barley. (From Håkansson, 1979.)

Weed density[a]	Weed delay (days)	Plant density of barley[a]			
		1	3	6	9

Barley weight in stands with weeds as a percentage of the weight in weed-free stands

2	0	68	90	95	97
	3	84	95	96	99
	6	88	97	99	99
6	0	57	84	87	97
	3	66	82	90	96
	6	77	88	90	98
18	0	43	72	86	89
	3	57	75	84	95
	6	70	88	90	97

Weed weight in stands with barley as a percentage of weight when free of barley

2	0	57	14	5	4
	3	30	10	4	2
	6	22	5	3	1
6	0	61	22	12	6
	3	47	17	8	4
	6	31	9	5	3
18	0	72	36	18	11
	3	56	25	15	8
	6	42	16	8	5

[a]Number of plants dm^{-2}.

Table 30. Effects of the relative time of emergence of weeds and cereal plants grown in mixtures in 12-l pots in two greenhouse experiments (1 and 2). Calculations based on the dry weights of above-ground plant parts (living + dead) 1 month after start of experiment. Differences between early and late emergence were 7 days. (From Håkansson, 1979.)

Time of emergence		Weed weight in stands with cereals as a percentage of weed weight in pure stands	
Weed plants	Cereal plants	1. *Sinapis arvensis* in barley	2. Annual weeds[a] in wheat
Early	Late	67	42
Early	Early	22	20
Late	Late	20	22
Late	Early	5	3

[a]*Capsella bursa-pastoris* dominating.

temperature was 12–14°C at a depth of 3 cm in the soil during the period from early sowing to late emergence of barley. This means temperatures that were 3–4°C higher than on average at sowing of spring cereals in the area (Rodskjer and Tuvesson, 1976). Further information on the experiment is presented with the results in Table 28 and Fig. 30.

In Tables 29 and 30 and Figs 31 and 32, treatments and results of three greenhouse pot experiments (in winter) are presented. The greenhouse temperatures varied from maxima of 15–20°C in the daytime (18 h of light ensured by fluorescent tubes) to minima of 10–12°C in the night. In one of the pot experiments (Table 29, Fig. 31) the time of weed emergence was varied by sowing weed seeds on different days. In

two experiments (Table 30, Fig. 32) times of emergence were varied both for the cereals (two-row barley and spring wheat, respectively) and for the weeds (annuals).

In pot experiments, influences of light from the sides certainly reduce the effect of differences in the relative time of emergence. Although probably underestimated for that reason, the effects of the relative time of emergence measured in the pot experiments were strong in favour of early emergence.

The small-plot field experiment described in Table 28 and Fig. 30 simulated conditions in a cereal field with an abundant annual weed flora. It shows that a delay of barley emergence by 3 days resulted in a doubling of the weed biomass per unit area. A delay by 12 days led to a strong dominance of weeds over barley.

According to the upper graphs in Fig. 30, the relationships between biomasses and the relative time of emergence are described by straight lines. Similar relationships are shown in Fig. 31, which represents one of the pot experiments. In field experiments in Canada (O'Donovan *et al.*, 1985), linear regressions satisfactorily described relationships between the relative time of emergence of *Avena fatua* and the yield reduction

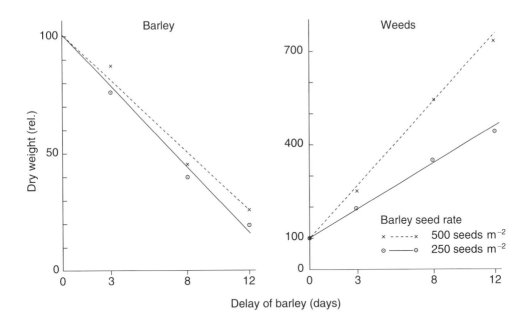

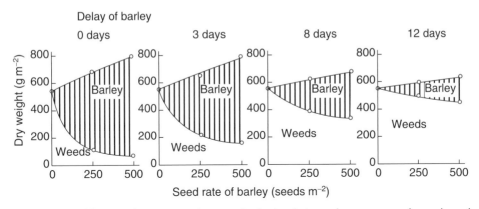

Fig. 30.   Effects of the time of emergence of two-row barley in relation to the emergence of annual weeds at two seed rates of barley. Dry weights of above-ground plant parts (alive + dead) of barley and weeds determined at the time of early milk ripeness of barley. Results from a small-plot field experiment presented further in Table 28.

in the crop (barley or wheat) caused by this weed. A delay of the weed emergence by one day relative to the crop lowered the yield reduction by 2–3% according to these regressions.

It is suggested, however, that the relationships discussed are not strictly linear. The graphs in Fig. 32 do not contradict this suggestion. At least at strongly different times of emergence, the relationships must

be non-linear (cf. Håkansson, 1991). Models representing sigmoidal relationships have been tested by Cousens *et al.* (1987) and appear reasonable. At the same time, however, the relationships can certainly be satisfactorily described by linear regressions at moderate differences.

The influence of a given change in the relative time of emergence depends on the temperature. To evaluate the influence of

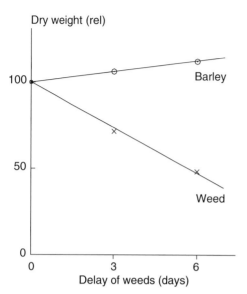

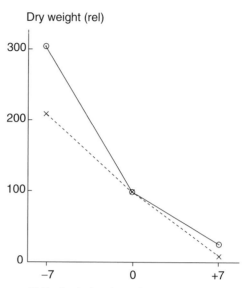

Fig. 31. Effects of the time of emergence of *Chenopodium album* in relation to the emergence of two-row barley in the greenhouse pot experiment presented further in Table 29. Relative values representing the dry weights of above-ground plant parts (alive + dead) per unit area at the termination of flowering in barley, representing means over four densities of barley and three of *C. album*.

Fig. 32. Effects of the relative time of emergence of annual weeds and cereal plants in mixed stands, otherwise illustrated in Table 30. Solid lines represent *Sinapis arvensis*, broken lines a weed flora dominated by *Capsella bursa-pastoris*. Relative dry weights of weeds are percentages of the weights of weeds sown at time '0', i.e. at the same time as barley in combinations 'Early–Early' or 'Late–Late' according to Table 30.

relative times of emergence in units of a more general validity, differences in time could be presented as temperature sums, 'degree-days' (cf. Håkansson, 1991: pp. 109–149).

Moreover, any factor influencing growth in a plant stand modifies the effect of the relative time of emergence. An important factor is plant density. At the higher seed rate of barley in the experiment presented in Table 28 and Fig. 30, the *relative* changes in the weed biomass were considerably greater and those of barley slightly smaller than at the lower seed rate. Table 29 illustrates joint influences of the relative time of emergence and the density of barley and weeds. By virtue of regression analyses, the values in the table indicate that, with increasing density of barley, the relative effect of delayed weed emergence became weaker on the barley plants and stronger on the weed plants, whereas increased weed plant density resulted in opposite effects. Similar effects were recorded in other pot experiments (Håkansson, 1991).

Table 22 illustrates joint effects of the time of emergence and the density in wheat, where every second plant emerged later than the others. An increased plant density led to an increased competitive pressure, particularly from the earlier plants. Although the density of the later plants was also increased, this caused a strengthened suppression of the delayed plants in favour of the earlier ones. Table 22 also shows that delayed plants suffered more in two-row barley than in wheat at an equal plant density (seed rate). This indicates a higher competitive pressure in barley than in wheat. When compared at the same plant densities, barley has mostly proven to be more competitive than wheat under the prevailing conditions (cf. Figs 23, 33 and 45 and Table 31).

Within each plant category distinguished in the experiments, the individual plants have varied more or less in growth and biomass production. This is certainly to a large extent a result of variation in the time of emergence among the individuals within the categories. Variation in the plant size was particularly pronounced among the weeds, the emergence of which has mostly extended over several days. Under field conditions, a small minority of the plants of a specified weed often produces the major part of the biomass of this weed. The dominant minority largely comprises plants emerged early in relation to the others, but variation in seed properties and soil microsites also contributes strongly. Consequences of heterogeneity in plant categories treated as units in mixed plant stands are discussed further on p. 136.

Competition Modifies the Response of Plants to Various Factors – Examples

The relative competitive power of different plants (species, subspecies, varieties, etc.) varies more or less strongly with the environment. A change in external conditions may affect different plant categories either in the same direction, but with different strength, or in a positive direction for some categories and a negative for the others. This will change their relative competitiveness.

When reactions to plant densities in a stand are used for characterizing changes in the relative competitiveness of different plants, it should be remembered that the relative time of emergence and spatial plant distribution are also very influential and should not be changed.

It may be possible to group plants reasonably well with respect to their 'average' relative competitiveness in specified situations as regards climate, soil, cultural conditions, etc. However, unforeseeable variations in weather and other factors make it difficult or impossible to predict the outcome of competition in a given field and year, even in a plant stand with a very well defined composition and structure in early growth. Thus, it is not only unforeseeable variation in environmental factors with well known influences that makes reliable

predictions difficult, but also many undefined factors, which often occur under field conditions.

In spite of these difficulties, knowledge of how various factors influence plants in different types of stand is necessary for important qualitative judgements. Some experimental results presented below will illustrate various environmental influences.

Response to nitrogen

Three spring cereals, two varieties of each, were sown on a loam soil moderately rich in clay near Uppsala (Table 31). The field was relatively rich in mineral nutrients even without fertilizer application. Before sowing, half of the experimental plots were treated with 75 kg nitrogen (N) ha^{-1} applied as calcium nitrate. Three seed rates occurred (2, 4 and 6 million seeds ha^{-1}).

Table 31 indicates that the nitrogen application caused a moderate grain yield increase, whereas the growth of weeds in most cases increased proportionally more. The weed flora almost exclusively consisted of annual species, the three most important of which are presented in Table 32. Although the weed plants responded differently to seed rates in the three cereal species, their responses to nitrogen are satisfactorily described by the means as in Table 31a. Wheat, and often also oats, are sown with somewhat higher seed rates than barley. Therefore, means for the two lower seed rates in barley and the two higher rates in wheat and oats are described in Table 31b, for comparisons with the means in Table 31a. (Analyses based on responses to the seed rate are discussed later.)

Regarding the reactions of the crop and weeds to nitrogen, there were sometimes differences between stands with different cereal varieties. However, the differences between varieties are significant only between Ingrid and Akka in barley (both for cereal and weed plants) and between Selma and Titus in oats (only for weeds). This applies both to comparisons over all the seed rates (Table 31a) and to comparisons where wheat and oats are represented by higher seed rates than barley (Table 31b).

Comparing cereal species, over varieties, there were significant differences in the response of weeds to nitrogen between stands of barley on one hand and stands of wheat on the other. At the higher nitrogen level (N_{75}), weeds were suppressed considerably more strongly in barley than in wheat (and in one of the cultivars of oats). It could be of interest to speculate on this.

Table 31.   Grain yield in spring-sown barley (two-rowed), wheat and oats and the dry weight of annual weeds (above-ground parts collected in early August, alive + dead) in an experiment with and without nitrogen application before sowing. N_0, no nitrogen applied; N_{75}, 75 kg ha^{-1} of nitrogen as calcium nitrate. A three-block field experiment at Uppsala in 1979 on a loam soil, rich in clay and well supplied with fertilizers in previous years.

Cereal species and variety	Grain yield (g m^{-2}) N_0	N_{75}	Dry weight of weeds (g m^{-2}) N_0	N_{75}	Relative weight (N_{75}/N_0) Grain	Weeds
(a) Means over seed rates of 2, 4 and 6 million seeds ha^{-1}						
Barley						
Ingrid	341	390	12.4	12.5	1.1	1.0
Akka	251	343	6.3	13.3	1.4	2.1
Wheat						
Drabant	255	315	4.6	28.3	1.2	6.2
Snabbe	232	297	6.0	27.4	1.3	4.6
Oats						
Selma	254	266	4.9	24.4	1.1	5.0
Titus	272	299	5.9	15.1	1.1	2.6
(b) Means over seed rates of 2 and 4 million seeds ha^{-1} in barley and 4 and 6 million seeds ha^{-1} in wheat and oats						
Barley						
Ingrid	334	384	16.6	15.2	1.1	0.9
Akka	252	336	8.1	14.5	1.3	1.8
Wheat						
Drabant	279	350	3.1	26.0	1.3	8.4
Snabbe	249	311	3.5	24.1	1.2	6.9
Oats						
Selma	281	282	3.8	25.0	1.0	6.6
Titus	289	305	2.9	8.3	1.1	2.9

Table 32.   Dry weight of above-ground parts of three annual weeds in two-row barley (Ingrid) and spring wheat (Drabant). The same field experiment as that dealt with in Table 31. N_0, no nitrogen applied; N_{75}, 75 kg ha^{-1} of nitrogen as calcium nitrate. *Che, Chenopodium album; Fum, Fumaria officinalis; Lam, Lamium purpureum.*

Cereal species	Dry weight of weeds (g m^{-2}) N_0 *Che*	*Fum*	*Lam*	N_{75} *Che*	*Fum*	*Lam*	Relative weight of weeds (N_{75}/N_0) *Che*	*Fum*	*Lam*
(a) Means over seed rates of 2, 4 and 6 million seeds ha^{-1}									
Barley	2.73	0.39	0.67	7.02	0.38	0.36	2.6	1.0	0.5
Wheat	2.39	0.18	0.27	12.71	4.30	2.82	5.3	23.5	10.4
(b) Means over seed rates of 2 and 4 million seeds ha^{-1} in barley and 4 and 6 million seeds ha^{-1} in wheat									
Barley	3.22	0.50	0.76	8.66	0.49	0.40	2.7	1.0	0.5
Wheat	1.55	0.07	0.24	11.11	2.24	2.89	7.2	34.5	12.3

Thus, the treatment with N_{75} is more similar to common fertilizer use in Scandinavia in the last few decades than is the N_0 treatment. The stronger suppression of weeds in barley at the N_{75} treatment is in accordance with the general opinion in Scandinavia that spring barley is more competitive than spring wheat. In treatment N_0, on the other hand, both wheat and oats suppressed the weeds more strongly than barley. Treatment N_0 (where the soil was still not extremely poor in N) corresponds to situations in areas with a more limited use of fertilizers.

It may be added that, in a pot experiment with mixed stands of two-row barley and spring wheat, an increased application of NPK, from low to comparatively high levels, gradually changed the relative competitiveness of the two species in favour of barley (Figs 34 and 45). In that experiment, the varieties used differed from those in the field experiment presented in Table 31. Although the two experiments were very different, their results harmonize well. It thus appears to be of interest to study barley and wheat systematically regarding changes in their competitive ability in response to fertilizer application, particularly with nitrogen.

In the stands of barley (Ingrid) and wheat (Drabant) in the field experiment, the three dominant weed species, which were fairly evenly distributed over the field, were weighed separately. Dry weights, shown in Table 32, indicate that the species, particularly *Fumaria officinalis* (*Fum*) and *Lamium purpureum* (*Lam*), were more strongly suppressed in wheat than in barley at N_0, contrary to the situation at N_{75}. The growth of *F. officinalis* and *L. purpureum* was unchanged, or decreased, by the nitrogen application in barley, but in wheat their growth increased dramatically. *Chenopodium album* (*Che*) increased with nitrogen application in both barley and wheat, but increased more in wheat than in barley.

Any alteration in the environment may be expected to cause selective effects on weeds. Because of extremely complex interactions between many factors influencing competition in a mixed plant stand, the influence of an individual factor, such as nitrogen, has to be measured under various conditions before it can be characterized reliably. Effects of nitrogen on the competition between cereals and weeds presented by Jørnsgård *et al.* (1996) illustrate this.

Growth of plants from seeds of different sizes placed at various depths in the soil

Table 33 describes differences in the early growth of red clover plants from seeds of different sizes placed at various depths in a homogeneous soil according to the table. Table 33 concerns the same greenhouse pot experiment as that described in Table 21 and Fig. 6, but is focused on influences of competition from another plant. Red clover

Table 33. Dry weight of above-ground plant parts of red clover (Rea) from seeds of different sizes (large, 3.18 mg; small, 1.72 mg) sown in soil at different depths.[a] Determinations 4 weeks after sowing. The same pot experiment as that dealt with in Table 21 and Fig. 6.

Sowing depth (cm)	Dry weight (g per 150 seeds sown)		Percentage dry weights	
	Large seeds	Small seeds	Large seeds	Small seeds
Pure stands				
0	0.04	0.01	(1)	(<1)
0.5	4.66	2.80	100	100
1	4.51	3.04	97	109
2	4.42	2.61	95	93
3	3.88	2.24	83	80
4	3.57	0.98	77	35
5	1.44	0.26	31	9
With barley				
0	0.09	0.01	(8)	(2)
0.5	1.18	0.67	100	100
1	1.07	0.54	91	80
2	0.97	0.36	82	54
3	0.81	0.29	69	43
4	0.41	0.14	35	21
5	0.33	0.02	28	3

[a]Sandy loam soil, homogeneous, with water contents sufficient for germination and plant growth at all depths, except in the most superficial layer.

was sown with 50 seeds dm^{-2} at each depth. It was sown either solely or together with barley sown with 16 seeds dm^{-2} at a depth of 2.5 cm. Seed rates were comparatively high in order to create competitive conditions from early periods of growth. Joint effects of seed size, depth position and competition could therefore be studied as early as 4 weeks after sowing.

According to Table 21, the emergence of clover plants decreased and the early mortality among the plants increased with a decreased seed size and an increased depth of the seeds in the soil, from 0.5 cm on. An increased depth also caused an increased delay in the emergence (Fig. 6). The delay was combined with an increased consumption of food reserves, particularly weakening plants from smaller seeds. (Cf. Tables 18 and 23, Figs 6 and 7.)

Table 33 shows that the competitive power of the clover plants decreased with a decreased seed size and an increased depth of the seeds in the soil, from 0.5 cm on. Competition from barley strengthened the negative effects of the smaller size and the greater depth position of the seeds described above.

Regrowth after breakage and burial of regenerative structures of perennial weeds with an example of nitrogen influences

When rhizomes and regenerative roots of perennial weeds are broken into fragments of various sizes and buried at different depths by soil cultivation, responses are analogous to those when seeds of various sizes are placed at different depths. Results from two experiments at Uppsala illustrate these conditions (Tables 34 and 35).

The results can be summarized and commented on as follows. With increased fragmentation, more buds were activated; more new shoots and roots thus developed. Each new shoot–root unit then has access to a gradually decreased amount of food reserves resulting in a gradually weakened growth.

With increased burial depth, a decreased number of shoots could emerge and continue

Table 34. Response to degree of fragmentation and depth of burial of rhizomes of *Elymus repens* in the spring. Comparison between plants developing in pure stands and in competition with spring-sown white mustard.[a]

Depth of burial (cm)	Rhizomes occurring in the autumn of the experimental year Length of planted fragments (cm)		
	18–36	9–12	4–6
(a) Dry weight in grams per gram of planted rhizome fragments			
Pure stands			
5	12.5	11.6	11.4
10	6.5	5.8	3.8
With mustard			
5	3.63	3.20	2.53
10	1.48	0.98	0.46
(b) Dry weight in stands with mustard as a percentage of that in pure stands			
5	29	28	22
10	23	17	12

[a]Rhizomes of *E. repens* were collected in the field in spring and immediately cut into fragments of different lengths and planted at two depths in a field with sandy soil at Uppsala. The same amount of rhizomes, by weight, was planted in each treatment. White mustard was sown at a normal seed rate in half the number of plots. Weights represent living rhizomes collected in the autumn. (From Håkansson, 1968a.)

growth. The inability to emerge increased with increased fragmentation of the rhizomes or roots.

At strong competition from crops, the initially weaker units became proportionally more suppressed than the stronger ones.

As illustrated in Table 34a and Table 35a, the competition in a crop stand can reduce the growth of the perennial weeds dramatically. In addition, competition strengthens negative effects of fragmentation and deep burial (Table 34b and Table 35b). From these results and a great number of others in Swedish experiments, the following can be stated:

> With competition from a crop, the biomasses produced by the weed do not only become smaller, but differences following different degrees of mechanical

Table 35.   Response to application of nitrogen (in calcium nitrate) and to fragmentation (F_{18} and $F_{4.5}$) and burial depth of regenerative roots[a] of *Sonchus arvensis* in the spring. Comparison of plants developed in pure stands and in competition with spring-sown barley.

N (g m^{-2})	Depth of burial (cm)	Without barley F_{18}	Without barley $F_{4.5}$	Competition with barley F_{18}	Competition with barley $F_{4.5}$
\multicolumn{6}{l}{(a) Dry weight in grams per gram of planted roots}					
3	5	97	115	4.1	3.2
	10	95	95	2.9	1.5
9	5	208	220	0.6	0.1
	10	150	149	0.0	0.0
\multicolumn{6}{l}{(b) Weight as a percentage of that of longer root fragments at the 5-cm depth}					
3	5	100	119	100	78
	10	98	98	71	37
9	5	100	106	100	17
	10	72	72	0	0

Regenerative roots in the autumn (column grouping header)

[a]Regenerative roots were collected in the field in spring and immediately cut into fragments of two lengths, 18 and 4.5 cm (F_{18} and $F_{4.5}$) and planted at two depths in a field with sandy soil at Uppsala. The same amount of roots, by weight, was planted in each treatment. Two-row barley was sown at a normal seed rate in half the number of plots. (From Håkansson and Wallgren, 1972b.)

disturbance also become proportionally greater.

When, after mechanical disturbance by soil tillage, a perennial weed is observed in competitive crops, its appearance, which is commonly looked upon as its response to the disturbance by the tillage, is always a result of joint effects of the mechanical disturbance and competition from the crop.

This is analogous to the interactive effects of herbicides and competition illustrated in Table 27.

The experiment with *Sonchus arvensis* included application of nitrogen at two levels (Table 35). In pure stands of the weed, the higher nitrogen level caused an increased biomass production at each degree of fragmentation and depth of burial. With competition from barley, however, *S. arvensis* was strongly disfavoured at the higher nitrogen level; at the same time, the crop was favoured. The negative effect of the high nitrogen level on *S. arvensis* increased with increased breakage and increased burial depth, i.e. with increased weakening of the initial growth of the weed by mechanical treatments.

The result shown in Table 35 is a challenging example of the complex interactions determining competitive outcomes. It is also challenging because *S. arvensis* is often regarded as a plant favoured by a high supply of nitrogen (e.g. Hanf, 1982). This characterization might have wide validity when the plant grows in less competitive crops, such as beets, potatoes or vegetables. Ellenberg (1974) classifies *S. arvensis* as 'indifferent' with respect to its demands for nitrogen.

The strengthened competitive suppression of *S. arvensis* at increased nitrogen supply may be conditional. The relative accessibility of mineral nutrients to different plants in a stand varies with the fertilizer location in the soil. In this experiment, the nitrogen fertilizer was applied to the superficial soil. This probably favoured barley more than the *S. arvensis* plants, particularly when these had to regenerate from the greater depth. This again highlights the influences of the fertilizer placement on the crop–weed interactions.

Response to temperature

Differences in the weed flora in areas with dissimilar temperature climates largely result from the different competitive abilities of diverse plants in the different climates.

A plant (species, genotype) that is only moderately weakened by a certain influence when growing in pure stands may be strongly suppressed in competition with plants that are better adapted to that influence. An example of this is given in Table 27c, illustrating the joint effects of a herbicide and competition from a crop on weeds.

An analogous interaction, now between temperature and competition, is illustrated in Table 36. The table describes results from an experiment with an annual weed,

Table 36. Effects of temperature, photoperiod and relative time of emergence on the outcome of competition between *Galinsoga ciliata* and two-row barley.[a]

Mean temp. light/ dark (°C)	Dry weight (g) of pure stands			Mixed stands/pure stands (%)				Galinsoga/barley (%)			
				Barley		Galinsoga		Pure stands		Mixed stands	
	Barley late	Barley early	Galinsoga	B_L	B_E	B_L	B_E	B_L	B_E	B_L	B_E
Light period 19 h, dark period 5 h											
14/9	9.87	12.65	3.21	98	101	13	6	32	25	4	1
24/19	6.42	7.79	3.79	90	93	41	22	59	49	27	11
Light period 13 h, dark period 11 h											
15/10	8.13	10.04	2.86	106	100	25	7	35	28	8	2
25/20	5.78	6.65	2.84	88	82	48	25	49	43	27	13

[a]Results from a growth-chamber pot experiment partly presented by Håkansson (1986). Barley sown with 64 seeds per 7 dm², either 'early' or 'late', which means either 7 days before or on the same day as planting of young seedlings of *Galinsoga* (28 plants dm⁻²). *Galinsoga* plants from seeds collected near Uppsala. Light intensity was about 10% of maximum daylight at Uppsala by midsummer. In each combination of temperature and photoperiod, all plants were weighed when barley sown early had five well-developed leaves. Figures represent dry weights of above-ground plant parts per 7 dm². B_L, barley late; B_E, barley early.

Galinsoga ciliata, and a crop, two-row barley. *G. ciliata* is a summer annual plant native to Central and South America, naturalized in Europe on waste and cultivated ground and spread to scattered places in Southern Scandinavia in the recent century (Holm *et al.*, 1977). In a growth-chamber experiment (described in Table 36), *G. ciliata*, from seeds collected at Uppsala, and two-row barley were grown in pure and mixed stands at two temperature levels and different light periods.

In pure stands, *G. ciliata* produced similar growth at the two temperatures, whereas barley grew better at the lower temperature than at the higher. In competition with barley, *G. ciliata* was favoured by the higher temperature. This can be largely ascribed to weaker competition from barley. Barley, as expected, exerted a stronger competition after early emergence (B_E) than after late emergence (B_L), an effect particularly pronounced at the lower temperature. Both plants produced lower growth at the shorter light period than at the longer, which was probably due more to the reduced amounts of light than to the photoperiod as such.

Table 36, indicates that barley suppressed *G. ciliata* proportionally much more than it was itself suppressed by *G. ciliata*.

Consequently, the reduction in barley due to competition from *G. ciliata* was much less influenced by their relative time of emergence than the reduction in *G. ciliata* due to competition from barley.

The amount of light reaching the plants in the pots in this growth chamber experiment was not very low when compared with light reaching the plants in a cereal stand in an average Swedish field situation. Sidelight reaching the plants in pots may have reduced rather than enhanced the temperature influences reported here.

The response of *G. ciliata* to competition at different temperatures may lead to some speculation about the occurrence and distribution of the plant. In its native areas of Latin America, this species, which is a C_3 plant, occurs both in hot climates at lower altitudes and in cold climates higher up in the mountains (Garcia *et al.*, 1975). This might point to a considerable genetic diversity, which, in turn, may be the reason why it has been able to naturalize in northern climates, where its distribution according to Mossberg *et al.* (1992) is still expanding. However, it certainly still has a limited competitive ability in its Scandinavian outposts. Thus, like its relative *G. parviflora*, it is found in this region only as a ruderal in places with thin vegetation or as a weed in weakly competitive crops, such as sugar beet, potatoes and vegetables.

Response to infections – example, discussion

Certain plant diseases invariably lead to a high mortality among infected individuals. However, infections by many agents may be non-lethal and may only influence the host plants very slightly under normal growth conditions. Influences of some of them may be lethal or reduce plant growth mainly in unfavourable growing situations. Plants that are weakened but not killed following an infection could be expected to be more strongly suppressed when growing in competition with vigorous plants – e.g. with healthy individuals of its own taxon – than when growing among equally infected plants of this taxon.

A greenhouse box experiment affirms this speculation (Table 37). In the experiment, plants of Italian rye-grass that were either non-infected or infected with a sublethal mosaic virus were grown in separate and mixed stands. In the mixed stands, every second plant was infected. Each box was enclosed by a white plastic frame surrounding the aerial shoots to a height of 25 cm, reducing the influence of side-light (for further details, see Table 37).

Table 37 shows that when healthy and infected plants were grown separately, the biomass of infected plants was of almost the same magnitude as that of healthy plants. However, in competition with healthy plants in the mixed stands, the infected plants became considerably suppressed. Here, the healthy plants effectively compensated for the weak growth of the infected ones. As a result, the average production was of the same magnitude in the mixed stands as in the stands with entirely healthy plants. Dry weights were about 9.5% of the fresh weights of the shoots in both healthy and infected plants, i.e. they were equal. No plants were dead, but two or three of 72 infected plants in the mixed stands were possibly dying.

Table 37. Influence of non-lethal infection of Italian rye-grass (*Lolium multiflorum*) with rye-grass mosaic virus under different competitive conditions. Dry weights of aerial shoots of healthy and infected plants determined 6 weeks after transplantation, in their late two-leaf stage, into boxes with soil in a greenhouse experiment.[a]

| | Stand type of plant categories | |
Plant category	Separate stand	Mixed stand
	Aerial shoots of 72 plants (g)	
Healthy	10.7	17.0
Infected	9.6	6.5
Average	10.2	11.8
	Infected relative to healthy	
Healthy	100	100
Infected	90	38
	Mixture relative to separate	
Healthy	100	159
Infected	100	68

[a]Soil surface in box: 40×40 cm^2. 72 plants per box, nine plants in each of eight rows. When in mixed stands, healthy and infected plants were alternated in the rows, 36 plants of each category in each of two boxes. Experiment in Nov.–Jan. Temp. 20–22°C in 18-h periods of daylight + artificial light, and 10–12°C in 6-h dark periods. Plants obtained in Nov. from B. Gerhardson, Department of Plant Pathology, Uppsala.

The competition between crops and weeds, and between the weeds, will certainly be of great interest for future investigations into effects of viruses, fungi, bacteria, herbivores, etc. attacking weeds. It is then of interest how and to what extent the effect of the attack by an agent is modified by factors such as air movements and humidity in the stand, soil conditions and, not least, the anatomy of the plant tissues. The shape of the tissues is strongly influenced by the supply of water and plant nutrients (not least nitrogen), the temperature, etc., and by the stage of plant development.

In attempts to find and develop *bioherbicides* as means of managing weeds, systematic studies of the joint effects of bioherbicides and competition in the crop stand are of great interest. The effect of parasites or herbivores attacking weeds selectively will often be increased by strong competition in a crop stand.

The 'system management approach for biological weed control' under development in recent years (Müller-Shärer and Frantzen, 1996; Frantzen et al., 2001) aims to utilize attacks of various agents, such as bacteria, on weeds, thereby reducing the competitiveness of the attacked weeds. A question that may be raised in this context is to what extent the weakening effects on the host plants of such agents could be increased by improvements in crop competition.

8

Weed Flora and Plant Adaptation to Environment and Competitive Conditions

Extremely Complex Causalities

When two plants in a mixed stand respond differently to a defined change in a specified factor, this is not necessarily caused solely by this factor. A changed supply of water or plant nutrients, particularly nitrogen, may, for instance, alter the shading conditions and the relative access of different plants to light. Altered access to light among the plants could thus be the immediate reason for a changed outcome of their competition, even when it was originally provoked by a changed supply of water or nutrients. Complex changes in the interactions, involving many factors, may be triggered by changes in one or a few factors under observation. This should be kept in mind when causalities behind competitive outcomes are to be evaluated.

Before abiotic factors in themselves become sufficiently extreme to prevent growth and reproduction of a plant, the establishment and persistence of this plant could be hindered by competition from other plants and/or by other biotic conditions. This competition and, for example, infections by certain parasites may thus be increased in the 'extreme' environment. Conditions that are extreme to a certain target plant may be less extreme or optimal to others, the competition from which may be adverse to the target plant. Fertilizer levels, drainage, irrigation, etc. in arable fields may strongly affect the composition of the weed flora by influencing the competitive conditions in the crop–weed stands, where the complex causalities may be hard to evaluate correctly.

Extensive alterations in agricultural and horticultural practices have thus strongly changed weed floras in the last century. Many alterations have largely affected the weed flora composition indirectly through their influence on the competitive conditions in the crop–weed stands.

Changes in the Weed Flora as Related to Increased Fertilizer Use – Some Scandinavian Examples

Many plants that were common weeds in past centuries have gradually decreased in abundance or almost disappeared as weeds on arable land during the 20th century. Important changes in the weed flora in the last half of that century are direct or indirect effects of herbicides. Others can be ascribed to changes in the soil tillage intensity. Many others, however, seem to be a result of changed competitive conditions in the crops. These changes have been largely ascribed to increased fertilizer use, not least to the increased nitrogen application. Although experiments confirming the causalities are lacking, or difficult to conduct strictly, many observations support the judgement that increased fertilizer use is a very important triggering factor.

Many observations indicate that plants that have declined as arable weeds respond positively to fertilizer in the absence of competition from crops, but negatively in dense crop stands. An example: *Rumex acetosella* has been described as an important weed in agricultural fields in large parts of Europe for centuries, particularly on light, acid soils. It declined dramatically in Swedish arable fields in the last century. With increased fertilizer application, this species obviously has difficulties in withstanding the competition in dense crop stands, but can be seen to prosper greatly in gaps of these stands.

Changes in Sweden during the 20th century can be characterized on the basis of literature from various periods. Literature from the 18th and 19th centuries to recent decades reviewed by Fogelfors (1979) indicates considerable changes. Among species that were abundant arable weeds in earlier times, but are now practically extinct in ordinarily fertilized fields, the following may be mentioned: *Arabidopsis thaliana*, *Bromus hordeaceus*, *Camelina microcarpa*, *Cardaminopsis arenosa*, *Melampyrum arvense*, *Rhinanthus angustifolius* and *Silene vulgaris*. Today they are mainly seen outside arable land, on poor ground among sparse vegetation.

Information from the early 20th century (e.g. Ferdinandsen, 1918; Bolin, 1922; Korsmo, 1930) and data from later field surveys (Laursen, 1967; Mukula *et al.*, 1969; Gummesson, 1975, 1979; Raatikainen *et al.*, 1978; Haas and Streibig, 1982; Erviö and Salonen, 1987; Andreasen, 1990; Hallgren, 1993, 1997; Salonen, 1993; Andreasen *et al.*, 1996; Fykse, 1999) allow broad descriptions of weed flora changes in the 20th century. Some examples are given below.

The summer annual *Spergula arvensis* is still an important weed but was much more common earlier, abundant particularly on acid soils (cf. Ellenberg, 1974). Its decrease partly results from herbicide control, but is doubtless also caused by changed competitive conditions in the crops due to liming and increased fertilizer use.

The rhizomatous perennial *Equisetum arvense* has decreased considerably from the 1940s. Research by Andersson (1997) has shown that nitrogen fertilizer application at the normal levels of today disadvantages

this plant in its competition with cereals, probably mainly due to increased shading in the cereal stands. *E. arvense* has certainly also been competitively disadvantaged by the drainage of periodically waterlogged fields, where it was previously one of the most important weeds.

Other species that were important weeds on moist and periodically waterlogged land are nowadays mainly seen in moist field margins and similar places with weak crop growth. Examples of such species are *Bidens tripartita* and *Filaginella uliginosa*. The decline of these species on arable land is probably an effect of increased crop competition due to drainage and increased fertilizer use.

In current high-input agriculture, the number of species occurring frequently and abundantly in the weed floras of today is much lower than in earlier low-input farming. One likely reason for this, besides the use of selective chemical weed control, is the changed competitive conditions caused by increased fertilizer use. The most important weeds of today are species that are favoured by, or endure, the changed conditions in principal crops, mainly cereals. Some weeds are obviously often more favoured by high fertilizer levels than the crops. Table 32 thus shows that *Chenopodium album*, *Fumaria officinalis* and *Lamium purpureum* can be much more favoured than spring wheat by nitrogen fertilization. Tables 31 and 32 indicate that the competitive ability in spring barley increased relative to that in spring wheat with increased nitrogen supply. None the less, *C. album* was also favoured considerably by nitrogen when in competition with barley (Table 32). It should be kept in mind, however, that the influence of a given factor can be changed by other factors.

Certain weeds that have difficulties in enduring the competition in closed cereal stands are important in crops with more open stands, or stands that close late, such as sugar beet, potatoes and vegetables. Examples of species with this behaviour under Swedish conditions are *Solanum nigrum*, *Sonchus asper*, *Sonchus oleraceus* and *Urtica urens*. They are obviously favoured by high fertilizer application in crops that close late. In such crops, even these plants may overgrow the crop if not efficiently controlled. However, they may in turn be overgrown by dense populations of other weeds, such as *C. album*, *Sinapis arvensis*, *Matricaria perforata* and even *Stellaria media*.

Deschampsia cespitosa, a tufted perennial grass, is one of the more common weeds on permanent grassland and in grass–clover leys on arable land, particularly in northern humid areas. It increases in number and size of tufts with the age of a ley (Tables 5 and 8). It can be a very troublesome weed, foremost on acid soils poor in nutrients (Hagsand and Thörn, 1960; Thörn, 1967). In the absence of effective competition from other vegetation, it responds positively to fertilizers. At the same time experience indicates that fertilizer use favours cultivated grasses at the expense of *D. cespitosa*. In experiments with leys for hay conducted by Hagsand and Anier (1978), an increasing nitrogen fertilizer application increasingly favoured the cultivated grasses at the expense of *D. cespitosa*. In recent decades, *D. cespitosa* has largely declined as a weed in arable leys, which is ascribed to increased fertilizer use in the first place, but also to decreased duration of the leys (see Table 5 and 8 regarding its increase with the age of the ley) and to decreased use of poorly drained land for cropping.

Competitive Conditions in Different Crops

The competitive conditions in a crop stand greatly depend on its plant density and spatial distribution and the time of emergence of the different plants of crop and weeds. The lower the plant density and the more uneven the plant distribution of the crop, the later its stand closes and the weaker its competitive power becomes.

In accordance with previous descriptions, we may distinguish two main categories of crops: (i) crops with stands that close late or remain relatively open until harvest; and (ii) crops that close early, i.e. crops rapidly developing a shading ground cover.

Crops grown with a wide plant spacing both within and between rows belong to category (i). Sugar beet and other beets, potatoes, most other root crops and vegetables belong to this category. Their stands may close with different degress of lateness and their competitive ability may differ, but in all of them, large yield losses due to weeds are unavoidable in the absence of direct weed control. Even relatively few weed plants escaping or surviving early control can grow and reproduce vigorously. These plants may cause problems at harvest and their seeds or vegetative propagules may strengthen their populations. Also weed plants that emerge late may grow and reproduce vigorously. Most crops of category (i) are grown for their vegetative production, but there are also crops grown for seed production. Many leguminous plants are such crops. They are either grown at a low plant density or a wide row spacing, or they grow slowly, or form weakly competitive stands for other reasons.

It is understood that direct weed control invariably has to be undertaken in category (i) crops discussed above. At low labour costs (in earlier times or in many areas of the world), hoeing and hand weeding are carried out. When effectively performed, the control in category (i) crops on the whole leads to a better weed suppression than the control in cereals without herbicides. In Scandinavia and many other areas, these crops, like black fallows, were considered 'rescue crops' with regard to weed suppression in a crop rotation. If mismanaged, however, these crops have an opposite influence. When weeds are controlled by herbicides, competition gives less interactive help in these crops than in cereals.

Certain crops with plants becoming tall and competitive later in the growth period, such as maize, have rather open stands for comparatively long periods in early growth. Early weed control is always of importance in these types of crop.

Small-grain cereals and oilseed *Brassica* and *Sinapis* crops are (mostly) grown at a relatively high plant density and a narrow row spacing leading to early stand closure. They are then typical category (ii) crops. *It is in crops of category (ii) that competition is a factor worthy of improvement and exploitation in integrated weed management. In these crops, it is thus meaningful to improve the stand build-up and breed for increased competitive ability in order to reduce the need for intensive direct weed control.*

Because the competitive ability of a plant varies with the environmental conditions (e.g. Tables 31, 32 and 36), it is not possible to rank crop species or varieties in definite hierarchies with respect to their competitiveness. However, a rough ranking of crops grown in ordinary stands and with varieties adapted to conditions prevailing in an area seems to be possible and is presented and discussed in the following. The ranking order, applied to Southern and Central Sweden, is based on general observations in field experiments and farmers' crops.

Spring cereals and oilseed crops may be characterized as follows: Two- and six-row barley are usually more competitive against weeds than are oats. In early stages, barley grows more rapidly than oats. Oats, on the other hand, usually exert effective shading during a longer period; weeds that grow actively during long periods therefore are disadvantaged in oats. Spring wheat usually suppresses weeds considerably less effectively. Differences between species and varieties largely depend on the rapidity of the canopy closure. Oats, however, sometimes exert a surprisingly strong competitive effect even on fast-growing weeds, though its stands do not close as rapidly as barley stands. On the basis of various reports, Seavers and Wright (1999) speculate on possible effects of allelopathic exudates in oats. Spring-sown rape and turnip rape exert about the same effect on weeds as spring wheat, but their stands and resulting competitive effects differ considerably. Due to faster early growth, turnip rape tends to check weeds more strongly than does rape (Fig. 23b; see also Håkansson, 1975b, 1979; Gummesson, 1981; Hallgren, 1990a,b, 1991b).

In stands of *autumn-sown cereals and oilseed crops* that have survived winter successfully, weeds emerging in the spring become much more suppressed than in comparable spring-sown crops. Winter-annual weeds emerged in the autumn have a good competitive position. Plants of these weeds often grow more vigorously in autumn-sown crops than do annual weeds emerged in spring-sown crops in the spring.

The competitive power of autumn-sown crops strongly depends on their winter survival. A well wintered stand of rye is considerably more competitive than a corresponding stand of wheat. The main reason is certainly that rye grows much faster than wheat in early spring, thereby producing an earlier ground cover. Good stands of winter barley and rye–wheat (*Triticale*) are also very competitive, particularly winter barley, which exhibits a rapid growth from early spring. Due to faster growth in the spring, winter turnip rape, on average, exerts a stronger competition than winter rape (e.g. Gummesson, 1981; Hallgren, 1991a,c).

In Scandinavia, winter oilseed *Brassica* crops are sown earlier than winter cereals. The former are sown in August, the latter in September/October. The seedbed preparation for the earlier sowing brings more winter-annual weed seeds to germinate than the preparation for later sowing (Fig. 1). Winter-annual weeds therefore often emerge more abundantly in the oilseed crops than in the cereals. The early emerged weed plants may become very competitive, certainly one reason why weeds in oilseed crops often grow very vigorously. However, due to varying moisture conditions in the seedbed at early sowing and to different winter survival of the autumn-sown oilseed crops, the conditions vary greatly.

Leys cut for fodder grown alternately with annual crops on arable land have been discussed previously with regard to the occurrence of weeds with different life forms (Tables 3–5). Perennial grasses and clover species are the main ley plants in Scandinavia, mostly grown in mixed stands. After sowing (usually undersowing in cereal crops) and wintering, the leys may be kept for 2–3 years (1 year and 5 years being extremes). Annual weeds have great difficulty in establishing in dense ley stands. However, such weeds, winter annuals in the first place, may grow vigorously in gaps and spots with weakened plants in the leys

(Fig. 2, Table 2), where they can produce considerable amounts of seed.

Weeds that are low-growing and, in addition, produce seeds germinating and establishing in different seasons, e.g. *Stellaria media* and *Lamium* spp., are particularly capable of developing and reproducing in leys. More tall-growing weeds, such as *Thlaspi arvense*, that develop and shed viable seeds before cutting in early summer, or between cutting occasions, can shed great numbers of seeds to the ground. Even in relatively uniform ley stands, the seeds produced by annual weeds should not be underrated. Their *relative* contribution to the soil seed bank and the population persistence is certainly important for annual weeds whose seed production is effectively prevented by control in cereals and other annual crops alternating with the leys.

Competitive Ability of Weeds

The importance of plant traits

As previously stressed, the competitive ability of a plant is conditional. At the same time, there are several plant traits that can be pointed out as favourable to the competitiveness of a plant in a given situation. With special reference to competition in plant stands of short duration, some examples of traits judged to be favourable are as follows:

- Rapid extension of roots and foliage in the early period of growth.
- Large seeds, or vigorous vegetative propagules, i.e. units containing great amounts of food reserves, thus enabling a vigorous growth of the young plants.
- Leaves developing at, or reaching to, high levels in a plant stand.
- Climbing shoots, favouring the exposure of leaves to light.
- Great phenotypic plasticity, enabling adaptation favouring the competitive position relative to neighbouring plants.
- Tolerance to shading: particularly favourable to low-growing plants.

Other properties or conditions favouring the competitive success of a plant in various situations are:

- Ability of the seeds in the soil seed bank to germinate in different seasons: This means a flexibility enabling new plants to establish in more varying situations than when germination is restricted to a short period of the year.
- Capacity to regenerate or regrow rapidly from vegetative plant structures after damage caused by tillage, cutting, etc.

Some annual weeds growing vigorously in competitive arable crops in modern agriculture are characterized below with respect to traits judged to favour their competitive success.

- *Avena fatua*, *Centaurea cyanus*, *Galeopsis tetrahit*, *Galeopsis speciosa*, *Polygonum lapathifolium*, *Polygonum persicaria* and *Sinapis arvensis* have large seeds and tall-growing shoots.
- *Bilderdykia convolvulus* and *Galium aparine* have large seeds and climbing shoots.
- *Chenopodium album* sets 'medium-sized' seeds. Its plants are extremely plastic, adaptable to very different conditions. They can reach a height of more than 2 m and produce huge amounts of seed under favourable conditions. After late emergence in a shading crop, the plants may become less than an inch high and yet set a few viable seeds (cf. Fogelfors, 1973; Holm *et al.*, 1977). Their leaf area is then large relative to their biomass (the dry matter in particular), which facilitates photosynthesis relative to respiration. Plants that are not delayed in relation to the crop can easily regulate their height in relation to the height of the crop stand. In studies of *C. album*, Röhrig and Stützel (2001a,b) show that its growth in height, its vertical leaf distribution and its leaf area vary, thereby optimizing its access to light.
- *Matricaria perforata* produces comparatively small seeds. Its seedlings are therefore rather weak, but when

starting their growth early in relation to neighbouring plants, they can reach a height of more than 1 m, branch heavily and produce numerous seeds even in a dense crop. Moreover, even in crop stands considered uniform, there are normally spots with weak plants, enabling seedlings of *M. perforata* to establish and grow vigorously even when not emerged earlier than the crop plants; this facilitates the establishment and reproduction of weed species with weak growth in early stages.

- The prostrate and low-growing plants of *Stellaria media* develop from small seeds. However, they are shade-tolerant and can survive and reproduce in strong shade (Fogelfors, 1973). Their height also varies to some extent with the character of the crop stand. When the light increases in maturing cereal stands or after harvest, these plants often grow explosively.

Reasons for variation often hard to identify in the field

Although individual traits favouring the competitive ability of a plant can be pointed out, their relative importance could change considerably owing to influences of environmental factors.

It is also hard to specify the relative time of emergence under ordinary field conditions, which is a problem due to its strong influence on the outcome of competition between plants. This problem is one reason why it is hard to define the relative competitiveness of plants on the basis of measurements in the field. Seeds with differing phenotype and in different positions with regard to the soil environment give rise to plants with different competitive power.

For reasons mentioned above, measurements of the relative competitiveness of a weed could be expected to vary from situation to situation. That they do so will be illustrated by results from some Swedish experiments. Weed species in the experiments are ranked on the basis of their

biomass reduction in response to increasing crop plant density (seed rate).

In greenhouse box experiments, 12 weed species were grown in a mixture with two-row barley sown at different seed rates (Fogelfors, 1973). The weeds studied were ranked as follows: *Matricaria perforata* ≥ *Viola arvensis* > *Thlaspi arvense* > *Capsella bursa-pastoris* ≥ *Lapsana communis* > *Stellaria media* ≥ *Veronica agrestis* ≥ *Sinapis arvensis* > *Myosotis arvensis* > *Chenopodium album* > *Lamium purpureum* > *Polygonum lapathifolium*.

Plants of weed species important in Swedish cereal fields were weighed in a great number of field experiments in spring cereals sown with different seed rates (Granström, 1962). The following species grew well in most experiments: *C. album*, *Lamium amplexicaule*, *Sinapis arvensis*, *Stellaria media* and *Viola arvensis*. However, their relative competitiveness varied between the experiments.

Based on data from a number of experiments with seed rates in spring cereals at Uppsala, the more abundant weeds were ranked (Håkansson, 1995d) as described below.

In a greenhouse experiment carried out in large boxes, three weeds uniformly distributed in the boxes were ranked in the following order: *Stellaria media* > *Polygonum lapathifolium* > *Chenopodium album*.

In three field experiments (1, 2 and 3) with spring cereals on clay soils near Uppsala, the more prominent and uniformly distributed weeds were given the following ranking:

1. *Bilderdykia convolvulus* > *Galeopsis* spp. > *Galium aparine* > *Stellaria media* > *Chenopodium album* >> *Sonchus asper*.
2. *Stellaria media* > *Galeopsis tetrahit* > *Chenopodium album* > *Polygonum lapathifolium* > *Brassica napus* ssp. *oleifera*.
3. *Galium aparine* ≥ *Lamium purpureum* >> *Chenopodium album* > *Fumaria officinalis*.

Ranking based on average values from 12 field experiments with seed rates in spring cereals presented by Andersson

(1987) turned out as follows: *Galeopsis* spp. ≥ *Fumaria officinalis* > *Stellaria media* > *Chenopodium album* > *Brassica napus* and/ or *rapa* ssp. *oleifera* > *Bilderdykia convolvulus* > *Polygonum aviculare* > *P. lapathifolium*. It should be noted that there was great variation between the 12 experiments.

Late emergence may be one reason for an occasional low ranking position of species that are considered very competitive.

The evaluation of the importance of given weed species in particular crops has to be based on broader observations of plant growth and reproduction in the crops under different conditions. Broad knowledge of their life history together with measurements of their competitive response and effect in particular crops are then crucial. On the basis of such knowledge, weed species may be classified into groups with regard to their typical behaviour in defined agricultural situations with an increased focus on their population dynamics.

Phenotypic Adaptation to Environments Exemplified by Weeds in Lawns

Plants may become exhausted by frequently repeated cutting if they have no parts of their foliage left after a cutting. Assimilating leaves remaining facilitate the recovery of the plants. For perennials, this particularly means the survival and renewal of their regenerative structures.

Typical lawn grasses are perennials with much of their foliage located near the ground surface. Proportionally large parts of their leaf area are therefore left even after cutting close to the ground. Moreover, these grasses develop new assimilating leaves from buds near the ground surface, thereby consuming minimum amounts of food reserves in maintaining or re-establishing states of positive net photosynthesis. In this way, they can withstand frequently repeated cutting, even when cutting each time deprives them of large parts of their foliage.

Some common dicotyledonous perennials considered as weeds when occurring in lawns persist in grass turfs because their leaves largely develop below the cutting level. These perennials therefore could be even less damaged by cutting than the lawn grasses. In response to the high intensity and the favourable spectral composition of the light accessible to them at repeated cutting, they develop their leaves close to the soil surface, which is a perfect adaptation to the conditions in a frequently cut lawn. A few examples will be given.

The growth habit of *Taraxacum officinale* in a frequently cut lawn on the whole differs from that in a dense stand of tall-growing plants, e.g. in a clover–grass ley that is mowed once or twice in the growing season. Thus, where a stand of tall, shading plants develops, the plants of *T. officinale* form longer and almost erect leaves with large surfaces relative to their contents of dry matter. This is an appropriate adaptation to such stands. In frequently cut lawns, the leaves of *T. officinale* form smaller areas in relation to their dry matter contents; these leaves, in addition, largely grow close to the ground and therefore to a great extent escape cutting and continue their assimilation in the absence of shading. Even typical lawn grasses may be deprived of a proportionally greater amount of assimilating leaves by cutting than *T. officinale*. By creating good light conditions, repeated short cutting therefore favours this species in its competition with the lawn grasses. Without special control efforts, plants of *T. officinale* can therefore easily grow in size in the lawn, at the same time as seedlings of the species can establish when the moisture conditions are suitable.

(The leaves of *T. officinale* sometimes do not grow as close to the ground in a lawn as could be expected from the above. This can happen under partial shade of trees, even where the light intensity is rather high in periods. The spectral composition of the light might favour a more erect growth there. However, such growth is sometimes observed even in cases where the light conditions should favour a more horizontal growth according to the simple description above. The growth habit of the leaves is certainly governed by many interacting factors.)

Plantago major is a stationary perennial developing rosettes of leaves close to the soil surface in places where they are not shaded, or where other vegetation is weak. Because its leaf rosettes largely remain undamaged by cutting, this plant is also a persistent weed in lawns. In addition, as it endures disturbance by trampling and soil compaction better than the lawn grasses, it is particularly competitive after such disturbance.

Bellis perennis is another perennial that adapts well to the conditions in lawns, where it develops rosettes of low-growing leaves. The plant also spreads by short, superficially creeping stem branches, from which the leaf rosettes often develop dense mats, sometimes completely crowding out the grasses.

Achillea millefolium is a rhizomatous perennial which can easily spread in lawns on many types of soil. When growing at sites where it is not cut, it develops tall flowering shoots. In a lawn where repeatedly cut, it produces shoots with low-growing leaves from short, superficial rhizome branches. The leaves largely escape cutting. By rapid branching near the soil surface, the rhizomes can give rise to patches of shoots forming dense covers of leaves, which sometimes crowd out the lawn grasses.

Many plants with long stolons or prostrate runners are weeds in lawns. Examples are *Ranunculus repens*, *Prunella vulgaris* and *Glechoma hederacea*. On ground where the surface soil is not dry for long periods in the growing season, such species often spread and root efficiently. They are favoured by frequent cutting close to the ground as their leaves largely grow near the soil surface under the light conditions created by the cutting. *Trifolium repens*, often considered an unwanted plant in lawns, behaves similarly.

In many areas, stoloniferous grasses are used as valuable lawn grasses. In cold temperate climates, *Agrostis stolonifera* is one of these. In warmer areas, *Cynodon dactylon* is a prominent lawn grass. In certain areas, other stoloniferous grasses persist as unwanted plants in the lawns, where they largely withstand cutting.

The rhizomatous grass *Elymus repens* is seldom associated with lawns. After lawn grasses have been sown in soil infested with vigorous rhizomes of *E. repens*, coarse and tall shoots of this grass emerge rapidly, becoming very conspicuous and contrasting to the slender seedlings of the lawn grasses. Vigorous rhizomes are effective means of plant regeneration after tillage. However, the development of such rhizomes as well as the regrowth after tillage is strongly assimilate-demanding. At repeated cutting, new shoots develop from buds located nearer and nearer to the soil surface. New rhizomes become gradually smaller; at the same time they develop close to the soil surface. After a period of repeated cutting of the lawn, surviving *E. repens* plants develop shoots with slender leaves, originating both from buds at the basal parts of aerial shoots and from apices of short rhizomes near the soil surface. These responses represent another means of adaptation to repeated cutting. Small amounts of assimilates are now used for rhizome production and the formation of new shoots draws slight amounts of food reserves. In this way, *E. repens* may survive for a long time in a lawn as inconspicuous plants, although it often seems to be crowded out within a few years of the lawn establishment.

Two examples of annual plants infesting lawns as unwanted weeds are the dicotyledonous plant *Polygonum aviculare* and the grass *Poa annua*.

Although morphologically rather variable, *P. aviculare* is always a typical summer annual in cold temperate areas. It germinates to a large extent in the very early spring (e.g. Martin, 1943; Courtney, 1968; Roberts and Ricketts, 1979). Its early plants can establish fairly easily among the lawn grasses. Whereas it develops more erect shoots in shading cereal stands, its stems and leaves grow close to the ground when they are almost unshaded in a frequently cut lawn. Its leaves therefore largely escape cutting. *P. aviculare* also tolerates trampling and soil compaction, which makes it a competitive plant not only in lawns but also in yards and tracks, etc., where most other vegetation becomes strongly suppressed. In such places it often grows together with *Plantago major* and *Poa annua*, which are also tolerant to trampling.

Poa annua is a very variable plant, often annual, but often also appearing as a short-lived perennial, which may grow together with the annual forms (e.g. Netland, 1985; Holm *et al.*, 1997). The perennial types are prostrate to semi-erect or semi-stoloniferous. The annual types may be characterized as facultative winter annuals. In full daylight, all types develop great parts of their leaves within levels of a few centimetres above ground; they therefore largely escape cutting in lawns. Many of their inflorescences also develop near the ground surface. Tillering, flowering and seed set occur throughout the growing season.

P. annua is ubiquitous in large parts of the temperate regions. Its seeds then occur in almost all open areas influenced by man. They can also germinate in almost any season. All these traits make the species a very invasive plant, rapidly colonizing bare soil or areas with weak vegetation, e.g. spots in lawns where the perennial grasses have been worn down. However, it is comparatively sensitive to drought. The presence of its seeds almost everywhere otherwise makes it a weed, or a potential weed, in many agricultural and horticultural crops. In crops such as cereals, it sometimes develops rather tall shoots, 25–30 cm in height, but it is mostly a weed of minor harm in cereals. However, in crops with more open stands, in flower beds, etc., it can rapidly produce very harmful covers.

9

Measurements of Competition and Competitiveness in Plant Stands of Short Duration

Introduction: Stands of Short and Long Duration

When plant stands (communities) develop freely over a period of years – becoming increasingly dominated by perennials in successional processes – the changes in their composition and production become more and more overshadowed by factors other than *early* plant densities. The growth and production of different plants in stands of short duration are, on the other hand, strongly dependent on the *initial* densities of the established plants throughout their period of growth. In arable grass–clover leys for fodder, which are kept for a few years, the initial densities of the early established plants (of perennials in particular) influence the stand composition and production over several years, although other factors become increasingly influential.

The following discussions focus on plant stands of short duration, mainly annual stands, where initial plant densities and other initial conditions determine the competitive conditions and, therefore, influence growth and production throughout the period of active plant growth. Interpretations of competitive effects in such stands are largely based on comparisons of plants in stands with different initial densities. The way of studying effects of *intraspecific* competition is to compare the performance of plants in monospecific stands at different densities. *Interspecific* competition between two plant categories may sometimes be

characterized by comparing one pure stand of each category and one mixed stand. However, for more thorough analyses, a greater variation of densities and density proportions is needed.

Regarding their qualitative influences on the relative competitiveness of different plants in a stand, changes in the spatial plant distribution, in the relative time of emergence, or in the environment could sometimes be characterized even in stands with unchanged initial densities. However, the strength of the effects of such changes is density-dependent. Deeper analyses therefore require experiments with varied densities.

Both plant densities and plant traits influence the competitive outcome in a mixed stand. *It is, however, not always readily seen that the influence of densities and traits can be easily confounded. Thus, differences in growth and production of plants competing in mixed stands may be mistaken for effects of plant traits in cases when they are, mainly or entirely, effects of plant densities. Experimental designs where density influences and effects of plant traits are mixed up can be detected by means of 'reference models', in which 'different' plant categories are represented by equal plants in uniform single-species stands.* Reference model calculations and their use are illustrated on pp. 138–144.

One problem in this context is that densities of plants with different traits cannot be described in comparable units without evaluations in special experiments.

Although this is an inconvenience, effects of densities can be used for characterizing the relative competitiveness of plants, e.g. by the 'Unit Production Ratio' (UPR).

Unit Production Ratio (UPR)

This ratio was first introduced under the name 'Relativt Produktionsindex' (= Relative Production Index, RPI, by Håkansson, 1975b). As it is based on the production per plant individual or some other well-defined unit of plants, it was changed to 'Unit Production Ratio' (UPR) in subsequent publications (e.g. Håkansson, 1983b, 1988b, 1991, 1997a,b).

A plant unit, i.e. a unit of plant individuals, has to be defined from case to case. It could be an emerged plant, a sown seed, a defined amount of seed (by weight), etc. The plant density can thus be expressed by the number of such units per unit area. The production per such a unit is termed the 'Unit Production' (UP). The UP of plant categories A, B, etc., is written UP_A, UP_B, etc.

The ratio of UP_A to UP_B is called the unit production ratio of A to B and is written UPR_{AB}. The corresponding ratio of B to A is thus UPR_{BA}. *UPR is only used for comparing the production of plant categories mixed in the same stands. Unless otherwise stated in the following, a plant unit means an individual (seed or emerged plant) in the initial period of plant stand development, and the plant density is the 'initial density' determined as the number of plant units per unit area.* 'Normal' death among the growing plants in a stand, leading to a decreased plant density, is considered a process among other processes influenced by the competition.

The Unit Production (UP) is a parallel to the term 'Reproductive Rate', used by de Wit (1960) and de Wit and van den Bergh (1965) expressing the number of seeds produced per sown seed. The meaning of UP is more open and must be specified from case to case. UP can be the production of specified plant parts, by weight or by number, per defined plant unit. It may be based on those plant parts that are of interest as harvested

products. In studies of more general principles concerning competition, the biomass of total plants or total above-ground plant parts may be used. When, on the other hand, the interest is focused on long-term population dynamics, the competitive influence on seeds and/or vegetative regenerative or perennating structures is of particular importance.

UPR in Measuring the Relative Competitiveness of Plants

It was observed by Håkansson (1975b) that the unit production ratio, UPR, then named RPI, of different plants in mixed stands changed with changed plant densities. When representing different species called A and B, UPR_{AB} changed when the density of A or B was changed. Changes in UPR_{AB} could be expected when A and B are plants with different traits. Because of such changes in UPR, a single value of this ratio is, in itself, no measure of the relative competitiveness of two plants. Neither is Wilson's (1986) ratio of the plant weight of weeds to crops, the 'Crop Equivalent' of weeds, which is analogous to UPR. Furthermore, 'Density Equivalents' such as those calculated by Berti and Zanin (1994) for different species will certainly largely vary with densities. However, the influence of density on UPR and corresponding ratios can be used for qualitative characterization of the *relative competitiveness* (*relative competitive ability*) of plants in mixed stands (Håkansson, 1975b).

When plants compete for largely the same resources, when in other words their niche differentiation is not too strong, the general competitive pressure increases with an increased density of *any* plant category in the mixed stand. Thus, the average growth rate of the plant units then decreases in all plant categories (cf. Håkansson, 1988a, 1991). However, when the *relative decrease* differs between diverse plants, this indicates dissimilarities in their relative competitiveness. Thus, if this decrease is relatively stronger in, for instance, plant category A than in B, UPR_{AB} decreases.

As an increased density of A and/or B means an increased competitive pressure in the mixed stand, an increase in UPR_{AB} logically means that category A is competitively superior to category B *in the prevailing environmental situation*, whereas a decrease in UPR_{AB} means that A is competitively inferior to B. If UPR_{AB} remains unchanged when the density of A or B is changed, A and B can be considered as equally competitive under the prevailing conditions. Plants other than A and B may also occur in the mixed stand. An increase in their density usually also means an increased competitive pressure in the stand, causing a change in UPR_{AB} in the same direction as a density increase in A or B (Håkansson, 1988b).

When A and B represent plants with different traits, UPR_{AB} usually differs from unity ($UPR_{AB} \neq 1$). This means that the production per initial plant unit of A and B is different ($UP_A \neq UP_B$). A deviation of UPR_{AB} from unity does not in itself mean that the relative competitiveness of A and B differs. Say, for example, that UPR_{AB} is 1.6 in a stand where A and B are mixed at a given number of emerged plants per unit area. This only says that the production per emerged plant of A corresponds to that per 1.6 emerged plants of B under the prevailing conditions. If UPR_{AB} proves to remain at a constant level, 1.6, when densities are varied, it is reasonable to consider A and B equally competitive in their mutual interference, differing only in their production per plant. Thus, when A and B are mixed at the density proportion 1/1.6, each of them stands for half of the production in the mixed stand. In other words, there is a *density equivalent* of 1.6 for B vs. A and 0.63 for A vs. B. (Under conditions represented in Fig. 33a, UPR of wheat vs. barley is thus almost constant, about 0.63; see below.)

However, a UPR value that differs strongly from unity mostly implies such a strong difference between the plants that they could be expected to be differently competitive. Changes in UPR with varying densities will reveal such differences. Before measurements of such changes are illustrated, changes in UPR caused by alterations in the environment are discussed below.

Thus, if A and B have different demands regarding their environment, UPR_{AB} may change with an alteration in the environment. Say that, in stands with unchanged densities of A and B, UPR_{AB} rises in response to an environmental alteration from situation s_1 to situation s_2. This means that the competitive ability of A has increased relative to B under otherwise unchanged conditions, but does not necessarily mean that A has become competitively superior to B. This question can be elucidated by comparing UPR_{AB} at different densities, say at a lower density, d_1, and at a higher, d_2. The following fictitious values of UPR_{AB} in environmental situations s_1 and s_2 illustrate this in two fictitious cases, *i* and *ii*:

	Case *i*		Case *ii*	
	d_1	d_2	d_1	d_2
s_1	0.8	0.4	1.0	0.5
s_2	1.6	1.2	2.4	3.2

In situation s_1, UPR_{AB} decreases with the density increase from d_1 to d_2 in both Case *i* and Case *ii*, which means that A is inferior to B in both cases.

In Case *i*, UPR_{AB} increases with the environmental change from situation s_1 to situation s_2, indicating that the competitive ability of A has increased relative to B with the changed environmental situation. However, as UPR_{AB} decreases with the density increase from d_1 to d_2 also in situation s_2, it is still competitively inferior to B. However, it is less inferior, which is indicated not only by the general increase in UPR_{AB} but also by the fact that the relative decrease in UPR_{AB} with increased density from d_1 to d_2 is smaller in s_2 (25%) than in s_1 (50%).

In Case *ii*, UPR_{AB} increases with the environmental change from situation s_1 to situation s_2; in situation s_2, it increases also with a density increase from d_1 to d_2. This means that the competitive position of A has not only increased relative to B, but that A has also become competitively superior to B.

The changes in the relative competitive power of A and B could have different causes. In both Case *i* and Case *ii*, the competitive power of A *relative to* B has increased with the change in the environmental situation from s_1 to s_2. The reason for this increase could be that A has been favoured at the same time as B has been either disadvantaged, uninfluenced or less favoured than A by the change in the environment. It could also be that A has been disadvantaged, but less so than B, etc. The reader can fill out the list of possible reasons for changes of UPR$_{AB}$ with changes in the environment.

Reasons why an alteration in the environment has led to a changed competitive power of A *relative to* B – as indicated by a change in UPR$_{AB}$ – might be interpreted through comparing the absolute production values of A and B at different densities in the mixed stands. However, the response of A and B to altered environmental conditions can be more reliably interpreted by comparisons of their response in the mixed stands and their response in added separate (pure) stands. Very many questions concerning plant competition and relative competitiveness can be answered reliably only in experiments including both separate and mixed stands at various densities of the plant categories studied.

Comparisons of UPR values over wide density ranges would probably reveal differences in relative competitiveness between most genotypically diverse plants, even when they are morphologically fairly similar – as are, for instance, wheat and barley. The strengths and weaknesses of diverse plants are not necessarily the same in different density ranges. The plastic responses of diverse plants may change differently with changed plant densities in different density ranges. This means that different plant parts may change differently in diverse plants. Therefore, measurements of the relative competitiveness based on the competitive outcome in mixed stands may differ depending on which organs are in focus for the evaluation of the relative competitiveness. These issues deserve systematic studies.

Even if there were similar differences in the relative competitiveness of two plants, A and B, in different density ranges, the relative change in UPR$_{AB}$ with a numerically equal density alteration would decrease when this

alteration occurred in a higher than in a lower density range. The simple reason is that a *numerically* equal density alteration means a *relatively* smaller alteration at higher densities than at lower. When the addition of a plant category C to a mixed stand of A and B results in an increase in the total density, this will usually have an effect on UPR$_{AB}$ analogous to the effect of an increased density in a stand of A plus B free of C.

However, if the added C is a plant that changes the environment in the plant stand considerably, this might also change the relative competitiveness of A and B. It is of interest to compare UPR$_{AB}$, UPR$_{AC}$ and UPR$_{BC}$ regarding their response to density changes in mixed stands with different combinations of A, B and C (see pp. 132, 133 and Fig. 33). With an increasing number of plants with different traits mixed in a stand, the interactions become increasingly complex. When the interest focuses on a few plants, changes in their relative competitiveness caused by changes in the plant composition in a stand can certainly be studied using UPR values for different pairs of plants in stands of varying composition.

It is widely experienced that environmental changes, including changed treatments, can change not only the total production but also the relative production of different plants in the stand. Changes in their relative production are results of changes in their relative competitive power (e.g. Tables 31–36). Through their morphology, through allelopathy, etc. many plants can cause such changes in the environment that the growth conditions and competitive power of other plants relative to each other are changed.

UPR is useful in stands where plant units can be distinctly defined. As stated above, it can be used for characterizing the relative competitiveness on the basis of its response to plant densities. UPR can also be used in studies of the competitive power of plants as influenced by their relative time of emergence and planting depth, seed size and spatial distribution – and by their external environment, including treatments with mechanical and chemical methods. As is shown in a later section, it can also be used for comparing the growth rate of plants in mixed stands. *UPR is useful mainly for characterizing plant relations in stands of short duration. However, the ratio is also useful for characterizing competitive conditions in older stands when these are composed of plant units that can be distinctly defined, such as trees and bushes, growing in uniform mixtures.*

The *initial* density of weeds is often difficult to determine. It may be only moderately influenced by the crop seed rate and even though there is usually some influence, this can be included among the influences of the crop seed rate on the weed biomass. All weeds, or specified weed species, on an area could then be looked upon as a unit in determining the UPR of these weeds to the crop at different seed rates. Relative changes in the UPR caused by the crop seed rate could then be used to characterize the competitive status of weed populations in relation to the crop.

The crop plant density (seed rate) influences the weeds to a different degree. An increased plant density thus has a selective effect on the weeds, even on genetically equal individuals, because their competitive position usually varies strongly due to variation in time of emergence, seed quality, microsite conditions, etc. Consequences of this are further discussed on p. 136.

Changes in UPR in response to changes of various factors are often preferably expressed in relative (percentage) numbers. These could then be easily compared by means of 'relative UPR' or 'percentage UPR' (i.e. percentage changes in UPR) as in Fig. 33b and d, discussed in the following section.

Use of UPR and Percentage (or Relative) UPR – Examples

Figure 33 illustrates changes of UPR and the 'Percentage UPR' (percentage changes of UPR) in response to changed densities of plants in mixed stands. The graphs are based on a pot experiment with spring wheat, *Triticum aestivum* (T), two-row barley, *Hordeum distichon* (H), and white mustard, *Sinapis alba* (S), in mixtures and pure stands at various densities. Pots containing a sandy loam fertilized with a complete fertilizer were placed outdoors and sown in the spring. Calculations presented in Fig. 33 are based on the dry weights of aerial shoots cut at the soil surface in early July when cereal culm elongation had ceased. The seeds sown are the plant units (see details in legend to figure).

Figure 33a and b shows that UPR_{TH} and Percentage UPR_{TH} were very slightly influenced by the densities (X_H, N_T or N_S). According to the above, this indicates that H and T (barley and wheat) were almost equally competitive under the conditions given. UPR_{TH} was almost constant, between 0.60 and 0.65. This means that in mixed stands one unit of wheat invariably produced the same amount of biomass as 0.60–0.65 (≈ 0.63) units (seeds sown) of barley, or that one unit of barley produced as much as about 1.6 units of wheat.

UPR_{SH} and Percentage UPR_{SH} – also shown in Fig. 33a and b – decreased with increasing density of H (X_H). In Fig. 33c and d it is shown that UPR_{HS} and Percentage UPR_{HS} increased with increasing X_S, as did also UPR_{TS} and Percentage UPR_{TS}. These responses indicate that both H and T were competitively superior to S. Calculations indicate that the proportion between UPR_{TS} and UPR_{HS} was almost uninfluenced by X_S, at values between 0.60 and 0.65.

Thus, within the density ranges now being considered, there was a nearly constant density equivalence of H vs. T, amounting to 0.60–0.65 in their interaction with S as well as in their mutual competition. In consequence, the curves in Fig. 33d, describing Percentage UPR_{TS} and Percentage UPR_{HS} as functions of X_S run nearly together.

It should be emphasized that a similarity such as that of UPR_{TH} to the proportion UPR_{TS}/UPR_{HS} should not be expected to occur unless all the plants studied are mixed in the same stands. This means, at the same time, that proportions characterizing the competitive power of two plant categories in relation to a third (as measured on the analogy of UPR_{TS}/UPR_{HS}) will not always exhibit the same proportion as that

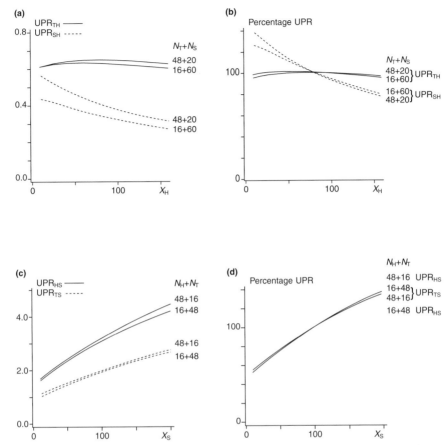

Fig. 33. Plant density effects on the unit production ratio (UPR) in mixed stands of two-row barley, *Hordeum distichon* (H), spring wheat, *Triticum aestivum* (T), and white mustard, *Sinapis alba* (S), in varying density combinations. Curves based on trivariate regression equations describing dry weights of above-ground shoots dm^{-2} as influenced by densities. Weights measured after termination of culm elongation in the cereals. N_H, N_T and N_S denote constant densities of H, T and S; X_H and X_S varying densities of H and S. Densities are numbers of seeds sown dm^{-2}. Results from an outdoor experiment in pots with a sandy loam soil fertilized with a moderate amount of a complete fertilizer, equal to 'Fert. 2' in Fig. 34. The experiment comprised 124 combinations of densities and species in four replications (Håkansson, 1988b; cf. 1997a).

characterizing their own relative competitiveness (as measured by UPR_{TH}).

Figure 34 describes the use of UPR in studying influences of the relative time of emergence (or germination) and the level of fertilizer on the relative competitive position of two plants. The graphs describe UPR_{TH} in mixed stands of spring wheat (T) and two-row barley (H) in an outdoor multifactorial experiment at Uppsala, in pots containing 5 l of a sandy loam soil (585 pots representing three replications of 195 treatments). A complete fertilizer was applied at three

levels corresponding to nitrogen supplies of 0.6, 1.35 and 2.1 g dm^{-2}. Barley was sown on the same day in all pots ('Day 0') at three densities (X_H = 33, 66 or 132 seeds dm^{-2}). Wheat (T) was sown at a constant density (30 seeds dm^{-2}) on four occasions. Seeds were sown in marked positions with a technique avoiding damage to previously sown plants of barley. Aerial shoot dry weights were determined on five occasions. (See further details in legend to Fig. 34.)

Figure 34 presents calculations based on dry weights dm^{-2} determined when culm

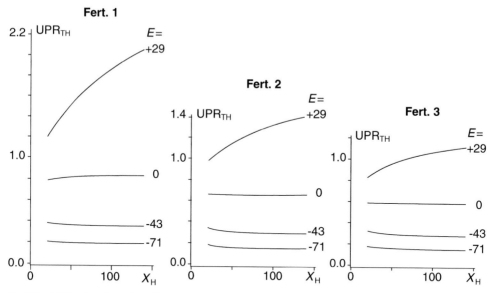

Fig. 34. UPR$_{TH}$, i.e. the unit production ratio of spring wheat, *Triticum aestivum* (T), to two-row barley, *Hordeum distichon* (H), described as a function of X_H (the number of seeds of H sown dm^{-2}) at a constant T density of 30 seeds dm^{-2} ($N_T = 30$). Influences of the relative time of germination of T and H ($\approx$ their relative time of emergence) at different levels of fertilizer. Start of germination of T occurred at four times relative to H: namely 7 days before, on the same day as, or 7 or 14 days after H. The differences in time are expressed in degree-days (°C.d, base 4°C) denoted $E = +29$, $E = 0$, $E = -43$ and $E = -71$, respectively. Fert. 1, Fert. 2 and Fert. 3 represent three fertilizer levels achieved by application of a complete fertilizer corresponding to nitrogen supplies of 0.6, 1.35 and 2.1 g dm^{-2}. Curves are based on regression equations describing the dry weights of above-ground shoots dm^{-2} as functions of X_H and E. Weights were measured after culm elongation had terminated in the cereal plants. Outdoor pot experiment at Uppsala. Soil and Fert. 2 correspond to those in the experiment dealt with in Fig. 33. (Further details, see Håkansson, 1991: pp. 109–149.)

elongation had ceased in all plants. The periods (E) between the sowing occasions of T and H are expressed by degree-days (°C.d). The curves describe UPR$_{TH}$ as functions of X_H (the density of H) at the four relative times of sowing T and H described by the E values. (Periods between times of sowing or starts of germination $\approx$ periods between times of emergence.) Curves representing cases when barley (H) and wheat (T) were sown on the same day ($E = 0$) are considered as reference curves.

Strongly differing levels of UPR$_{TH}$ at differing E values indicate a great impact of the relative time of germination/emergence on the outcome of the competition between T and H (cf. Tables 28–30 and Figs 30–32).

At $E = 0$ (T sown at the same time as H), UPR$_{TH}$ was very slightly influenced by the density of H (X_H). It appears to be almost constant at the two higher levels of fertilizer use

(Fert. 2, Fert. 3), but may tend to increase at the lower level (Fert. 1). This appears more clearly in Fig. 45 (p. 153), where UPR$_{TH}$ is described on the basis of regression equations representing biomass changes over time. (These equations allow a somewhat more reliable description of density influences at $E = 0$ than do the equations behind the curves in Fig. 34, because they cover a long period of growth with many data at $E = 0$.) In Fig. 45, UPR$_{TH}$ at Fert. 1 is always higher at a higher H density than at a lower (called N_H here). This points at a greater competitiveness of T relative to H at the lower fertilizer level. At the highest level, Fert. 3, no density influence on UPR$_{TH}$ was indicated, either in Fig. 45 or in Fig. 34. At Fert. 2, the curves suggest an intermediate density influence.

The analyses illustrated in Figs 34 and 45 – accomplished in different ways – thus point in the same direction, invariably inti-mating that the competitiveness of T (wheat) relative to H (barley) decreased with increas-ing fertilizer level. Although results of pot experiments do not reliably characterize the relative competitiveness of plants in field stands, it is worthy of note that the results described here are in harmony with results of the field experiment presented in Tables 31 and 32. These tables thus showed that spring wheat at a low level of nitrogen fertil-izer suppressed weeds more strongly than spring barley, whereas the opposite was evident at a higher level (cf. pp. 111–113).

The dry matter production of aerial shoots of T (wheat) sown in pure stands at $E = 0$ increased with increased fertilizer rates. Dry weights determined at the same time as comparable dry weights underlying the curves at $E = 0$ in Fig. 34 were thus: 82, 135 and 158 g dm^{-2} at Fert. 1, Fert. 2 and Fert. 3, respectively. The curves at $E = 0$ in Fig. 34 show that UPR$_{TH}$ decreased with increasing fertilizer application, i.e. that the production of T per sown seed decreased relative to that of H. The curves at $E = 0$ indicate a decrease in UPR$_{TH}$ from about 0.8 at Fert. 1 to about 0.65 at Fert. 2 and 0.60 at Fert. 3. (Fert. 2 resembles the fertilizer levels in the experiment dealt with in Fig. 33, where UPR$_{TH}$ was 0.60–0.65 and only slightly influenced by density.)

When T emerged about a week earlier than H, i.e. at $E = +29$, the general level of UPR$_{TH}$ was higher than at $E = 0$, indicating that the competitive position of T relative to H had been improved. The UPR$_{TH}$ curves at $E = +29$ at the same time increased with increasing X_H, indicating a competitive superiority of T to H when T emerged earlier than H. The degree of superiority decreased with increasing fertility level which favoured H over T.

With increasing delay of T in relation to H, from $E = -43$ to $E = -71$, UPR$_{TH}$, expectedly, decreased below the levels at $E = 0$. There was no marked difference between Fert. 1, Fert. 2 and Fert. 3 regarding the relative decrease in UPR$_{TH}$. A stronger decrease could have been expected as a result of an increased fertilizer application, which favours H over T, according to the

curves at $E = 0$ and $E = +29$. The reason why the relative decrease in the UPR$_{TH}$ did not change markedly can only be speculated. Thus, when T was delayed, H could capture more of the applied fertilizer than when T emerged simultaneously or earlier. Being more tolerant to low nutrient levels, T might have lost less in its *relative* competitiveness than H could gain through increased capturing of nutrients at higher levels of fertilizer application.

Many of the effects illustrated in Fig. 34 can be somewhat more easily illustrated by combinations of curves of UPR$_{TH}$ and Percentage UPR$_{TH}$ as in Fig. 33.

When plants A and B have different demands for some mineral nutrient or other resource, or tolerate resource shortages differently, their relative competitiveness will always change with an altered access to the resource. Then, both the general level of UPR$_{AB}$ and its response to plant density change. In most cases, the relative competitive effect of A and B on a third plant category, C, changes – probably mostly in the same direction as the change in their relative competitive power (as measured by UPR$_{AB}$), but not necessarily as a general rule.

The changes under discussion can be studied by comparing UPR$_{AB}$ with the rela-tionships between UPR$_{AC}$ and UPR$_{BC}$ repre-senting different densities and proportions of A, B and C in mixed stands. This was exemplified in Fig. 33, where UPR$_{TH}$ and the ratio of UPR$_{TS}$ to UPR$_{HS}$ are compared.

As is illustrated in other sections, comparisons aimed at describing relative competitiveness can be made in other ways. Comparisons of A and B with regard to their effect on C must be done in separate stands of A + C and B + C, respectively, and should therefore be done with great care. There is then a greater risk of confounding influences of plant traits and densities on the com-petitive outcome than when the plants are mixed in the same stands (see reference models, pp. 138–139).

The influence of fertilization on the relative competitiveness of barley and wheat – as characterized on the basis of their effect on weeds (Tables 31 and 32) – is seen in comparisons of stands of types A + C and B + C. It is in harmony with the influences characterized on the basis of UPR$_{TH}$ in Figs 34 and 45. Although the effects shown in Tables 31 and 32 are measured in separate stands, they are reasonably informative, as they have been measured under the same environmental conditions and density influences have also been considered.

Frantíc (1994) presented a comprehensive study on competition between maize and two weeds, *Chenopodium suecicum* and *Amaranthus retroflexus*, in different combinations. Among many interesting results, he showed that the competitive ratio (CR) of *C. suecicum* to *A. retroflexus* gradually increased with increased total plant densities. The ratio increased from values below unity at low densities to values far above unity at higher densities.

Frantíc's conclusion was that the former species was the inferior competitor at low densities, but became the superior competitor at higher densities. Because of the increase in CR with increased plant density (= increased competitive pressure), the conclusion should instead be that *C. suecicum* was more competitive than *A. retroflexus* under the prevailing conditions. CR (according to Willey and Rao, 1980) is presented on pp. 142–144. UPR would have shown an increase similar to that of CR, which means that *C. suecicum* was more competitive than *A. retroflexus*. Changes in the simpler UPR at different densities (sometimes even as few as two) allow competition analyses in multifactorial experiments aimed at characterizing the relative competitiveness and its dependence on various factors.

Selective Effects of Competition within Heterogeneous Components in Mixed Stands

Let plant categories A, B and C be the components in a mixed stand. Theoretically, the individuals in a given category could be considered completely identical. In reality, however, each category consists of more or less differing plants. Even when all plants of a specified category represent the same strain or variety of a species, they usually diverge genetically to some extent. Apparently similar plants may therefore respond differently to external influences. However, even when a crop, call it category A, is represented by genetically identical plants, these may be experiencing different growth conditions due to varying 'quality', varying depth location and varying conditions in the microsites of the seeds sown. Category B could be weed plants representing one species discerned in the weed flora or a mixture of species. As can be understood from previous discussions, even individuals within the same species normally represent a very heterogeneous group.

This heterogeneity normally results in a biomass dominance of a minority of the weed plants in a stand. As stressed by many authors (e.g. Hallgren, 1974a), this dominance may be won by plants of a few species and usually also by a minority of the individuals within the species.

Effects of the heterogeneity under discussion must be considered when the results of competition experiments are evaluated. Competition exerts selective effects, which are strengthened when densities are increased.

An increased density thus leads to an increased dominance of the more competitive plants, whether these plants are compared with other plants within the same category or with plants representing other categories distinguished in a mixed stand. When two plant categories differ in their degree of heterogeneity, their response to density may lead to difficulties in interpreting the relative competitiveness of the categories correctly. UPR, for example, may be influenced by a differently strong selective effect within the two categories. Within ranges of higher and lower densities, UPR may even change in different directions (cf. Håkansson, 1988b: p. 157). All relationships or ratios aimed at comparing the growth of different categories, seen as entireties, will be influenced by differences in their degree of heterogeneity.

The variation in the size of the plants within the categories discerned in mixed stands deserves systematic studies.

Accomplished *in addition to* the broader measurements treating plant categories as units, such studies will facilitate correct interpretations of the effects behind the outcome of competition in different situations.

Determination of Production–Density Relationships in Mixed Stands

A great number of models considering two plant categories (say crops and weeds) have been developed from simple equations of type Eqn 1 (p. 88). For a plant category A (e.g. a crop), this equation can be written $Y_A = X_A \cdot (a + bX_A)^{-1}$, where Y_A is the production per unit area and X_A the plant density.

Watkinson (1981), Spitters (1983a,b), Firbank and Watkinson (1985), Cousens (1985b) and others have tested models describing the production of two plant categories, called A and B here, in mixed stands. In the simplest case, such models could be characterized by the following example describing the production per unit area of A mixed with B: $Y_A = X_A \cdot (a + bX_A + cX_B)^{-1}$. The production of B could be written $Y_B = X_B \cdot (d + eX_A + fX_B)^{-1}$.

If $c/b = f/e$, plants of A and B are replaceable in proportions given by these ratios; and these ratios are, in addition, independent of the densities of A and B. Using Connolly's (1987) terminology, there is a constant *substitution rate* of A vs. B, i.e. a rate that is independent of the A and B densities.

However, it has been observed in various situations that c/b and f/e are different. Vleeshouwers *et al.* (1989) interpret a deviation of the ratio $(c/b)/(f/e)$ from unity as a result of niche differentiation. Many models have been proposed with the assumption that the substitution rate is not constant, i.e it does not only depend on environment, but also on densities. On the basis of experiments and calculations of UPR, first called RPI, Håkansson (1975b) suggested that such proportions and related ratios representing different plant species mostly depend on densities; the ratios being approximately constant only when growth habits and morphology of the species are very similar

(as shown above for wheat and barley in certain situations).

The denominators $(a + bX)$ of single-species models of type Eqn 1 are described by straight lines, and the models $(Y = X \cdot (a + bX)^{-1}$ by hyperbolic curves. Modifications of single-species models described by parabolic curves (curves with a maximum) have been used by Bleasdale and Nelder (1960), Holliday (1960b), Willey and Heath (1969) and others (Håkansson, 1975b, 1983b, etc.).

In developing models for mixed stands, Spitters (1983a) found no substantial improvement in adding an 'interaction term' $(X_A X_B$, multiplied by a constant) in the denominator. Models allowing for deviations from linearity by using exponents have been presented by Watkinson (1981), Spitters (1983b), Firbank and Watkinson (1985) and others.

It has been proposed (e.g. Vleeshouwers *et al.*, 1989) that the plant category that exhibits the greater substitution rate of two categories in mixed stands should be considered the superior competitor. However, *the relative size of the individuals of two plants in a mixed stand is not a good measure of their relative competitiveness.* The simple reason is that *this relative size mostly changes with densities, i.e. with the competitive pressure*, which has been illustrated above by the changes in UPR (Håkansson, 1975b, 1983b, 1988b). The changes in UPR$_{AB}$ (UPR of A to B) with changed densities are usually (see above) more reliable descriptions of the relative competitiveness. Risks of misinterpretation (e.g. due to heterogeneity as described above) burden any ratio or expression aimed at characterizing relative competitiveness.

UPR has the advantage of not being bound to specific models. In its practical use, UPR of A to B can often be calculated directly on the basis of biomasses recorded from a few stands of A + B representing different plant densities. In the analyses of more complex situations, e.g. those treated in Figs 33 and 34, UPR is preferably based on regression equations well fitted to experimental data.

From Fig. 18 (p. 92) and associated discussions it is understood that there is no simple equation suited as a standard for any production–density relationship. However, within narrower density ranges, simple standard equations can mostly be used.

In analysing the suitability of various expressions for the interpretation of competitive effects and relative competitiveness of plants in mixed stands, production–density equations giving good fits to different plants and densities are needed. On the basis of experimental data, such equations have been produced (Håkansson, 1983b: p. 63, 1988b: p. 18, 1997a). All of them can be described as extensions of Eqns 1, 2 or 3. To achieve good fits to data from experiments covering wide density ranges of different plant categories, terms indicating interactions between densities of the different categories had to be included in the denominator polynomials. In equations describing growth as influenced by densities, terms representing density–time interactions were also included (Håkansson, 1991, 1997b).

These equations are empirical regression equations for *ad hoc* use. They confirm the above statement that no simple standard equation can describe any production–density relationship. As used hitherto, they are the basic equations underlying a number of graphs aimed at characterizing competition from various aspects – such as graphs in this book. Certain equations include, in addition to densities, either row spacing (e.g. Håkansson, 1975b) or relative time of emergence (Håkansson, 1991) as independent variables.

Reference Models

'Reference models' are based on competitive conditions in stands of identical plants. These plants are assumed to be a mixture of different plant categories (A, B, etc.) uniformly distributed in the stands, now fictitiously considered to be mixed stands. Reference models have been used to detect and logically explain general characteristics of relationships and ratios describing the competition in mixed plant stands. They have special use in the identification

of density influences, when these are easily mistaken for effects of the traits of dissimilar plants.

The reference models are expressions (indices or relationships) derived from equations describing the production–density relationship in uniform pure stands of a single plant category (e.g. Eqns 1, 2 or 3, pp. 88–89; Fig. 13). Regression equations fitted to experimental data have been used (Håkansson, 1975b, 1983b 1988b, 1997a, etc.).

Equations where both density and time (after germination or emergence) are the independent variables have been used in deriving reference models considering biomass changes over time (Håkansson, 1991, 1997b).

The 'reference model technique' was first used to describe and explain characteristics of relationships between weed biomass and crop density and between crop production and weed density (Håkansson, 1975b). There were conflicting opinions about the characteristics of these two relationships. The shape of these relationships could be characterized and logically discussed on the basis of reference models (e.g. Fig. 20).

Reference models are preferably presented graphically. In most cases, an equation of type Eqn 1 can be used as the basic production–density equation. When wide density ranges, including maximum production, are to be covered, it is more correct to use equations of types Eqn 2 or Eqn 3 (p. 89). Equations of the three types under discussion can be written:

$$Y = X \cdot [f(X)]^{-1} \tag{4}$$

where Y is the production per unit area and X the initial density. Now, assume that the plant stand is composed of two categories of equal plants, A and B, uniformly mixed at densities X_A and X_B, where $X_A + X_B = X$. If the total production is $Y_{A+B} = Y_A + Y_B = Y$, Eqn 4 can be written:

$$Y_{A+B} = (X_A + X_B) \cdot [f(X_A + X_B)]^{-1} \tag{5}$$

The production of A and B when mixed at different densities is:

$$Y_A = X_A \cdot [f(X_A + X_B)]^{-1} \tag{5a}$$

$$Y_B = X_B \cdot [f(X_A + X_B)]^{-1} \tag{5b}$$

The pure stand can, of course, be considered to be a composition of more than two plant categories, say three, A, B and C. Their total production is then:

$Y_{A+B+C} = Y_A + Y_B + Y_C = Y$, and Eqn 4 can be written:

$$Y_{A+B+C} = (X_A + X_B + X_C) \cdot$$
$$[f(X_A + X_B + X_C)]^{-1} \qquad (6)$$

The production of A, B and C evenly mixed can be presented by the following equations, analogous to Eqns 5a and 5b:

$$Y_A = X_A \cdot [f(X_A + X_B + X_C)]^{-1} \qquad (6a)$$

$$Y_B = X_B \cdot [f(X_A + X_B + X_C)]^{-1} \qquad (6b)$$

$$Y_C = X_C \cdot [f(X_A + X_B + X_C)]^{-1} \qquad (6c)$$

Using different combinations of Y_A, Y_B and Y_C described as functions of $(X_A + X_B + X_C)$, or using expressions derived from them, we can determine biomass–density relations, calculate biomass losses in one category caused by competition from the other(s), or compute the production or production ratios at optional levels of densities X_A, X_B and X_C. We can also determine densities resulting in optional production levels. The production of A, B or C could be calculated as a function of the density of one or two categories, the densities of the other(s) being constant. Constant densities have been written N instead of X, the values of N being inserted in the topical equations.

When a reference model expression varies with densities of the plant categories fictitiously discerned in monospecific stands, it is a variation solely in response to the density variation. Thus, as categories A, B, etc. consist of identical plants, the response of the expression to densities is not mixed up with responses to differences in plant traits. When an expression derived as a reference model varies with densities, an analogous expression (ratio, index, etc.) cannot be used (straight off) to characterize the relative competitiveness of different plants. Experiments aimed at characterizing the relative competitive ability of different plants must allow calculations of expressions where influences of plant traits and densities are not confounded. The use of reference models is illustrated in the following.

Reference Models in Evaluating Relationships and Indices Aimed at Characterizing Relative Competitiveness – Examples

According to the reasoning in previous sections, the general competitive pressure in a mixed stand increases with increasing density in all or some of the plant categories in the stand. Furthermore, when two plant categories in this stand are compared, the one that exhibits the strongest biomass reduction per plant unit with the increased density is considered the inferior competitor. If we accept this, we can characterize the relative competitiveness of the two plant categories by means of expressions (relationships, indices) that describe their response to increased densities. An expression suitable for this purpose must not change with changed densities for reasons other than differences in the competitiveness of the plants.

According to the above section, reference models where plant categories A and B are identical plants can be used to unmask misleading density influences. The use of reference models to test the suitability of UPR$_{AB}$ for characterizing the relative competitiveness of two plant categories, A and B, is illustrated first.

Thus, where A and B are represented by identical plants, the response of UPR$_{AB}$ to density can characterize the relative competitiveness of A and B only in experimental arrangements where UPR$_{AB}$ = UPR$_{BA}$ = 1 at any density of A or B. This only applies to situations where the unit production of A and B (UP$_A$ and UP$_B$) is equal, regardless of densities. As shown by Håkansson (1988b: p. 142) this is the case only when A and B are mixed in the same stands. The UPR graphs in Figs 33, 34, 44 and 45 therefore represent pairs of plants mixed in the same stands.

From the production–density equation of a monospecific plant stand where the plants are fictitiously grouped into more than two categories – A, B, C, etc. – UPR$_{AB}$, UPR$_{AC}$ and UPR$_{BC}$, etc. can be calculated as reference models. In this case, all categories are uniformly mixed in the same stand and all UPRs become unity (i.e. UPR$_{AB}$ = UPR$_{AC}$

= UPR$_B$, etc. = 1), irrespective of densities. This means that UPR, in principle, can be used with the aim of measuring density effects caused by plant traits without being confounded with other density effects. If the plant pairs to be compared using UPR occur in different stands, UPR varies with densities for reasons other than different competitiveness of the plants (Håkansson, 1988b: p. 142). There are thus risks of confusing effects of non-comparable densities and effects of different plant traits.

However, the use of two-component stands can sometimes be acceptable in combination with special studies of density influences. At the same time, it should be noted that it is often very hard to define 'equivalent densities' of plants with different traits regarding their relative impact on growth and competition. It is hard even in specified unchanged environments, because the (possible) equivalence shifts with the general density level in the plant stand. It is also hard because the effects of given densities vary with the environment. In trying to find comparable densities for two plants with different traits, we realize that these relationships usually change both with the general density level and with the environmental conditions, and certainly also with the addition of other plants to the stand (Håkansson, 1988b). Therefore, experiments aimed at a reasonable ranking of plant species as to their relative competitiveness under defined environmental conditions should be based on measurements in stands with different composition, densities and density proportions as regards plant species.

It is basically impossible to separate the relative competitiveness of plants in terms of 'competitive response' and 'competitive effect' (as defined by Goldberg and Landa, 1991). The outcome of competition between plants, as characterized by UPR, is always a combination of effects and responses. However, having that in mind, we can, of course use the term 'effect' when we describe the influence of a plant category, A, on another category, B, and the term 'response' when we describe the reaction of B to impacts of A.

Let us now study the relative competitiveness of two weeds, call them B and C, by comparing their response to competition from a crop, A. The focus is now on the biomass reduction in B and C when the A density (X_A) increases at constant densities of B and C (N_B, N_C). Let Y_B(rel) and Y_C(rel) – described as functions of X_A – denote the production of B and C (weeds) per unit area as a percentage of their production in stands free of crop A (i.e. at $X_A = 0$). In Fig. 35, the curves of Y_H(rel), Y_T(rel) and Y_S(rel) represent such functions.

Before discussing these curves, we may look at the Y_B(rel) curves in Fig. 20a (p. 94). These are *reference model* curves illustrating the biomass reduction in a plant category called B with an increasing density of A (X_A). *Figure 20 shows that* Y_B*(rel) decreases more strongly with increasing* X_A *at a lower* N_B *than at a higher. This phenomenon should be considered when the intention is to evaluate the relative competitiveness of dissimilar plants (called B and C) by comparing their biomass reduction in response to an increased* X_A.

A differently strong decrease in Y_B(rel) and Y_C(rel) with increasing X_A could be expected to illustrate a different ability to withstand competition from A. However, according to the reference model curves in Fig. 20a, the decrease of Y_B(rel) depends on N_B. Therefore, a differently strong decrease in Y_B(rel) and Y_C(rel) could be the result of 'non-equivalent' levels of densities N_B and N_C. If we intend to evaluate the relative competitiveness of different weeds by comparing their biomass reduction at an increased crop density, we have to avoid confusing effects of non-equivalent densities.

The question is whether, or to what extent, this is possible. We know from the above that comparable densities of plants with different traits are hard or impossible to determine. Such changes are indicated by the density influences on UPR. However, reference model and other analyses by Håkansson (1988b: pp. 130–133) suggest that these density problems are at least partly overcome when B and C are mixed in the same A stands. The curves in Fig. 35 illustrate this.

These curves are based on results from a pot experiment where the A, B and C plants

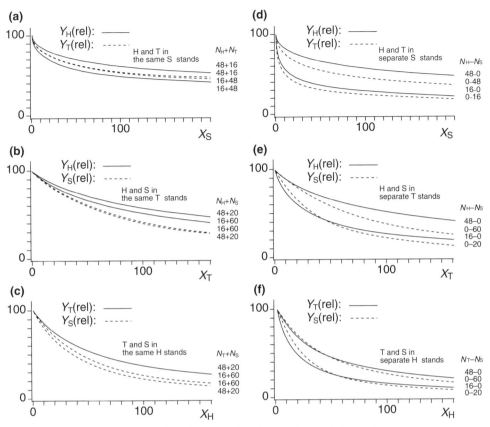

Fig. 35. Relationships of type Y_B(rel) and Y_C(rel) as functions of X_A aimed at assessing the relative competitiveness of plant categories B and C in response to category A. Curves based on the same experimental material and regression equations as the UPR curves in Fig. 33. They illustrate risks of disturbing density influences in experiments and calculations aimed at assessing relative competitiveness (see text). Plant categories A, B and C are represented by two-row barley, *Hordeum distichon* (H), spring wheat, *Triticum aestivum* (T) and white mustard, *Sinapis alba* (S), in different orders. (See Håkansson, 1988b: pp. 130–133, 1997a.)

were different species; the same experiment as that underlying the UPR graphs in Fig. 33 with plants H (*Hordeum distichon*, barley), T (*Triticum aestivum*, wheat) and S (*Sinapis alba*, white mustard). In Fig. 35, the different plants are compared in pairs regarding their response to densities. The paired plants are looked upon as categories B and C occurring at constant densities. The curves describe their decrease with an increased density of the third category, looked upon as A. The H, T and S plants, alternately, represent categories A, B and C.

The left-hand graphs (Fig. 35a, b and c) describe the decrease in plant categories B and C at an increased density of A

when they are mixed in the same stands (A + B + C). The right-hand graphs (Fig. 35d, e and f) describe the analogous decrease in B and C, but now in separate A stands (B + A and C + A).

Figure 35d–f shows that densities N_B and N_C strongly affect the biomass decrease in B and C in response to an increased A density. This effect corresponds to the effect of N_B in the reference model in Fig. 20a. The effects of N_B and N_C overshadow influences of possible differences between B and C in their relative competitiveness. The B or C curves presented in Fig. 35a, b and c are influenced to a much lesser extent by densities N_B and N_C. Their courses are, instead, 'species-oriented', less influenced by densities N_B and N_C.

Thus, with the B and C plants growing together, divergences in the general courses of the Y_B(rel) and Y_C(rel) curves can be interpreted as largely caused by different relative competitiveness. Graphs in Fig. 35b and c thus show that both H (barley) and T (wheat) were competitively superior to S (white mustard) under the given conditions. Figure 35a, on the other hand, suggests that H and T were rather equally competitive. This agrees with the conclusions drawn using UPR in Fig. 33.

The relative competitiveness of plants in specified environments may be ranked on the basis of changes in UPR, Percentage UPR or Y(rel) in response to density changes (Fig. 33 and Fig. 35a–c). By comparing specified pairs of plants in stands with different composition or densities caused by additional plants, influences of the stand composition may be detected. These aspects of experimental design have been discussed previously (Håkansson, 1988b: pp. 130–147, 1997a: pp. 66–70).

The 'competitive effect' on a crop A caused by weeds B and C, respectively – or, in other words, the 'response' of A to competition from either B or C – cannot be measured when B and C are mixed with A in the same stands. As seen in Fig. 35d–f, it is then hard to avoid confounding density influences.

Problems in comparing the relative ability of two plants, B and C, to reduce the production in a stand of another plant, A, have been investigated by Håkansson (1988b: pp. 90–95 and 147–151, 1997a: pp. 70–71). This relative ability was called the *relative aggressiveness* of B and C to A. It was described by the *Unit Reduction Ratio* of B to C with respect to the effects of these plants on the production in A, and was abbreviated to URR_{BC} in A.

Because competitive effects of different plants have to be compared using measurements in separate stands, interpretations of URR_{BC} in A are disturbed by inadequate density influences. This was illustrated in reference model calculations by Håkansson (1988b: p. 92).

Accurate evaluations of the relative aggressiveness of plants by means of URR are thus hard to attain. For various practical uses, however, URR values – e.g. for comparing different weeds – can be calculated at ordinary ('standard') densities of crops and different weed densities. However, judgements based on such comparisons always have to be interpreted very cautiously. This applies to any ratio aimed at characterizing weeds with regard to their relative yield reducing capacity, e.g. ratios presented by Wilson and Wright (1990).

The *Relative Crowding Coefficient, k*, introduced by de Wit (1960), and the *Competitive Ratio*, CR, proposed by Willey and Rao (1980) have been analysed using the reference model technique (Håkansson, 1988b: pp. 51–85, 1997a: pp. 65–66). It was shown that assessments of the relative competitiveness of two plant categories, A and B, by these indices easily lead to wrong conclusions. The indices are based on 'replacement series', i.e. series of separate stands. They therefore strongly depend on the densities in the pure stands of plants A and B, which determine the densities in the replacement series throughout. Density influences on the results in replacement experiments have been illustrated in other ways by Inouye and Schaffer (1981) and Connolly (1986).

In reference models, the A and B plants are, by definition, equal in each separate stand. Therefore, expressions such as k_{AB}, CR_{AB} (k or CR of A vs. B) and RYT_{A+B} (RYT of A plus B (RYT = *Relative Yield Total*)) should be unity (one) regardless of densities. If they vary with densities, confounding density influences must be considered when they are aimed at characterizing properties of A and B, e.g. their relative competitiveness. Now, reference model analyses show that the expressions diverge from unity as soon as the densities of A and B are not equal. This indicates that, in experiments where A and B are plants with different traits, effects of 'non-equivalent' densities and effects of plant traits are confounded.

Figure 36 shows examples of reference model calculations. In Fig. 36a, it is seen that

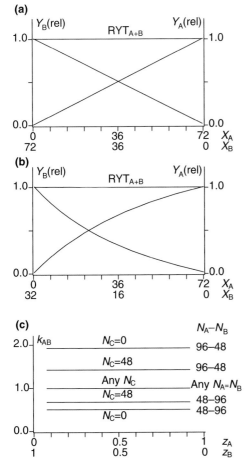

Fig. 36. Examples of reference model calculations illustrating density influences in replacement-series. (a, b) Relative biomass production per unit area – Y_A(rel) and Y_B(rel) – of two plant categories, A and B, mixed in different proportions. Densities X_A and X_B (number of plants dm^{-2}) approach pure-stand levels which are either equal as in (a), i.e. at $N_A = N_B = 72$, or different as in (b), where $N_A = 72$ and $N_B = 32$. In the former case, k and CR are always unity, but differ from unity in the latter case. Diagram (b) shows that RYT is less influenced by densities and (c) that k_{AB} (i.e. k of A versus B) depends on the pure stand densities of A and B (N_A and N_B) and on the density (N_C) of a plant category C intermixed in all stands of the replacement series, including the otherwise pure stands of of A and B. Furthermore, z_A and z_B denote relative densities when N_A and N_B in pure stands are given as unity (1) regardless of their absolute values. (Examples from Håkansson, 1988b.)

when the densities of A and B in the pure stands are equal (here 72), all curves are linear. This means that expressions k, CR and RYT are unity (one) throughout. Figure 36b illustrates a situation where the densities of A and B in the pure stands are different (72 and 32, respectively), which results in curved lines.

Calculations exemplified in Fig. 36c, illustrate that when the densities of A and B in the pure stands are equal ($N_A = N_B$), then k_{AB} is unity at any proportion of A and B in the mixed stands of a replacement series. This also applies when a third plant category, C, is added at any constant density (N_C) to all stands of the replacement series of A and B. However, as soon as $N_A \neq N_B$, k not only depends on densities N_A and N_B, but also on N_C (see further Håkansson, 1988b: pp. 51–85).

Figure 36b illustrates that RYT, in comparison to k or CR, differs only slightly when the densities of A and B in pure stands are different (see further Fig. 46). This agrees with the results of other methods of analyses presented by Connolly (1986).

The findings now discussed indirectly indicate that there are difficulties in interpreting indices derived from replacement series involving plants with different traits. Thus, without systematic studies of the influence of different densities of these plants, we do not know at what levels their densities are comparable, or to what extent they influence the ratios derived from the data of a replacement arrangement. An additional difficulty is that the influence of specified densities depends on the environment. Reference model calculations (as in Fig. 36c) also indirectly indicate that when two plants with different traits – e.g. plants of two crops, A and B – are compared in a replacement series, the presence of a third category of plants, C (e.g. weeds) may be influential.

CR$_{AB}$ varies almost identically to k_{AB} (minor differences, see Håkansson, 1988b: pp. 53–60). Thus, $k_{AB} > 1$ or CR$_{AB} > 1$ does not necessarily rank A as competitively superior to B in a given environment. However, when an environmental change results in a change in k_{AB} or CR$_{AB}$ in a replacement series where A and B are different plants, this reliably describes the direction in which the relative competitive ability of A and B is changed. For instance, an increase in k_{AB} or CR_{AB} means that the competitiveness of A relative to B has increased. However, this is no answer to the question of which category is the superior or inferior competitor in any of the environments compared. This question can be answered by comparing the responses to density changes in mixed stands of A and B. An increase in k_{AB} or CR_{AB} with a changed environment may mean that A, from having been inferior to B, has become either superior or less inferior, or that A, from having been superior to B has become an even stronger competitor in relation to B.

The relative competitiveness of A and B in the different environments could be determined more easily and more reliably by studying changes in UPR_{AB}, according to Fig. 33, or by comparing Y_A(rel) and Y_B(rel), according to Fig. 35a–c, in stands with similarly changed densities in A and/or B in the different environments.

Plant Growth as Influenced by Density and Relative Time of Emergence

Plant competition should, when possible, be studied over time, to monitor its influence on plant growth and development. In introductory studies of general principles, it seems natural to measure biomasses, if the following measurement is restricted to aerial shoots. Biomasses are given as dry weights, which are more easily determined without errors than are fresh weights.

For many purposes, however, fresh weights are required in addition to the dry weights. Fresh weights often reflect better than dry weights the extension of plants or plant parts in a stand. The ability of a plant to occupy space in competition with other plants is certainly facilitated by its ability to extend by tissues with temporarily very low contents of dry matter.

The plastic modifications of morphology and anatomy sometimes must be studied in detail. Various plant parts should then be characterized separately.

The graphs in Figs 37–46 represent results from two pot experiments. They are based on dry weights of aerial plant parts determined at different times during growth. In each experiment, two species were studied in pure and mixed stands at various densities and, in one experiment, also after different relative times of emergence. The experiments are presented in detail as Experiments 1 and 4 by Håkansson (1991). They are named Experiments 1 and 2 here. In Experiment 1, the plants were two-row barley, *Hordeum distichon* (H), and white mustard, *Sinapis alba* (S). In Experiment 2, they were two-row barley (H) and a spring-sown variety of wheat, *Triticum aestivum* (T).

Non-linear regression equations describe the dry weights (Y) of aerial shoots (g dm^{-2}). Dry weights, i.e. the dependent variables, are thus Y_H and Y_S for H and S in Experiment 1 (Håkansson, 1991: Eqns 31 and 32), and Y_H and Y_T for H and T in Experiment 2 (Håkansson, 1991: Eqns 45 and 48). Independent variables are time and densities. In Experiment 1, time is expressed in days after emergence (D); in Experiment 2, it is given as the sum of degree-days after sowing (°C.d; base 4°C). When used as independent variables in the equations, densities are denoted X (X_H, X_S, or X_T), stating numbers of seeds sown dm^{-2}. Constant densities are denoted N (N_H, N_S, or N_T). The relative time of emergence, called E, is defined in the legend to Fig. 40.

The regression equations are represented by the curves, the 'growth curves', in parts a and b of Figs 37, 39 and 40. The growth curves describe dry weights of aerial shoots (Y) as functions of D (days) or °C.d (degree-days). The graphs in Figs 37–46 are described in detail in the legends to the figures. Underlying values are presented with the curves in Fig. 37a and b, and Fig. 40a and b, as means of replications in the experiments.

Growth rates (GR) at given times (D or °C.d) are presented in Figs 37, 39 and 40. They are, in principle, differential coefficients (derivatives) of the equations behind the corresponding growth curves. The calculated GR values are presented as functions of time. The GR curves describe the dry-weight increase per day in different plants at given densities or relative times of emergence. Their equations are related to the 'growth curve equations' as follows.

If time (e.g. D = days after emergence) is the independent variable and densities and relative times of emergence are constant, the complex equations behind the growth curves can be simply discussed as

$$Y = f(D)$$

GR has been computed from such equations as follows:

$$GR = 100 \cdot [f(D + 0.01) - f(D)] \qquad (7)$$

Using $f(D + 0.01)$ instead of $f[D + 1)$ in the equation and multiplying by 100, GR values that are practically identical with differential coefficients have been easily computed.

Figure 37 describes pure stands of H and S. Their growth is characterized in Fig. 37 by the growth curves, the growth rate curves (which describe the dry weight increase in H and S dm^{-2} day^{-1}, GR$_H$ and GR$_S$, respectively) and the other curves as follows:

- In an early period of growth, the weight per unit area increased, as expected, more rapidly at a higher plant density than at a lower.
- In the very first period after emergence, plant weight and growth per unit area were almost proportional to density. This is better illustrated by the $Y(rel)_H$ and $Y(rel)_S$ curves in Fig. 38 than by the curves in Fig. 37.
- Figure 37e and f shows that the growth rate (GR) per plant was restrained both earlier and more markedly (by competition) at a higher plant density than at a lower.
- With increased density, the GR maximum therefore fell earlier, both per unit area and per plant (Fig. 37c–f). The time of the GR maximum is probably the time of optimum leaf area index (LAI) or close to that time. GR$_S$ reached maximum earlier than GR$_H$ (Fig. 37c–f). This is probably a result of an earlier and stronger mutual shading among the leaves in S than in H.
- In the period following their maximum, both GR$_H$ and GR$_S$ decreased more

rapidly at higher plant densities than at lower. This could be interpreted as a negative effect of the stronger and more rapidly increasing competitive pressure with increasing plant density.

The curves in Fig. 38a and b, which represent H or S in pure stands, are analogous to the 'growth curves' in Fig. 37a and b. The curves in Fig. 38c and d, on the other hand, represent stands where the total plant density has been increased by addition of S and H plants, respectively, at constant densities ($N_S = 36$, $N_H = 36$). A comparison of the lower diagrams with the upper in Fig. 38 illustrates the influence of the added plants on the growth of the target plants. With decreasing density of the target plants (from 144 to 9 dm^{-2}), the added plants caused an increasingly strong relative change in the total density in the stand. They therefore exerted a gradually stronger competitive influence on the target plants.

Figure 38 also shows that, in very early stages, the target plant biomasses are proportional to densities, not only in pure stands, but also in stands with other plants added. Competitive effects appear gradually earlier with increasing plant density, which is shown by the course of the curves. These are first horizontal and parallel, but with the progress of time they converge, and earlier so with increasing plant density. With further progress of time, the levels of the curves representing different densities become proportionally less different. At the same time, the differences remain stronger when the target plants are subjected to competition from added plants than when they grow in pure stands. This corresponds to the previously illustrated fact that weeds reduce the production of an annual crop relatively more at a lower crop plant density than at a higher density (Figs 20 and 22).

The relative growth rate (RGR) of young plants decreased with the progress of time, first very rapidly, then more slowly (e.g. Håkansson, 1991: Figs 6, 9 and 72). In very early stages, before competitive effects appear, the decrease seems to be independent of densities. This is understood from the curves of 'relative' RGR, described as

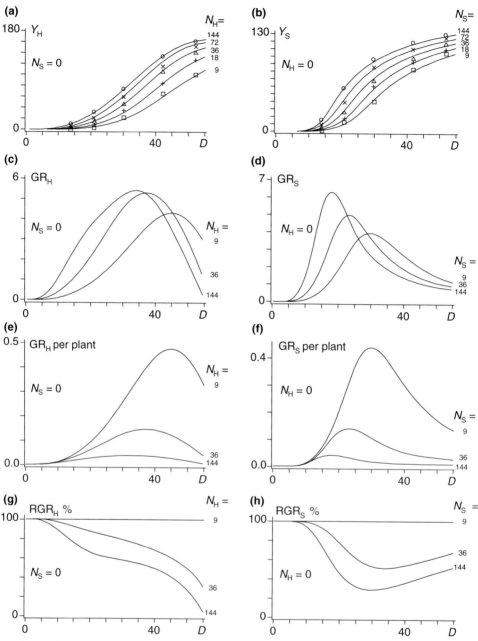

Fig. 37. (a, b) 'Growth curves', representing aerial shoot dry weights (Y_H and Y_S, g dm^{-2}) in pure stands of two-row barley, *Hordeum distichon* (H), and white mustard, *Sinapis alba* (S), respectively, presented as functions of time (D, days after emergence) at five constant densities, N_H or N_S (number of seeds of H or S sown dm^{-2}). (c, d) Corresponding curves of the growth rate per unit area, GR$_H$ or GR$_S$ (d.wt., g dm^{-2} day^{-1}), at three densities. (e, f) Corresponding curves of the growth rate per 'initial plant unit', GR$_H$ or GR$_S$ (d.wt., g seed sown day^{-1}). (g, h) Corresponding curves of the relative growth rate, RGR$_H$ or RGR$_S$, presented as percentages of the level at $N_H = 9$ and $N_S = 9$, respectively. All curves are based on trivariate regression equations (Eqns 31 and 32 in Håkansson, 1991) describing Y_H and Y_S as functions of densities X_H and X_S and time D. Outdoor pot experiment at Uppsala. *Note that vertical scales differ between diagrams.*

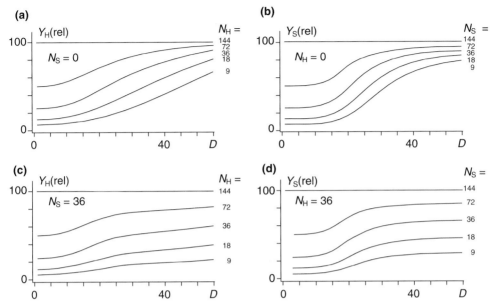

Fig. 38. (a, b) Curves corresponding to the growth curves in diagrams a and b of Fig. 37, now presented as percentages of levels at $N_H = 144$ and $N_S = 144$, respectively. (c, d) The corresponding curves of H grown in mixture with S at $N_S = 36$ and of S grown in mixture with H at $N_H = 36$.

'RGR percentage', in Fig. 37g and h, and Fig. 39e and f. These curves run together, horizontally, in the early stages of growth, at the same time as their underlying RGR curves decrease with an equal rapidity. When competitive effects set in – earlier at higher densities than at lower – this leads to a divergence of the curves of RGR and RGR percentage representing different densities. Differences in the course of the curves in more advanced periods of growth reflect differences between the species in their behaviour when approaching and entering the generative phase. (These differences, including those caused by experimental errors, cannot be measured or explained with certainty in the present studies.)

Figures 39 and 40 illustrate the growth per unit area of a plant species (category A: say a crop sown at a constant density) when influenced by intermixed plants of another category (B : say a weed) established at different densities (Fig. 39) or emerged at different times (Fig. 40). All curves represent plants of category A. Categories A and B were represented by species H, S and T as

is understood from the diagrams. In Fig. 40, time is given in degree-days (°C.d) – so also the difference in time of emergence (E). See details in legends to Figs 39 and 40.

Parts a and b in Figs 39 and 40 indicate that the growth in A was, as expected, more strongly reduced at an increased B density or an earlier emergence of B relative to A.

Parts c to f in Figs 39 and 40 describe effects of B on GR_A and RGR_A–percentage. At a higher B density or an earlier B emergence, B made GR_A pass its maximum earlier and RGR_A (as understood from RGR_A–percentage) become earlier influenced by competition. These effects correspond to the effects of increased density in pure stands of A shown in Fig. 37e–h.

In Figs 41 and 42, the divergences of the growth curves described in Figs 39 and 40 are illustrated by curves describing relative divergences. Say that all curves describe biomasses of a plant category A as influenced by intermixed plants of category B. The divergences are caused by differences in the density or the relative time of emergence of the category B plants. An increased density

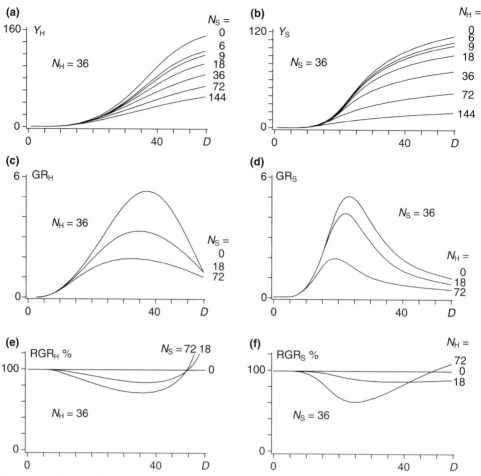

Fig. 39. (a, b) 'Growth curves' representing aerial shoot dry weights (Y_H and Y_S, g dm^{-2}) of two-row barley, *Hordeum distichon* (H), and white mustard, *Sinapis alba* (S), respectively, presented as functions of time (D, days after emergence) at various constant densities, N_H or N_S (number of seeds of H or S sown dm^{-2}). (c, d) Corresponding curves of the growth rate per unit area, GR$_H$ or GR$_S$ (d.wt., g dm^{-2} day^{-1}) at three levels of N_S and N_H, respectively. (e, f) Corresponding curves of the relative growth rate, RGR$_H$ or RGR$_S$, presented as percentages of the level at $N_S = 0$ and $N_H = 0$, respectively. Curves based on the same regression equations as those in Fig. 37. *Note that vertical scales differ between diagrams.*

or an earlier emergence of the category B plants causes an earlier and stronger divergence of the biomass production in category A. This divergence is elicited and increases during a comparatively short period of strong growth, which occurs before stages of GR maxima (cf. parts c and d in Figs 37, 39 and 40). Figures 41 and 42 show that the proportions of the biomasses of A in the different competitive situations were then fairly stable. Other experiments show a similar stability (Håkansson, 1983b: Figs 1

and 2, 1991). In later sections, a corresponding stability in other relations is demonstrated.

If proportions of the type now described will prove stable over a long period of time, this facilitates studies of competitive effects. Many competitive effects can then turn out similarly on the basis of measurements made at optional times within the period of relative stability. To what extent such stable relationships occur therefore deserves further research.

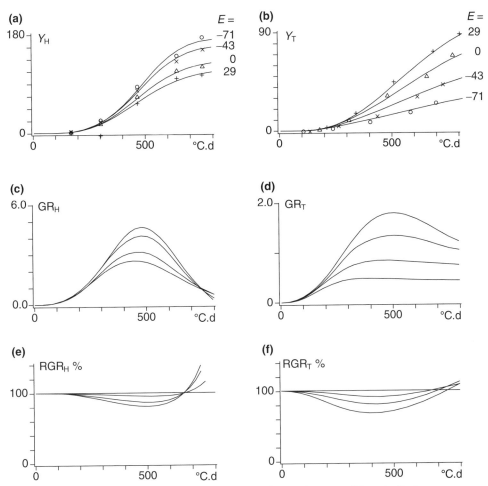

Fig. 40. (a, b) 'Growth curves', representing aerial shoot dry weights (Y_H and Y_T, g dm^{-2}) of two-row barley, *Hordeum distichon* (H), and wheat, *Triticum aestivum* (T), as functions of time expressed by the number of degree-days from sowing (°C.d, base 4°C). *E*, time of emergence of T relative to H, expressed in degree-days (+29, ±0, −43 or −71°C.d, representing situations where T was sown 7 days before, on the same day as, 7 days after and 14 days after H, respectively). Species H and T were mixed at constant densities ($N_H = 33$ and $N_T = 30$, i.e. sown with 33 and 30 seeds dm^{-2}, respectively). (c, d) Corresponding curves of the growth rate per unit area, GR$_H$ and GR$_T$ (d.wt., g dm^{-2} day^{-1}) at the four *E* values presented in a and b. (e, f) Corresponding curves of the relative growth rate (RGR$_H$ and RGR$_T$) expressed as percentages of the level at $E = -71$ and $E = +29$, respectively. Curves are based on trivariate regression equations describing dry weights of aerial shoots as functions of the H density, the time after sowing and the relative time of sowing, all times being expressed in degree-days. Weights determined after culm elongation had ceased in all plants. Outdoor pot experiment at Uppsala (Håkansson, 1991: pp. 109–149 and Eqns 45 and 48). *Note that vertical scales differ between diagrams.*

Growth Reduction Rate (RR)

Growth rate (GR) describes the biomass increase per unit of time, representing plants either per unit area or per plant unit (individual, etc.). An analogous expression, which could be called the growth *Reduction Rate* (RR), has been derived from equations describing growth in comparable pure and mixed stands (Håkansson, 1991). RR illustrates the degree of biomass reduction per unit of time per unit of area in a

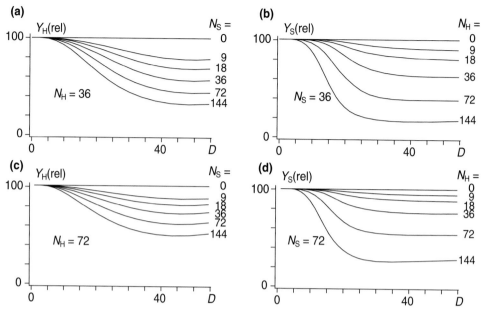

Fig. 41. (a, b) Relative aerial shoot dry weights, Y_H(rel) and Y_S(rel), per unit area of two-row barley, *Hordeum distichon* (H), and white mustard, *Sinapis alba* (S), sown at constant densities ($N_H = 36$ and $N_S = 36$, i.e. both sown with 36 seeds dm⁻²). Graphs correspond to the growth curves in Fig. 39 and are based on the same equations. Horizontal lines represent pure stands of H and S. The others, concerning mixtures, describe weights as percentages of those in the pure stands. (c, d) Analogous curves of Y_H(rel) and Y_S(rel) at densities $N_H = 72$ and $N_S = 72$, respectively.

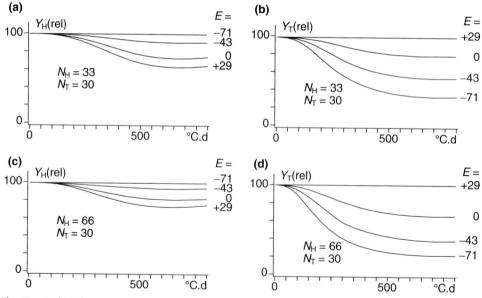

Fig. 42. (a, b) Relative aerial shoot dry weights, Y_H(rel) and Y_T(rel), per unit area of two-row barley, *Hordeum distichon* (H), and wheat, *Triticum aestivum* (T), sown in mixtures at constant densities ($N_H = 33$ and $N_T = 30$, i.e. sown with 33 and 30 seeds dm⁻², respectively). Graphs correspond to the growth curves in Fig. 40 and are based on the same equations. T emerged at different times relative to H, the difference (E) being expressed by degree-days (°C.d) as in Fig. 40. (c, d) Analogous curves of Y_H(rel) and Y_S(rel) at $N_H = 66$.

plant category (A) caused by competition from intermixed plants of another category (B). This reduction rate in A caused by B may be written 'RR$_A$ caused by B' (Fig. 43). The reduction rate is calculated as GR$_A$ in an A stand free of B ($N_B = 0$) minus GR$_A$ in a corresponding stand where B is intermixed. The densities of A and B are constant (N_A, N_B) at specified levels. RR$_A$ caused by B is derived from one and the same equation describing the growth of A at different levels of N_B (including $N_B = 0$).

Figure 43 illustrates RR$_H$ caused by S (*Sinapis alba*) and RR$_S$ caused by H (*Hordeum distichon*). The constant densities (N_H, N_S) are given in the legend to the figure. The curves are based on the equations underlying the growth curves of H and S in Fig. 39 (cf. Håkansson, 1991: pp. 59–60).

When comparing the curves of RR$_A$ in Fig. 43 with those of GR in Fig. 39, it can be seen that the maxima of RR$_A$ fall near those of GR$_A$. It can further be seen that RR$_A$ passes

its maximum earlier at higher N_B than at lower. Additional calculations show that an increased density in A also gets the maximum biomass reduction in A per unit of time to fall earlier, which parallels the density influence on the GR$_A$ maximum.

Changes in UPR and Percentage UPR Over Time – Comparing Growth Rates and Competitive Conditions in Mixed Stands

UPR is a quotient suitable not only for comparing different plants regarding their relative competitiveness, but also for comparing their growth rhythm in mixed stands. Analogous comparisons of plants in pure stands can, of course, also be made with results differing more or less from those obtained in mixed stands (Håkansson, 1991: p. 52). The use of UPR and percentage UPR for comparisons in mixed stands will now be illustrated.

The graphs in Fig. 44 compare species H and S on the basis of results from Experiment 1. They correspond to the graphs in Figs 37 and 39 and are calculated from the same regression equations (Eqns 31 and 32 in Håkansson, 1991). The two left-hand graphs (a and c) describe changes in UPR$_{SH}$ with time at different constant densities (N_H and N_S). The corresponding right-hand graphs (b and d) show percentage changes in UPR$_{SH}$ with changed densities (UPR$_{SH}$ = 100 at $N_S = 36$).

The left-hand graphs in Fig. 44 show that UPR$_{SH}$ increases in the first 15–25 days of growth. During this period, the growth rate was thus higher in S than in H. The length of this period depends on densities, on N_H as well as on N_S. The GR curves in Fig. 39 show that H grew intensively during a longer period than S. The UPR$_{SH}$ curves in Fig. 44 thus indicate that the growth rate of S decreased relative to that of H after the relationship had passed a maximum. The maximum was passed earlier and was lower and the following decrease more pronounced at higher densities than at lower, of both S and H (compare Fig. 44a and c).

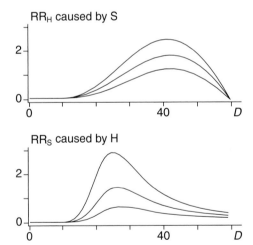

Fig. 43. Reduction rates in H and S (RR$_H$ and RR$_S$, by aerial shoot dry weight, g dm^{-2} day^{-1}) – functions of time (*D*, days after emergence). Upper diagram: Reduction rate in H (RR$_H$) at $N_H = 72$ caused by plants of S intermixed at densities $N_S = 18$, $N_S = 36$ or $N_S = 72$ (lower, middle and upper curves, respectively). Lower diagram: Reduction rate in S (RR$_S$) at $N_S = 72$ caused by plants of H intermixed at densities $N_H = 18$, $N_H = 36$ or $N_H = 72$ (lower, middle and upper curves, respectively). Densities = numbers of seeds sown dm^{-2}. Curves based on the same equations as those in Figs 37–39.

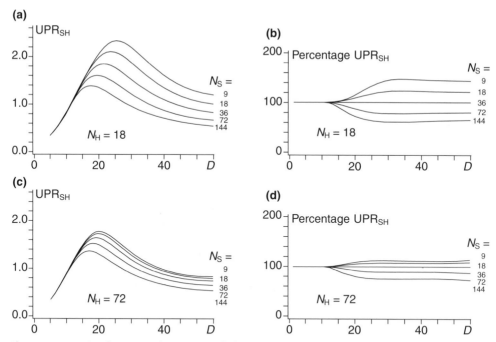

Fig. 44. UPR$_{SH}$, i.e. the unit production ratio of white mustard, *Sinapis alba* (S), to two-row barley, *Hordeum distichon* (H), and the corresponding percentage UPR$_{SH}$ – as functions of time (*D*, days after emergence). Different combinations of constant initial densities of H (N_H) and S (N_S) represented by the number of seeds sown dm^{-2}. Curves based on the same equations as those in Figs 37–39.

During the first 10–15 days, the UPR$_{SH}$ curves in Fig. 44 run together regardless of densities. This is the period before effects of competition are recorded and is therefore shorter at higher densities than at lower. In this early period, Percentage UPR$_{SH}$ is constant, represented by a common horizontal line (level 100 in Fig. 44b and d).

The UPR$_{SH}$ curves both diverge, pass maximum and begin to decline earlier at higher densities than at lower (N_H and/or N_S). The lowering of the levels of the declining curves with increasing N_S or N_H indicates that S was competitively inferior to H (cf. Figs 33 and 35).

Let us now compare the divergence of the Percentage UPR$_{SH}$ curves in Fig. 44. A period of increasing divergence means a period with intensified competitive pressure. After that period, the curves remain at fairly stable levels. The constant levels of the curves indicate a period with more stable competitive conditions (cf. *Y*(rel) in Figs 41 and 42, *k* and RYT in Fig. 46).

Figure 45 illustrates UPR$_{TH}$ over time (sum of degree-days, °C.d). The graphs represent Experiment 2. (See graphs in Fig. 34, which illustrate the influence of the relative time of emergence of H and T at an advanced stage of growth in the same experiment.) Figure 45 describes UPR$_{TH}$ in mixed stands of species T and H sown on the same day (treatment E_0) at three levels of fertilization (Fert. 1, Fert. 2, Fert. 3), at three H densities (N_H) and at a constant density of T ($N_T = 30$). The curves are calculated from regression equations describing aerial shoot dry weights of T and H per unit area over time (see legend to Fig. 34). Calculations representing the first period of growth are uncertain and therefore excluded.

The rather moderate changes in UPR$_{TH}$ over time shown in Fig. 45 indicate that the growth rhythm of H (*Hordeum distichon*, barley) and T (*Triticum aestivum*, wheat) differed only slightly within the study period. This contrasts with the rhythm of barley (H) and white mustard (S), which differed considerably according to Fig. 44.

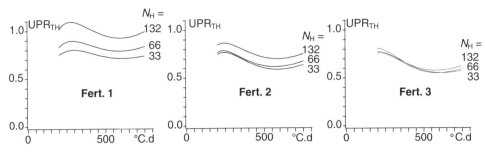

Fig. 45. UPR$_{TH}$, i.e. the unit production ratio of wheat, *Triticum aestivum* (T), to two-row barley, *Hordeum distichon* (H) – as functions of time. Time expressed as degree-days from sowing (°C.d, base 4°C). One degree-day was, on average, 8.6°C in the study period. Initial H densities (N_H) were 33, 66 or 132 seeds sown dm^{-2}. The T density (N_T) was 30 throughout. The three diagrams represent the same fertilizer levels (Fert. 1, Fert. 2 and Fert. 3) as those in Fig. 34 in treatments without an arranged difference in the relative time of emergence (i.e. $E = 0$ in Fig. 34). Calculations according to Håkansson (1991: Fig. 56).

In an advanced period of growth, UPR$_{TH}$ is about 0.8 at the lowest fertilizer level (Fert. 1). However, UPR$_{TH}$ decreases with an increased fertilizer level and is about 0.6 at the highest level (Fert. 3). This is in accordance with the graphs at E_0 and Fert. 3 in Fig. 34. Thus, the graphs in both Fig. 34 and Fig. 45 (based on different regression equations) indicate that a raised fertilizer level increased the production less strongly in T than in H.

As previously commented on, the UPR$_{TH}$ curves in Fig. 45 intimate that spring wheat (T) was more competitive than barley (H) at the lower fertilizer level (Fert. 1) than at higher levels. Thus, whereas the UPR$_{TH}$ curves run at increasingly high levels with increasing N_H at Fert. 1, their divergence decreases with increased fertilizer levels. This is an example of environmental influences on the relative competitiveness; further examples being shown in Tables 31, 32, 35 and 36.

Other examples of the use of UPR for determining not only the relative competitiveness of two plants but also their growth rhythm have been presented by Dock Gustavsson (1989).

The Relative Crowding Coefficient (*k*) and the Relative Yield Total (RYT) Observed Over Time

Figure 46 describes changes in *k* and RYT (as defined by de Wit, 1960) over time when

the densities in the pure stands of the plants are changed in a replacement series. The categories are, again, species H and S in Experiment 1. Indices k_{SH} and RYT$_{H+S}$ (i.e. *k* of S vs. H and RYT of H+S, as defined in association with Fig. 36) are calculated on the basis of the same regression equations as the curves in Figs 37–39, 41, 43 and 44. N_H and N_S are the densities in the pure stands of H and S in replacement series. The indices represent mixtures of H and S at densities $0.5 \cdot N_H$ and $0.5 \cdot N_S$.

Figure 46 shows that, in the initial period of growth, k_{SH} was unity, irrespective of densities, which means that competition was not yet apparent. With further growth, competitive influences became more and more evident. This is indicated by the divergence of the k_{SH} curves representing different combinations of N_H and N_S. (Note: different scales and different N_S values in Fig. 46a and b.) After a period of increasing divergence, the curves exhibit more stable differences. This is analogous to the behaviour of Y(rel) in Figs 41 and 42, and percentage UPR in Fig. 44.

Figure 46, like Fig. 36, shows that index k_{SH} varies strongly with densities. Thus, *k* is not well suited for assessments of the relative competitiveness of plants. As was stated previously, the same applies to CR (the Competitive Ratio according to Willey and Rao, 1980).

As shown by reference model calculations exemplified in Fig. 36, RYT$_{H+S}$ is only

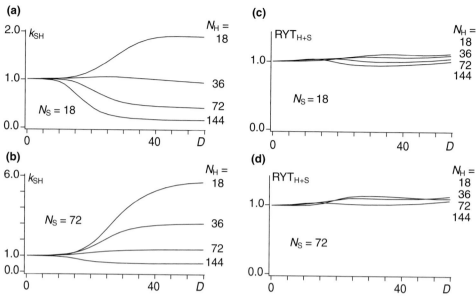

Fig. 46. The relative crowding coefficient of S versus H (k_{SH}) and the relative yield total of H and S (RYT_{H+S}) as functions of time (D, days after emergence). Experiment and symbols as in Figs 37–39. (Calculations based on equations referred to in the legend to Fig. 37.) N_H and N_S are the pure-stand densities in replacement arrangements, and the densities of H and S in mixtures were 50% of these pure-stand densities. *Note that vertical scales differ between diagrams a and b.*

slightly influenced by densities (cf. Connolly, 1986). RYT is therefore fairly well suited for comparing the total production of different plants in mixed and pure stands. RYT, like the closely related LER (the Land Equivalent Ratio), is thus well suited for comparing the production of crops grown in mixture (intercropping) and in monocultures.

'Critical Periods' of Weed Control in a Crop Stand

Figures 41–44 illustrate that different periods can be discerned during the progress of growth in a plant stand: (i) an early period when influences of competition cannot be measured as effects of plant density; (ii) a subsequent period when competition becomes increasingly apparent through rapidly enhanced density effects; and (iii) a period when competitive influences on the growth of different plant categories in a mixed stand seem to change only slightly with the progress of time.

The phenomena described above are obviously related to *critical periods* concerning times of weed control. Thus, many weed control experiments have focused on: (i) the minimum length of the period when the crop has to be kept free of weeds so as to avoid an unacceptable yield loss; and (ii) the maximum length of the period during which weeds can be allowed to grow uncontrolled without causing a considerable reduction in the crop yield. It is the period between the last point of time acceptable for starting control and the point when continued control is unnecessary that has been called the *critical period*.

Nieto (1960) and Nieto *et al.* (1968) defined and studied critical periods, regarding competition and weed control in maize and beans. An example illustrating how yields in common bean can be influenced by different weeding strategies is shown in Fig. 47, which is based on field experiments in Nicaragua (Aleman, 1989).

Zimdahl (1980, 1988) refers to a great number of papers on experiments concerning critical periods. The experiments show that critical periods differ with crop and weed types, with their density and spatial plant distribution, with temperature, etc. The experiments have been carried out mainly in crops grown with a large row spacing or crops with stands that close late for other reasons, exerting a weak competition during early periods of growth. Examples of such crops are common bean, soybeans, maize, grain sorghum, beets, potatoes and various vegetables. However, critical periods have also been studied and discussed concerning more competitive crops (e.g. Black and Dyson, 1997; Blackshaw and Harker, 1998).

It may be suggested that weeds could be allowed to grow with the crop for a somewhat longer period than the period when competitive effects are not observed according to Figs 41, 42, 44 and 46. The end of that period is the latest time that allows the

crop to compensate acceptably for a lowered early growth. The influence of weed plants recovering or emerging after the crop has been kept 'weed-free' during an early period depends on many factors. In a strongly competitive crop, weeds that emerge or recover after the period of repeated control will have little influence even when succeeding a short 'weed-free' period. In a weakly competitive crop, on the other hand, these weeds may grow vigorously and cause a considerable yield loss even after a longer 'weed-free' period. Optimum times of starting or finishing weeding also depend on the degree of injury to the crop that is caused by weeding carried out in different stages of crop development.

Early Predictions of Yield Loss Due to Weeds

Studies of competition in crop–weed stands have largely focused on influences of densities on the competitive outcome. However, the biomass ratio of plants A to plants B in mixed stands and the percentage biomass reduction in A caused by B were compared by Håkansson (1983b, 1984a). In reference models, this ratio proved to be strictly linear and uninfluenced by densities. Based on the relationships found, the value of expressions reflecting biomass ratios of weed plants to crop plants in young stands of annual crops for predicting the yield loss in the mature crop stands was discussed. A disadvantage of plant densities over biomasses is that a certain density may represent extremely different amounts of biomass, certainly representing different competitive situations.

The use of ground cover, or leaf area, of weeds relative to crops in young stands for predicting yield losses has been studied in different experiments. Experiments in the Netherlands (e.g. Kropff, 1988; Kropff and Spitters, 1991; Kropff and van Laar, 1993) and England (Van Acker et al., 1997b) may be mentioned in particular. These experiments have indicated that ratios based on leaf areas or measures of ground cover are

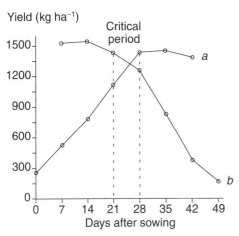

Fig. 47. Seed yield in common bean (*Phaseolus vulgaris*) after different weedings. Average of four field experiments in Nicaragua reported by Aleman (1989). Curve *a*: yield after a differently long period with repeated hand-weeding keeping the stands largely weed-free. Curve *b*: yield after a differently long period of undisturbed weed growth before the first of repeated weedings was performed to keep the bean stand largely weed-free until harvest. Bean seed rate was 40 seeds m^{-2}. Weed flora typical of the area, dominated by the composite species *Melampodium divaricatum*, *Melanstera aspera* and *Bidens pilosa* (names according to Garcia et al., 1975).

better suited than plant densities for predicting yield losses.

The question of early predictions of
yield losses in a plant category A caused
by intermixed plants called category B has
been studied further by Håkansson (1991, cf.
1997a). Various ratios have been calculated
using either biomasses of plants A and B (Y_A
and Y_B) or growth rates of these plants per
unit area (GR_A and GR_B) determined at different times in early periods of plant growth.
These ratios have been compared with the
percentage biomass loss in A – called R_A(rel)
– caused by B in a late period of plant development. In the early period following plant
emergence, both Y and GR reflect leaf areas
fairly well. Using reference models, different ratios based on Y and GR were tested
(Håkansson, 1991, 1997a). The only ratios
that could conceivably be used for prediction were $Y_B/(Y_A + Y_B)$ and $GR_B/(GR_A +
GR_B)$. These ratios are analogous to the leaf
cover ratios studied by Kropff (1988).

Such a ratio is useful for predicting the
future relative yield loss, R_A(rel), only if its
proportion to R_A(rel) is largely independent
of plant densities, relative times of emergence or environmental conditions. In addition, this proportion should preferably be
unity and constant over a period of time. If
not, it is very difficult in field practice to
define the suitable times for reliable predictions. The desirable, perfect situation can be
described as follows:

$$\frac{R_A \text{ (rel) in a specified late stage of stand development}}{Y_B/(Y_A + Y_B) \text{ as a function of time in earlier stages}} = 1$$

or

$$\frac{R_A \text{ (rel) in a specified late stage of stand development}}{GR_B/(GR_A + GR_B) \text{ as a function of time in earlier stages}} = 1$$

Biomasses or growth rates cannot be
easily determined in field practice for predicting yield losses. They are used for tests
here as they are easily determined in the
experiments and as they are, at the same
time, fairly well correlated with leaf areas in
stands of very young plants. The usefulness
of leaf area, or the like, in predicting future
yield loss can therefore be reasonably well
evaluated on the basis of biomasses.

Proportions, or quotients, described
above have been calculated using regression equations describing biomasses against
time. They are based on experimental data
(Håkansson, 1991: Figs 28–35, 45–48, 89, 90,
cf. 1997b: pp. 89–92) and are briefly written
as follows in diagrams:

$$R_A(\text{rel}) \cdot (Y_A + Y_B)/Y_B \text{ or}$$
$$R_A(\text{rel}) \cdot (GR_A + GR_B)/GR_B$$

in which Y and GR vary with time.
The entire quotients are thus presented as
functions of time.

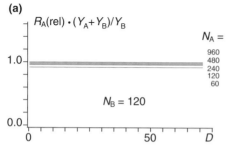

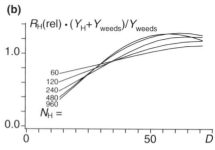

Fig. 48. The quotient of R_A(rel) – which is the relative loss of production in plant category A caused by
B in a specified late stage of stand development – divided by $Y_B/(Y_A + Y_B)$ varying as a function of time
in earlier stages. For detailed explanations, see text. (a) Reference model calculations, where A and B are
equal plants of two-row barley (H). (b) A situation where plant category A was barley (H) and category B
a population of weeds dominated by *Chenopodium album*. Production values represent aerial shoot dry
weights m^{-2}. Densities (*N*) are number of seeds sown m^{-2}. *D*, days after emergence of H. (From Håkansson,
1991: pp. 94–99.)

Results are exemplified in Fig. 48, describing quotients based on a box experiment with two-row barley (H) grown at different densities in weed-free stands (Fig. 48a) or together with weeds (Fig. 48b).

Figure 48a illustrates the result of *reference model calculations* based on a regression equation describing the aerial shoot dry matter of pure stands of barley against density of barley and time (D = days after the emergence of H). Plants A and B are now identical plants, fictitiously considered to be different categories in mixture. Both N_A and N_B thus denote constant densities of H. The quotient described by the lines in Fig. 48a varies very slightly, and close to unity, although the density, N_A, varies strongly. The quotient is also independent of time. Reference model quotients containing GR instead of Y also respond very slightly to density and are independent of time. Such quotients representing different plants from different experiments show the same result (Håkansson, 1991).

Reference models thus invariably prove that the quotients, whether they contain Y or GR, are independent of time and depend only slightly on densities, N_B or N_A. The question is then whether there is a real potential for approximate predictions of the percentage yield loss by assessing some parameters reflecting biomasses of crops and weeds in early periods of plant growth.

In an experiment where plant categories A and B were spring wheat and barley, i.e. species with rather similar stature, the quotient varied moderately near unity with variation in density, fertilizer level and time of 'yield loss predictions' based on Y or GR (Håkansson, 1991: p. 125). Other calculations show, however, that the relative time of emergence can exert a strong influence even when plants A and B are very similar (Håkansson, 1991: pp. 150–164, cf. 1997b).

Figure 48b illustrates a quotient when plants A and B were barley (H) and weeds (mainly *Chenopodium album*), respectively. This quotient varies strongly both with the H density, N_H, and with the time of 'yield loss prediction'.

In experiments where plant categories A and B were barley and white mustard, the variation was less strong but still considerable (Håkansson, 1991: pp. 66–82, 1997b). Where plants A and B are more different, variations greater than those shown in Fig. 48b, could be expected.

In an experiment studying the influence of densities and the relative time of emergence on the quotients under discussion, the quotient had a great influence even when densities modified the quotient more moderately (Håkansson, 1991).

The impression based on the various experiments is that predictions of the yield loss based on early observations are very hazardous. Even when the weed closely resembles the crop, e.g. *Avena fatua* in cereals, its emergence at different times will lower the reliability of a prediction of the yield loss. In addition, there is usually more than one important weed species in a crop, some of which differ strongly from the crop in their growth habit.

Results of field experiments presented by Lotz *et al.* (1996) support the idea that predictions of yield losses caused by weeds may be based on leaf areas or the like (reflecting biomasses), rather than on plant densities. At the same time, however, an inspection of the dispersion of the data presented indicates that an early prediction of the yield loss in a crop in a specified field and year is far from reliable. Difficulties in predicting yield losses have also been illustrated by Morin *et al.* (1993).

Whether the weeds observed in a young crop stand ought to be controlled or not should be decided after considering many aspects. The decision should not only be based on concerns for the present crop with regard to quantitative yield, product quality, harvesting problems, etc. It should also, and often mainly, be based on judgements regarding weed problems in subsequent crops. Good knowledge of the biology of the weeds, including a good understanding of their response to crops and possible measurements in subsequent years, are a prerequisite for rational decisions in a cropping-system perspective.

10

Soil Tillage Effects on Weeds

_____

Introduction

Uncultivated land brought into cultivation

Some type of operation to break the vegetation cover is usually one of the initial measures in preparing virgin land for cropping. Most plants in the vegetation of uncultivated land are sensitive to soil cultivation. These plants are mostly considerably restricted even by rather light tillage followed by competition and/or weeding in a subsequent crop. Such plants therefore disappear more or less rapidly in areas brought into permanent cultivation.

When arable crops are grown on really virgin land, very few plant species from the original vegetation survive as persistent weeds. Among plants appearing as weeds from the outset, some may derive from scattered dispersal units spread by birds, etc. before or after the area was broken for cultivation. In the first years, these weeds usually occur as scattered plants, but some of them may reproduce rapidly. However, the typical arable weeds are largely introduced by the growers, directly or indirectly, e.g. through unclean seed, agricultural implements and tools, clothes and domestic animals.

In typical shifting cultivation, land is cropped and tilled with simple tools for a few years alternating with longer periods of non-cultivation. Species that become dominant during these periods often, or mostly, represent plant types that

are sensitive to the disturbances during the periods of cropping. The decomposition of their dead tissues in the first few years of cropping results in a release of mineral nutrients useful for the crops. Until they have been sufficiently broken down, some of them may cause more or less severe physical or mechanical problems, _in_ or _on_ the soil. The rate of decomposition and plant nutrient release is increased by burning, which is a frequent introductory measure when the area is again brought into cultivation.

Some seeds or other reproductive units of typical arable weeds survive in the soil in periods between the years of cropping even when these periods last for decades. They may be comparatively few, resulting in moderate numbers of typical arable weeds in the first year(s) of cropping. Typical arable weeds are primarily annuals. Perennials vary strongly in species composition and abundance. Species persistent as arable weeds usually increase rapidly. Together with plants from dispersal units continuously introduced with the cropping activities, they thus cause a rapid increase in the weed populations in a few years.

In non-permanent farming, such as shifting cultivation, still practised in parts of the world, the field is usually abandoned after a few years of cropping, because of a combination of nutrient deficiencies and weed problems. If the field is retained for continued cultivation, the need for effective weed control increases rapidly.

©CAB _International_ 2003. _Weeds and Weed Management on Arable Land:_
an Ecological Approach (S. Håkansson)

Soil tillage is among the most important factors determining the absolute and relative abundance of weeds with different traits in a given cropping system. Tillage effects on weeds with different traits are therefore examined in the following, with the focus on tillage in modern cropping systems in permanent arable fields.

Selective effects of tillage

The immediate purpose of a tillage operation is often something other than weed management. Nevertheless, soil tillage always affects the weed flora in some way. Weed species (genotypes) with different traits are affected differently, in other words, selectively.

The strength and selectivity of tillage effects on weeds differ with the type of operation and, very strongly, with the time of the year. An example of differences in the selectivity pattern caused by the timing of soil operations is that summer and winter annual weeds appear in different proportions in spring- and autumn-sown crops. (Chapter 5, Tables 3 and 5–13). The main reason is that seedbed preparation influences dormancy and germination of these weeds differently in spring and in autumn (Fig. 1).

The response of growing weed plants to specified tillage depends on their life form, stage of development, etc., and may be strongly modified by external conditions such as moisture and temperature situations. Burial, breakage and other effects of tillage may result in restricted growth and death. The direct effects of tillage are strengthened by plant competition when a competitive crop is rapidly established after tillage. The reason is that weakened plants suffer from competition *relatively* more than uninjured, more vigorous plants.

Under certain conditions, the long-term effect of mechanical disturbance on growing weed plants may be increased populations. For instance, creeping perennials usually respond to breakage by producing an increased number of shoots. Although these shoots are initially weaker than the fewer shoots developed from undisturbed plants, the greater number of them sometimes results in an increased biomass production. Tillage may also improve the growth conditions in the soil, outweighing the negative effects of injuries to the plants. However, such favouring effects on weeds can be reversed by different measures, for instance, repeated tillage or chemical control, which suppress previously weakened plants more strongly than they suppress undisturbed plants. In addition, competition in crops counteracts the growth of weakened plants relatively more strongly than the growth of undisturbed or less weakened plants.

No more tillage than necessary

Almost all types of weeds in a weed community can be effectively controlled or strongly restricted by tillage repeated throughout a growing season. This was possible on those fallows that were regularly used in alternation with cereals in earlier times. Frequently, however, these fallows were not only aimed at controlling weeds and were therefore poorly treated with this respect. In the late 19th and the first half of the 20th century, 'black fallows' were commonly used in many areas and usually effectively tilled with the aim of controlling weeds. For many reasons, however, intensive tillage is avoided today, and attempts to reduce even moderately intense tillage are in progress.

Intense tillage increases soil erosion, mineralization and leaching of plant nutrients and other substances. In temperate areas, this may not lead to obvious problems in a short-term view, but the negative effects accumulate with time. Where the precipitation is high, tillage resulting in bare soil strongly increases the leaching of water-soluble substances.

In warm and humid climates, even moderate tillage may cause devastating erosion and leaching. In areas where precipitation, topography and texture make the soils particularly disposed to gully erosion, tillage can rapidly cause disastrous effects. In such areas, the control of weeds and other vegetation by herbicides has become a way of reducing soil deterioration. Here, a responsible choice and application of herbicides is often a great environmental advantage over tillage. Avoiding, or strongly minimizing, soil tillage is necessary in these areas.

On certain light soils in particular, tillage must be restricted so as to prevent wind erosion. In the dry prairie regions of North America, for instance, zero tillage (no-tillage, no-till) and various forms of reduced tillage, largely 'non-turning tillage', are widely practised. In that way, continuous soil covers of plants and/or plant residues are maintained.

Attempts to minimize tillage are worthwhile in all agricultural areas also because intense tillage and passes of implements in the fields lead to an adverse soil structure deterioration (e.g. compaction) and to high costs and other drawbacks associated with energy consumption.

Replacing chemical weed control by intensified tillage is no way of creating sustainable cropping systems.

Instead, we should combine and adjust tillage and other measures into weed management systems (integrated weed management) considering both long-term and short-term effects on weeds, resource economy and impacts on the environment and soil productivity. This should be kept in mind in cases when discussions in the following are restricted to the mere effects of tillage on weeds.

Different forms of reduced tillage, including 'zero tillage', are discussed in the last part of this chapter. The other parts deal with effects of tillage used in more conventional cropping systems, particularly those typical of Northern Europe. They focus on principles regarding effects of mechanical injuries and changes in the depth location of various plant organs in the soil. Influences of competition in crop stands, which interact with and thereby strengthen the tillage effects, are repeatedly stressed.

Tillage Effects on Weed Seeds in the Soil Seed Bank

Seed dormancy and factors affecting dormancy and germination of seeds in the soil seed bank of weeds were described from various aspects in Chapter 6. Referring to that chapter, some aspects centred on effects of tillage are stressed or added below.

Increased aeration and other impacts of tillage stimulate the activity of aerobic organisms in the soil, including the seeds in the soil seed bank. Dormancy-breaking processes and germination are thus increased among these seeds. At the same time, tillage increases the mortality among the seeds; not only because it intensifies the attacks of noxious aerobic microorganisms, but also because it hastens biochemical processes in the seeds that may lead to death following germination or other changes in the seeds. Strong moisture and temperature variations in superficial soil layers trigger death of seeds caused by germination as well as by other processes.

The germination readiness, or degree of dormancy, of the seeds in the soil seed bank varies seasonally, as summarized in Fig. 1 concerning typical summer and winter annual weeds in Southern Scandinavia. In humid temperate areas, this variation is largely caused by the temperature variation, although it is modified by other factors (Baskin and Baskin, 1998; Vleeshouwers and Bouwmeester, 2001). If germination is not impeded by drought, tillage can trigger a *quantitatively* strong germination increase among given seeds, particularly in seasons when they have their lowest degree of dormancy. However, tillage often causes an increase even in periods when the germination readiness is low, although this increase is quantitatively lower.

A particularly interesting phenomenon found in Swedish experiments (Håkansson, 1983a, 1992) is that seeds of facultative winter annual weeds germinate to a comparatively low extent in summer and autumn in the absence of physical disturbance of the soil. Strong germination in the later part of the growing season mainly follows after disturbance of the soil through tillage or the like (Table 2). Roberts and Potter (1980) report that showers of rain can provoke an increased germination.

Disturbance of the soil increases the soil aeration, which enhances the gas exchange between seeds and their surrounding soil, thus enhancing the supply of oxygen to the seeds and the removal of carbon dioxide from their tissues. The diffusion and removal of toxic metabolites from seeds may be facilitated both by mechanical disturbance and by water supply (cf. Benvenuti, 1995). It is likely that heavy showers of rain induce germination of winter annuals in late summer and autumn, not only due to the water supply as such, but largely also to increased supply of oxygen and to removal of carbon dioxide and other substances inhibiting germination.

Seeds transferred from deeper to shallower soil layers may thus germinate more easily owing to better gas exchange, more fluctuating temperatures and, for some of them, the influence of light. As can be understood from the above, there are also several reasons for an increased mortality.

Seed coats become scarified to some extent by soil particles and machinery. In combination with the general aeration of the soil, this may trigger germination of the seeds by facilitating their uptake of oxygen and emission of carbon dioxide. In seeds with coats that are impermeable to water (hard seeds), scarification will also allow the water uptake that is necessary for germination. Microbial activity weakening the seed coverings is certainly also hastened.

A few experimental results are presented below, exemplifying tillage effects on weed seeds.

In early English experiments (Brenchley and Warington, 1936, 1945) the effect of repeated soil cultivation on the weed seed bank was studied on fallows. The number of living seeds after repeated tillage during one season was compared with the number in undisturbed soil covered with vegetation and protected from addition of fresh weed seeds. Germination and death caused a decrease in the seed number in all treatments. In fallows, however, the number decreased to about 50% of that in undisturbed soil, as averaged over weed species and cultivation intensities. The decrease varied greatly between species. The number of seeds decreased more at higher than at lower tillage intensity, and more in moist soil than in dry soil.

In American experiments, Bibbey (1935) compared the seedling emergence of two summer annual weeds in field plots cultivated in early spring with the emergence in uncultivated plots. The weeds were *Avena fatua* and *Sinapis arvensis*. Their emergence in cultivated plots relative to that in uncultivated plots was about 500 and 200%, respectively.

Roberts and Dawkins (1967) studied the annual decrease in the number of seeds in the weed seed bank in untreated soil and soil tilled twice or four times a year over 6 years. Weed plants were removed after emergence so as to prevent fresh seeds adding to the seed bank. The percentage annual decrease in the seed number was approximately the same during each of the 6 years and could therefore be represented by averages. Average percentages of emerged weed seedlings and seeds that had died without seedling emergence within a year were as follows:

	Emerged seedlings (%)	Dead seeds (%)	Decrease of seed number (%)
Untreated soil	1	21	22
Tillage 2 times	7	23	30
Tillage 4 times	9	27	36

The percentage annual reduction in the seed number thus increased with the tillage intensity. Few seedlings emerged relative to the number of seeds that had died. Some of the latter seeds had certainly germinated without resulting in seedling emergence. The experiment involved several weed species. Their response to the soil operations differed but was similar in principle.

The response to tillage described here represents populations of seeds distributed over different depths in the soil. However, the ability of the seeds to produce new plants differs strongly with the depth. This was illustrated in Fig. 5 (p. 63), describing the establishment of plants from seeds of various sizes placed at different depths. In order to study the pure depth effect, moisture and other factors were adequate for germination and plant growth at all depths and, in addition, seeds of cultivated plants were used, as they were non-dormant (except a minor percentage of hard seeds in red clover). (For conditions and responses very close to the soil surface, see p. 64.) The optimum depth range proved to be narrower for smaller than for larger seeds: furthermore, plant establishment decreased more rapidly for the smaller seeds with a depth increase below the optimum.

In principle, the same applies to weed seeds with pronounced innate dormancy. However, these seeds germinate to a lesser extent and survive in dormant states for longer periods at greater depths in the soil than at shallower. Their decreased readiness to germinate at greater depths counteracts the waste of seeds caused by failure of seedling emergence.

The behaviour of seeds of *Chenopodium album*, *Matricaria perforata* and *Sinapis arvensis* was studied in experiments similar to those presented in Fig. 5. Considerable proportions of these seeds remained dormant in the soil at the end of the experiment. Although exact counts could not be made, it was obvious that the number of dormant seeds increased considerably with increasing depth.

In a German field experiment presented by Koch (1969), the number of germinated seeds of annual weeds (mainly *Sinapis arvensis*, *Thlaspi arvense*, *Chenopodium album* and *Lamium purpureum*) was counted in three soil layers at different depths on the eighth day after soil cultivation in spring. The results were as follows:

Depth (cm)	Number m^{-2}
0–5	2437
5–10	69
>10	0

Seed germination drastically decreased with increasing depth below 5 cm. According to judgements based on observations 2 weeks later, slightly more than 50% of the germinated seeds were represented by emerged plants; the rest had died. An extensive mortality among young seedlings of weed species can always be expected under field conditions.

Large seeds usually show both a greater readiness to germinate at greater depths and a greater capacity for shoot emergence from these depths than small seeds. Many observations indicate that the relatively large seeds of *Avena fatua* and *Galium aparine* can germinate and establish plants from depths exceeding 15 cm.

The principles exemplified above are important as a basis for discussions on seedbed preparation. First, however, tillage effects on seedlings in different stages of development are dealt with.

Tillage Effects on Seedlings

The response of seedlings to soil tillage depends on many factors:

- Character of the tillage operation.
- Weather and soil conditions.
- Size and morphology of seedlings as determined by the genotype and the properties of the seeds (e.g. by the contents of food reserves in the seeds).
- Stage of development (plant size, amount of food reserves).

In characterizing the response of plants from weed seeds to tillage, it seems appropriate to distinguish those of *annuals* (and biennials) and *perennials*. The former do not develop perennating vegetative structures with a high capacity for regeneration. Plants of the latter category develop such structures after an initial period of growth.

Seedlings of annuals

Young plants of annual species can be killed to a large extent even by relatively light soil cultivation. Effects of shallow cultivation have been studied by many researchers (e.g. Habel, 1954, 1957; Kees, 1962; Koch, 1964; Meyler and Rühling, 1966; Rasmussen, 1992; Rydberg, 1993, 1995). The knowledge collected may be summarized and interpreted as follows.

The strongest effect of cultivation on emerged seedlings results from their covering with soil. Small plants with low contents of food reserves are frequently killed when covered with a soil layer of a few millimetres in thickness. As larger plants regrow more easily than smaller plants, the effect of thin soil covers is strongly selective. Below a thicker and more complete soil cover, more plants are killed and, in addition, this effect is less selective.

Seedlings are particularly sensitive to tillage when they have reached their compensation point (the stage of minimum amounts of energy reserves; cf. Fig. 4). Those that have originally emerged from greater depths are more deprived of food reserves (cf. Fig. 7) and therefore more sensitive to tillage than those from shallower depths.

The ability of the plants to re-emerge and establish after having been covered with soil decreases with increased injury to their tissues in combination with their covering with soil. Injuries in the form of broken roots and destroyed shoot apices are particularly severe. Plants from seeds with low contents of food reserves are, on the whole, more easily killed than those from seeds containing more reserves. However, the anatomy of the seedlings also strongly determines their sensitivity to tillage.

If the soil cover is thin or incomplete and the roots still remain in contact with moist soil, the plants frequently survive cultivation. However, injuries and growth delay reduce their competitive power.

Plants are uprooted to different degrees by different harrowing. Those that have been entirely pulled up onto the soil surface die rapidly if the soil and weather are dry. However, if only a single root remains covered with moist soil and the water stress is not too strong, even such plants can survive.

Immediately after germination, the small seedlings from weed and crop seeds are comparatively tolerant to tillage, and more so in moist than in dry soil. Young seedlings of grasses, in particular, are often only slightly injured by soil cultivation and can then grow and emerge, although delayed, if not obstructed by drought or too thick a soil cover. Seedlings of dicotyledonous plants may pass their most sensitive stage, the compensation point, when their first foliage leaves begin to expand, or even earlier, depending on the shape of the plant, the light and temperature conditions, the depth location of the seeds in the soil, etc.

With continuing growth after the compensation point, the plants become increasingly able to recover from mechanical disturbance. When they have developed flowers, but long before they have mature seeds, plants of many species can develop viable seeds even after burial at depths preventing re-emergence. This is often experienced in the field and experimentally documented. Immature seeds often do not become dormant and therefore germinate or die earlier than fully mature seeds (see Baskin and Baskin, 1998).

Seedlings of perennials

In their early stages, seedlings of perennial species perform similarly to those of annuals. Their response to mechanical disturbance also changes similarly in their early growth. However, soon after passing their compensation point, seedlings of perennials develop vegetative structures that are increasingly regenerative. After a period of continued growth and accumulation of food reserves, these structures give the plants the specific characteristics of perennials. This general description is based on observations of seedlings of many perennial species at Uppsala.

Many perennials known to be important weeds in a global perspective, e.g. *Cyperus rotundus*, set very few seeds (Holm *et al.*, 1977). Others largely rely on seeds in addition to perennating vegetative structures. In *Elymus repens*, the seed set varies greatly between fields and years (see further p. 239), but the reproduction by seeds is important, particularly as a way of recovery when the occurrence of rhizomes has been strongly reduced by control measures (e.g. Melander, 1988).

Studies by Kephart (1923) and Arny (1927) indicate that young seedlings of *Elymus repens* are easily killed by soil tillage until they have formed rhizomes. In greenhouse experiments at Uppsala under temperature and light conditions simulating conditions in Swedish summers (Håkansson, 1970), plants from seeds of this species were buried horizontally in moist soil at various stages of development. Even after burial at the shallowest depth in the experiment, 1.5 cm, the plants invariably died after burial in early stages up to their five-leaf stage. At that stage, rhizome branches about 1 cm in length had developed. When plants with six well-developed leaves and rhizomes up to 3–4 cm in length were buried, shoots emerged from most of those that were buried at the 1.5-cm depth and from a few buried at 4 cm. Individuals with emerged shoots survived and resumed their rhizome production.

Similar experiments were carried out with seedlings of *Sonchus arvensis* (Håkansson and Wallgren, 1972a). Young seedlings had only thin vertical roots without regenerative capacity. When the plants had developed four foliage leaves exceeding about 4 cm in length, some roots had increased in thickness, the central vertical root first. Root parts with a diameter of 1.5–2 mm proved to be regenerative. Plants with six to seven foliage leaves exceeding 4 cm in length, had developed some horizontal, partly thickened roots with regenerative capacity. Excellent descriptions of *Cirsium arvense* (e.g. Lund and Rostrup, 1901; Åslander, 1933) indicate similar early plant development from seeds of this species.

Tillage Effects on Perennial Plants with Fully Developed Vegetative Structures

General statements

When seedlings have passed their juvenile stages, they develop vegetative structures capable of regeneration. Their growth and development then follow specific patterns. In temperate, humid climates where warmer growing seasons alternate with distinctly colder winters, the plants exhibit a regular seasonal variation in their initiation and growth of new roots and shoots, their development and growth of regenerative perennating structures and their flowering and seed set. Even though, in certain species, green shoots survive winter, there is normally a flush in the development of new green shoots and active roots in the spring.

In most species, the initial development of new shoots and roots following tillage later in the spring or in the summer largely corresponds to the earlier spring activities in undisturbed plants. *When active shoots and roots have been destroyed by soil tillage, new roots and shoots immediately start to develop and grow, if not obstructed by drought, by lack of food reserves or by innate dormancy in the regenerative plant structures.* Perennating structures covered with soil seldom die of drought, but drought delays and may weaken the growth. In many perennials, the regenerative structures become dormant in late summer or early autumn. Then, new shoots will not develop after tillage until dormancy has been broken through the lower autumn and winter temperatures. In the field, they will not start a rapid growth again until the temperature has become sufficiently high in the following spring.

In undisturbed plants, new aerial shoots are initiated mainly from structures near the ground surface. In this way, minimum amounts of food reserves are required for their early growth. This explains why many perennials that exhibit a weak capacity to regenerate after burial by tillage become competitive and persistent weeds in arable fields where no tillage is carried out during a period of years, as in leys. Because regrowth after cutting or grazing requires smaller amounts of food reserves than regrowth after tillage, these perennials persist in leys.

Table 5 presents examples of perennial weeds that produce a vigorous growth in leys but are of minor importance as weeds in annual crops established after conventional tillage. These perennials are thus 'ley weeds'. Perennials that are important weeds in the annual crops in conventional tillage have a high capacity to regenerate after burial. Some of them are not well adapted to the conditions in leys for cutting or in grazed areas, although they usually survive for many years under those conditions.

Perennial plants may be grouped into two categories with respect to their capacity to regenerate after tillage, as follows.

Plants with low tolerance to tillage

Most of the stationary (non-creeping) perennials belong to this group. In addition, creeping perennials with above-ground prostrate shoots or runners (stolons) and also many plants with underground creeping shoots (rhizomes) have a weak capacity to endure burial by tillage. These plants seldom appear with vigorous growth in competitive annual crops in current cropping systems with conventional soil cultivation (Tables 3 and 5). Plants from surviving parts of vegetative structures, such as taproots, often occur, but mostly as scattered individuals. In annual crops, the species under discussion may develop many plants from seeds, but the seedlings seldom manage to produce vigorous growth in competitive crops such as cereals.

As stated above, many of them are tolerant to mowing or grazing and are competitive in leys, pastures and meadows. The majority of such plants in a first-year ley have established as seedlings in the preceding annual crop (mainly spring cereals) among the young ley plants undersown there. They may have produced a weak growth in that crop, but having survived winter, they can start a vigorous growth in the ley. These plants increase in size with increasing age of the ley. They may also increase in number owing to establishment of seedlings in gaps or spots with weak ley plants.

Many perennials that normally produce a weak growth in competitive annual crops on tilled soil can grow vigorously and become troublesome in weakly competitive crops such as vegetables. Swedish examples

are *Taraxacum officinale* and *Rumex crispus* with taproots, *Ranunculus repens* with stolons, and *Plantago* spp. with short subterranean stems. Taproots of *T. officinale* can send shoots to the soil surface even after deep burial. Petersen (1944) reported emergence from depths of up to 18 cm, although the mortality among the taproots was then high and the emerged plants were weak. However, even such weakened plants can recover and grow vigorously in crops exerting a weak competition.

Plants with higher tolerance to tillage

This category of plants is represented by creeping perennials with strongly regenerative rhizomes (e.g. *Elymus repens*, *Agrostis gigantea*, *Tussilago farfara* and *Equisetum arvense*) and thickened plagiotropic roots (e.g. *Cirsium arvense* and *Sonchus arvensis*). These plants are comparatively tolerant to ordinary soil tillage and are sufficiently competitive to grow vigorously and propagate in many types of crop (Table 5). They are therefore important weeds in most types of crop in a crop sequence.

Some of these perennials may persist differently under different chemical conditions in the soil. This may be exemplified with *Rumex acetosella*, which was of much greater importance as a cereal weed in Scandinavia in earlier times than today. Though one reason for that may be that tillage was less effective in the past, a more important reason was probably that the plant withstood competition better on soils with lower nutrient contents. Vigorous plants of this species can still be seen in cereals on acid and weakly fertilized soil. According to Ellenberg (1974), the plant is light-requiring and prefers soils that are acidic and poor in nitrogen. In situations where the cereal crop is heavily fertilized with nitrogen, even a species such as *Sonchus arvensis* may have difficulties in recovering after breakage and deep burial delaying its shoot emergence (Table 35 and associated text). *Elymus repens*, on the other hand, seems to be favoured by high fertilizer levels (of N) on very different types of soil (pp. 236–242).

Factors affecting the response of growing plants to tillage – an overview

Some factors determining the response of a perennial plant to tillage are listed as follows:

- Regenerative characteristics of the plant
  - Life form and specific traits of the species
  - Stage of plant development
  - Modifications of the regenerative structures caused by conditions during their formation
- Type of tillage influence
  - Depth of burial of plant structures
  - Type and degree of injuries to plant structures
- Environmental conditions

The main environmental factors influencing the plant response to tillage are temperature and moisture. They modify this response to a varying degree. They cause direct injuries to the plants mainly at extreme levels. At moderate levels, they affect the speed of the life processes, including the response of the plants to mechanical disturbance.

Frost and drought in winter are particularly injurious when affecting plant parts that have, according to their nature, developed at greater depths but have been brought to the soil surface by tillage. These injuries become most extensive in the absence of a snow cover; even incompletely uprooted plant parts may then be severely harmed by freezing. At temperatures constantly above zero in soil and air, drought mainly harms completely uprooted plant parts. At higher temperatures, the regenerative organs of weeds such as *Elymus repens* may be killed in a few days when placed on the soil surface, even when wind is moderate and air humidity rather high (e.g. Grümmer, 1963; Håkansson and Jonsson, 1970; Permin, 1973a). Certain regenerative organs, e.g. many bulbs (Håkansson, 1963a), are so well protected against drying that they survive on the soil surface for months.

Even though most vegetative structures are killed in a few days or a couple of weeks when completely uprooted, frequent observations indicate that strong structures may survive for weeks even in a very dry soil, where the water tension is low when compared with that in dry air (Håkansson and Jonsson, 1970). However, it is not easy to effect an extensive complete uprooting of perennating organs; therefore tillage designed for partial uprooting seems worthwhile mainly in areas where regularly recurring periods of drought are expected.

In cooler and more humid areas with irregular, rather short, periods of dry weather, as in Scandinavia, soil cultivation intended to control perennial weeds is, and should mostly be, focused on reducing stored food reserves needed for regrowth. Then, effective breaking and burial by tillage should mainly aim at stimulating regrowth – which, in its early period, means a consumption of food reserves. During unforeseeable, and usually short, periods of severe drought, buried regenerative structures only die of drought to a minor extent. Instead they slow up their life processes, including their consumption of stored reserves. Drought hence slows up the intended weakening of the plants by tillage (studies on *Elymus repens*: e.g. Bylterud, 1965; Håkansson and Jonsson, 1970).

However, even in temperate humid areas, weakening and killing by drying should not be underrated as complementary effects in certain situations and when implements suited for bringing parts of the plant structures to the soil surface are available. *Drying in combination with frost in the winter can cause a considerable mortality among plant parts on or near the soil surface. This should be kept in mind, although the following discussions are focused on controlling perennial weeds by lowering their contents of stored reserves.*

When the situation is suitable for field operations, assimilating green shoots should be buried or destroyed by tillage at the earliest, so as to stop photosynthesis rapidly and impede the translocation of assimilates to the regenerative organs. These organs should, at the same time, be broken up or wounded as much as possible. An increased breakage brings an increased number of new shoots and roots to develop. This process of regeneration either starts immediately and proceeds rapidly after tillage or – in cases of dormancy, low temperature or drought – becomes delayed or proceeds slowly.

The consumption of stored reserves in the early period of regrowth thus increases with the increased number of buds activated as a result of increased breakage and wounding. The structures are then also exposed to increased microbial attacks. Furthermore, the more units of new shoots and roots that are forced to develop, the weaker they are in their early stages, simply because the amount of food reserves available to each unit decreases. More shoots, or 'plant units', therefore die before and after emergence, and the surviving units become increasingly sensitive to subsequent tillage and/or crop competition.

The effect of breakage increases with an increased depth of burial brought about by ploughing. Joint effects of breakage and burial depth are described in more detail later on. Times of soil cultivation are also discussed then.

It is mainly when the underground regenerative structures are distributed within shallow depths, as in *Elymus repens*, that ploughing can increase the average depth of these structures considerably. An increased average depth of the regenerative structures is attainable to a lower degree when the plants have deep-growing structures, such as *Cirsium arvense*, *Convolvulus arvensis*, *Equisetum arvense* and *Tussilago farfara*. Only the mean depth of the shallower parts of their regenerative structures is increased by ploughing. However, when the shallow parts have been broken by some soil operation before ploughing, the emergence and establishment of shoots from them is restricted. Structures located at shallow depths are particularly ready for rapid sprouting. Their breakage and burial to deeper levels are therefore considerably suppressive. This effect probably usually exceeds the favourable effect of the movement of structures

from deeper positions in the topsoil to shallower. In field experiments, some suppression is thus obtained as a joint effect of breakage and burial on deep-growing perennials, although it is often weaker than the effect on shallow-growing perennials (e.g. Permin, 1961b).

A single tillage operation sometimes only slightly suppresses perennials with great regenerative capacity; it can even increase their populations. If it has led to an increased number of established shoots, these may, although initially weak, give rise to a greater biomass production than the fewer, but stronger, shoots of comparable undisturbed plants. Such a favourable effect on growth is counteracted by repeated tillage destroying the new shoots. Tillage should preferably be repeated when regrowth has consumed food reserves for a period. It should be repeated at the latest when new aerial shoots have become sufficiently large to allow the initiation of new regenerative organs. At this stage, the plant contains minimum amounts of food reserves (Figs 49, 52, 53, 59 and 60). However, the shoots do not emerge simultaneously, which leads to a subsequent variation in shoot size. The optimum timing of tillage must therefore be a compromise, considering this variation. Effects of repeated tillage are discussed in the following. It should then be stressed again that intense tillage should be avoided as far as possible.

Perennials with rhizomes or plagiotropic roots

Soil tillage directly affects growing plants in the following two ways:

- by breaking and wounding the plant structures, and
- by changing their depth in the soil.

Tillage may also affect the plants indirectly by changing the structure and moisture conditions in the soil.

The response of plants to mechanical disturbance at different stages of development is described below, along with the growth and development of undisturbed plants. Descriptions focus on those creeping perennials that have been character-ized as plants with higher tolerance to tillage.

Undisturbed growth and response to a single soil operation

Figure 49 describes the dry-weight changes in different plant parts in the early period of plant growth from perennating vegetative structures. The characterization is based on

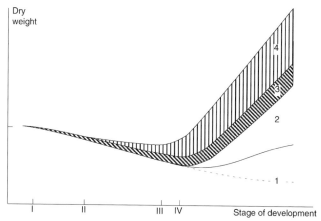

Fig. 49. The dry weight of different plant parts in the period of early growth from regenerative structures of creeping perennial plants. Principles including three grasses (*Elymus repens*, *Agrostis gigantea* and *Holcus mollis*) and two composite plants (*Cirsium arvense* and *Sonchus arvensis*) according to studies in Sweden. Growth may have started either in spring or, following tillage, later in the growing season, unless prevented by physiological dormancy of the regenerative structures. (The roots of *Sonchus arvensis* are dormant during late summer and autumn.)

Plant parts:
1. Original regenerative structures (rhizomes in the grass species; thickened roots in the others).
2. New regenerative structures (rhizomes or thickened roots).
3. Vertical shoots not yet emerged plus underground bases of aerial shoots.
4. Aerial shoots.

Stages:
I.   Initial growth of new shoots from the regenerative structures.
II. Emergence of the first aerial shoots.
III. Dry-weight minimum of total plant (thin roots excluded: see text).
IV. Dry-weight minimum of underground regenerative structures. Initial development of new regenerative structures. In the grasses, this means activation of certain underground buds giving rise to rhizomes. In the com-posite plants, it means the start of secondary growth in thickness in some of the originally thin roots. In *Cirsium arvense*, older roots die more or less rapidly when new thickened roots develop (hinted at by the broken line).

The time taken to reach the various stages depends on environment, mainly on temperature, provided that moisture and light conditions are satisfactory. The first development of new regenerative structures (stage IV) is closely related to the number of leaves of the primary aerial shoot under specified environmen-tal conditions. Leaf numbers given below represent unshaded plants at normal temperature variation from 'mid-spring' to 'mid-autumn' and represent leaves with visible laminae exceeding about 4 cm in length:

Elymus repens	3–4
Agrostis gigantea	3–5
Holcus mollis	4–5
Cirsium arvense	4–7
Sonchus arvensis	5–7

experiments with *Elymus repens* and *Sonchus arvensis* (Håkansson, 1967, 1969d,e), *Agrostis gigantea* and *Holcus mollis* (Håkansson and Wallgren, 1976), and *Cirsium arvense* (Kvist and Håkansson, 1985; Dock Gustavsson, 1997). Figures 50 and 51 illustrate plants of *E. repens* and *S. arvensis* in different stages.

Vertical underground bases of orthotropic aerial shoots of creeping perennials are fairly regenerative compared with the plagiotropic rhizomes or thickened roots. This was observed in *E. repens* (Håkansson, 1969c), in *S. arvensis* (Håkansson and

Wallgren, 1972a) and in *C. arvense*. However, while still soft, the orthotropic bases of young shoots of *S. arvensis* and *C. arvense* die without apparent regenerative activities.

The graph in Fig. 49 has many similarities to the graph of seedlings in Fig. 4 (p. 60). Both graphs show that the minimum dry-weight of underground plant structures, looked upon as entireties, is passed somewhat later than the minimum of the total plant. This is logical because the net translocation of organic material to the underground plant parts is negative until the aerial shoots are sufficiently large to produce a surplus of assimilates sufficient for a positive downward translocation. In the perennial species now being discussed, the initiation of regenerative organs starts as soon as the supply of assimilates to the underground structures compensates for the consumption caused by respiration

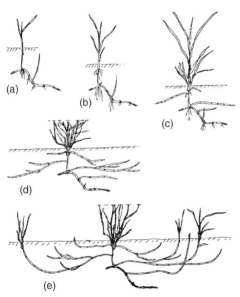

Fig. 50. *Elymus repens.* Undisturbed development and growth starting in spring from a rhizome fragment and proceeding in the absence of strong interspecific or intraspecific competition. (a) Early 3-leaf stage. (b) Primary shoot in its early 4-leaf stage: its first lateral tiller has emerged and new rhizomes have just started developing from buds below soil surface. (c, d) Stages in late spring and early summer when new tillers and rhizomes are rapidly increasing in number, length and degree of branching; new rhizomes may now also develop from buds on older rhizome branches. (e) Stage in late summer or autumn when many rhizome apices are bending upwards, some of them having developed aerial shoots. Note that, in a dense plant stand with mutually shading shoots, the development of new rhizomes starts later and the rhizome production becomes lower than intimated in this figure.

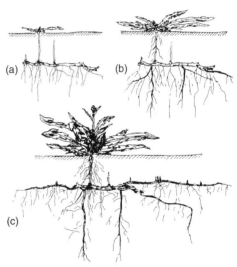

Fig. 51. *Sonchus arvensis.* Undisturbed development and growth starting in spring from a root fragment and proceeding in the absence of strong interspecific or intraspecific competition. (a) A primary shoot in its early rosette stage: a few leaves developed; all new roots still thin. (b) Rosette with 6–7 leaves: some of the new roots have started a secondary growth in thickness. (c) Stem elongation has started in the aerial shoot: inflorescence buds appearing; some new roots are now 2–4 mm thick, several shoots from which are extending towards soil surface; like the thickened roots, many underground stem bases are now regenerative.

and formation of aerial shoots. In the experiments of Kvist and Håkansson (1985), the same was found in *Tussilago farfara*, whereas in *Equisetum arvense* new rhizomes seemed to be initiated earlier, but they did not start rapid growth until the dry-weight minimum under discussion had been passed.

The thin roots of the plants have been hard to collect and weigh quantitatively. However, approximate weight values and some special measurements (Håkansson, 1969d) indicate that the statements given about dry-weight minima of total plants and underground structures apply also when these roots are considered.

The first regenerative organs develop from underground structures that are well connected with the new aerial shoots on whose photosynthesis they are relying. They start to develop as soon as these new shoots have reached a stage that can be satisfactorily defined by their number of leaves under specified conditions. Leaf numbers given in the legend to Fig. 49 represent unshaded plants in full daylight and within temperature variations normal from 'mid-spring' to 'mid-autumn' in a Scandinavian field. These numbers thus represent conditions when soil tillage can be carried out. The relationship between the leaf number of an aerial shoot and the development from the underground structures connected to this shoot is valid not only for a detached plant unit producing a single shoot; it is also valid for the relationship between the development of each aerial shoot and its adjacent underground parts of an unbroken system of rhizomes or regenerative roots.

In shading crop stands, the shoots, naturally, must develop more leaves (larger leaf areas) than stated in Fig. 49 until net photosynthesis becomes positive and new regenerative structures develop. This has been illustrated in *E. repens* (Håkansson, 1969d).

The initiation and further development and growth of regenerative organs means different processes in rhizomatous plants and plants with thickened plagiotropic roots, which is described in the legend to Fig. 49.

The original rhizomes and thickened roots lose stored reserves, leading to a dry-weight reduction in the period of early growth of new shoots. After net photosynthesis has become positive, their dry weights increase due to replenishment of assimilates (Fig. 49) – except for the original regenerative roots of *Cirsium arvense*. These die fairly extensively after new thickened roots have developed, but parts of them may survive for some period (Dock Gustavsson, 1997). However, the thickened roots of *C. arvense* normally die, largely or completely, after having functioned as mother roots of new shoots in their second year of life (Lund and Rostrup, 1901; Åslander, 1933). This is hinted at by the broken line in Fig. 49.

Seeds, in comparison, die during seedling development (Fig. 4). Similarly to seeds, bulbs of species such as *Allium vineale* die after their depletion when new bulbs develop. Like the new rhizomes and thickened roots of creeping perennials, new bulbs start a rapid growth immediately after net photosynthesis has become positive (Håkansson, 1963a).

Experiments with the perennial species discussed here indicate that their regenerative ability (capacity to regrow after mechanical disturbance) is particularly low at the time when new regenerative structures are initiated and in a short period after that (Håkansson, 1963a, 1967, 1969e; Dock Gustavsson, 1997). Figure 52 describes undisturbed growth of *Elymus repens* plants; Fig. 53 presents experimental results regarding the regenerative ability of the plants on different dates. This ability decreases with decreased amounts of substances needed for the formation of new plant structures. After a period of further growth, an increasing rate of positive net photosynthesis results in an increased regenerative capacity. This is low until the new rhizomes have developed 'stable' tissues, i.e. it is low even in the first period of positive net photosynthesis.

Figure 53 illustrates that reestablishment became increasingly difficult with increasing burial depth – particularly obvious after burial of the plants in their weakest stages. In *E. repens*, the weakness lasted over a long period of May. In

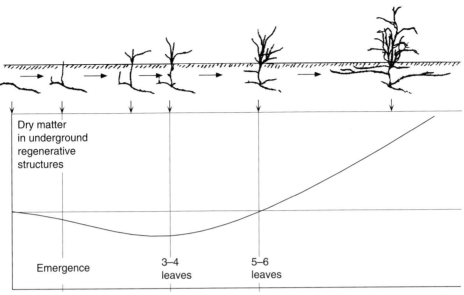

Fig. 52. Upper drawings describe undisturbed growth from a rhizome fragment of *Elymus repens*. When the primary aerial shoot has developed 3–4 leaves and the development of new rhizomes has just started (stage IV in Fig. 49), the regenerative underground stems are passing the stage of minimum dry weight. In a period around this stage, the plant is especially sensitive to soil tillage.

Sonchus arvensis, the first growth in thickness among originally thin roots – i.e. the 'development of regenerative organs' – was observed on 25 May at Uppsala (Håkansson, 1969e); the minimum regenerative capacity was recorded after burial in late May and early June.

Within certain limits, increased fragmentation of rhizomes or thickened roots activates an increased number of buds to start growth. Figure 54 illustrates the shoot emergence from rhizomes of *Elymus repens* cut into pieces of different lengths and buried at two depths, 3 or 7 cm, in a pot experiment (Håkansson, 1968a). Repeated watering prevented shortage of water at the shallower depth, where, under field conditions, drought often delays growth (although seldom kills the plants). At the 3-cm depth, few shoots died before emergence; the number of emerged shoots was nearly proportional to the number of activated rhizome buds. At the 7-cm depth, the mortality was much higher and was the main reason for a lowered emergence. The mortality was particularly high among weak shoots developed from the shortest rhizome pieces, which

were only 2.5 cm long. At the 7-cm depth, fewer shoots therefore emerged from these pieces than from the 5-cm pieces. The results are in harmony with effects of rhizome length and depth of planting presented by Vengris (1962).

Figures 55–57 show principles based on observations of shoots from rhizomes of *E. repens* cut into fragments of different lengths and buried at different depths in field experiments (Håkansson, 1974).

Figure 55 describes relationships based on experimental data using probit analyses according to Finney (1952). Depths shallower than 5 cm were excluded in these analyses because of irregularities created by variations in moisture. Broken lines roughly characterize typical situations on, or very near, the soil surface. Figure 57 illustrates responses to fragmentation and depth of burial of *E. repens* rhizomes.

Percentage numbers given in the legend to Fig. 55 present the degree of survival of buried rhizome pieces as determined in late autumn 1 year after burial. Pieces placed on the soil surface (depth 0) were entirely dead even though a few of them had produced

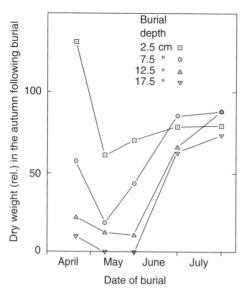

Fig. 53. The regenerative ability of *Elymus repens* plants in different stages of development, as characterized by the dry weight of rhizomes in the autumn following burial of plants at different depths on different dates earlier in the growing season. Dry weights are presented as ratios to the dry weight of rhizomes planted in the previous autumn and from which the experimental plants originated. Primary aerial shoots had 3–4 leaves on 10 May and 5–6 leaves on 31 May, i.e. on dates when the plants proved particularly sensitive to burial. (From Håkansson, 1967.)

established plants. The surviving parts of the buried pieces decreased with their decreasing length and, at depths from 2.5–10 cm to about 20 cm, with increasing depth of burial (see next paragraph). Parts characterized as living were represented by tissues that were judged to be alive and, in addition, harboured at least one bud. The mortality among rhizome pieces buried below the soil surface was certainly largely caused by depletion of energy reserves. Death and moulding of the tissues, due to attacks by fungi and/or to other adverse influences, mostly started and extended from the ends of the pieces. This is likely to be part of the reason why the mortality increased with decreasing length of the rhizome pieces.

According to assertions from many experienced people, deep burial results in

'conservation' of the rhizomes. My own observations support this, as do the percentage values in the legend to Fig. 55. These thus suggest a higher degree of survival at depths of 25–30 cm than at 17.5–20 cm. Several rhizome pieces from the greatest depths were alive without having developed aerial shoots supplying them with assimilates. Survival at great depths may be favoured by restricted gas exchange, which retards energy-consuming processes. Perhaps it is also a result of reduced attacks by fungi, etc. At ordinary ploughing, only few rhizomes are, in their entirety, buried at depths exceeding 20–25 cm.

Experiments and observations, overall, indicate that breaking and wounding the rhizomes of *Elymus repens* followed by ordinary ploughing not only restricts the regrowth; burial to deeper layers in the topsoil also reduces the quantity of old rhizomes that are viable and capable of producing regrowth when brought from deeper to shallower soil layers by the next ploughing.

The vitality of older rhizomes (rhizome fragments originally planted) was studied in a special experiment (Håkansson, 1968a). In late autumn, fragments with intact structures judged to be alive (harbouring at least one bud) were excised from rhizome pieces that had been planted in the previous autumn. Similar fragments were excised from young rhizomes of plants originating from the older rhizomes. All fragments were immediately buried at a depth of 5 cm in a sandy loam so as to compare both their winter survival and their regenerative capacity. *Older rhizome fragments with intact structures exhibited a somewhat reduced winter survival and, in the following growing season, they produced plants with a biomass amounting to about two-thirds of the biomass of the plants from comparable young rhizome fragments.* According to my own general observations, winter survival varies greatly, particularly among older rhizomes on the soil surface or in superficial soil layers.

When rhizomes or regenerative roots are placed on, or *very* near, the soil surface, early bud activities may be strong in rainy weather, but halt in dry weather. Their

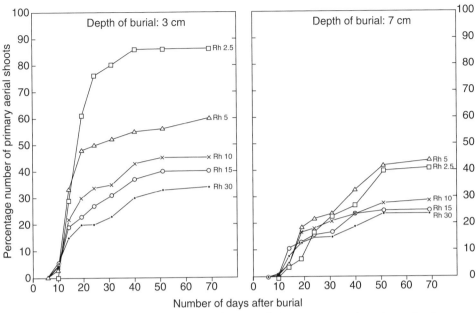

Fig. 54. Numbers of primary aerial shoots from rhizomes of *Elymus repens* cut into pieces of various lengths and buried at depths of 3 or 7 cm. They are expressed as percentages of the number of buds on the buried rhizomes. These had been cut into lengths of 2.5, 5, 10, 15 or 30 cm (Rh 2.5, etc.). Greenhouse pot experiment. (From Håkansson, 1968a.)

mortality and production of established plants therefore varies greatly, particularly when they have been placed in their entirety *on* the soil surface. When placed on, or very near, the soil surface, short rhizome fragments die more rapidly and extensively than longer fragments without establishing new plants. When rhizomes are placed at depths approaching the optimum, fragmentation into rather short pieces leads to an increased bud activation that results in an increased number of established shoots.

Where an increased shoot number per unit area is caused by increased fragmentation of rhizomes or regenerative roots, the individual shoots become weaker. The quantity of new regenerative structures produced per unit area therefore sometimes differs only slightly with the number of emerged shoots. In a competitive crop, an increased number of aerial shoots caused by fragmentation of the regenerative structures usually means a lower production of new regenerative structures (cf. Tables 34 and 35). The joint effect of extended fragmentation and burial depth is a lowered number of

emerged shoots in relation to the number of activated buds and, at great depths, also a lowered absolute number of aerial shoots. The production of new regenerative organs decreases proportionally more than the growth of aerial shoots.

Tables 34 and 35 and Figs 53–57 present effects of burial where plant structures have been spread horizontally at different depths. Important conclusions on principles can indeed be drawn on the basis of these experiments. However, tillage distributes plant structures at various angles across different depths. An experiment carried out at Uppsala with burial of rhizomes of *E. repens* at different angles (Håkansson, 1968b) is presented next (Tables 38 and 39).

Rhizome branches of *Elymus repens* were dug up either in late September or in early May. From unbroken branches of a length of 40–50 cm, 33-cm long pieces were then immediately excised and buried in a field with sandy loam. The 33-cm pieces were either left whole or cut into three 11-cm parts.

Percentage number of rhizome fragments
with established shoots

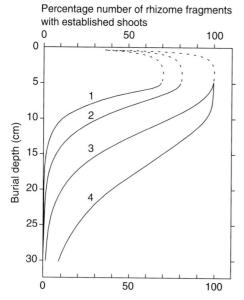

Fig. 55.   Percentage numbers of rhizome fragments of *Elymus repens* giving rise to at least one established aerial shoot. Fragments of various lengths cut from rhizomes and buried, horizontally, at different depths in late autumn in a field with sandy loam at Uppsala (Håkansson, 1968b). Determinations in the year following burial. Lengths and dry weights of rhizome fragments: 1, 4 cm, 50 mg; 2, 8 cm, 103 mg; 3, 16 cm, 220 mg; 4, 32 cm, 496 mg. Values presented below (from Håkansson, 1969b) represent parts of the buried rhizome fragments (1–4) classified as alive in late autumn 1 year after burial. Values describe dry weight as a percentage of the dry weight at burial:

Depth (cm)	1	2	3	4
0	0	0	0	0
2.5–5	1	30	41	56
7.5–10	0	31	43	59
12.5–15	2	2	22	45
17.5–20	2	0	2	28
25–30	0	6	21	31

The rhizomes were buried within square plots of 0.5 m² surrounded by wooden frames, either horizontally, at depths of 2.5, 12.5 and 22.5 cm, or at an angle of about 35° from the horizontal plane. Of the 33-cm pieces, 11 were buried in each plot (Table 38). When buried at an angle, the 33-cm long pieces were placed with their ends 2.5 cm and 22.5 cm below the soil surface, i.e. their mean depth was 12.5 cm. At horizontal burial, the three 11-cm parts of the divided 33-cm pieces were kept in the positions they would have been in without division. When buried at an angle, only one of the 11-cm parts was planted (see Table 39).

Dry weights of aerial shoots and new rhizomes were measured in mid-July. Relative values, comparing treatments, were essentially similar after burial in September and burial in May (see Håkansson, 1968b) and are therefore averaged in Tables 38 and 39.

Table 38 represents burial of whole 33-cm rhizome pieces. Horizontal burial at the shallowest depth, 2.5 cm, resulted in the highest production of aerial shoots and new rhizomes. Angled rhizomes with the terminal ends upwards showed a lower production, but this was much higher than the production after horizontal burial at 12.5 cm, which represents the mean depth of the angled rhizomes. With the terminal ends pointing downwards, the production was about half of that with these ends pointing upwards, but still consistently higher than after horizontal burial at 12.5 cm.

In the lower part of Table 38, the average production from horizontal rhizomes at the three depths is compared with the production from angled rhizomes, comprising those with their terminal ends pointing upwards and those with these ends downwards. The mean depth of the different categories was the same, 12.5 cm. The production from the angled rhizomes was higher than the average production from rhizomes buried horizontally.

The buried rhizome pieces either included the terminal tips (apices) of the dug up branches or were cut out 4 cm behind these tips. (Some of these tips are broken to some extent at tillage in the field. Due to 'immature' tissues, *separated* apical parts mostly have a low regenerative value.) The production from rhizomes with and without apices was, in principle, similar and is therefore presented by means in Table 38. However, there were interesting differences in the number and position of aerial shoots.

Shallow burial

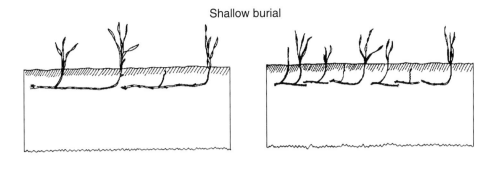

Deep burial

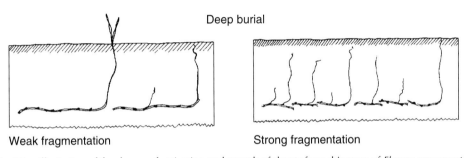

Weak fragmentation   Strong fragmentation

Fig. 56.   Illustration of the degree of activation and growth of shoots from rhizomes of *Elymus repens* cut into fragments of different lengths and buried at different depths in soil – below the surface layer where growth is often affected by drought (cf. Fig. 57).

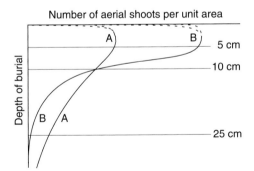

Fig. 57.   Establishment of shoots from rhizomes of *Elymus repens* buried at different depths after (A) moderate and (B) strong fragmentation. Principles summarized on the basis of various field experiments at Uppsala (Håkansson, 1967, 1968a,b, 1974; cf. Fig. 56).

When rhizome pieces without apices were buried, the number of primary aerial shoots persevering after emergence was three times as high from horizontal pieces at the 2.5-cm depth as from angled pieces whose terminal ends were pointing upwards. The latter pieces, in turn, produced twice as many primary aerial shoots as those whose terminal ends were pointing downwards. At horizontal burial, aerial shoots developed from buds in different positions along the buried rhizomes, although with some predominance of shoots from buds near the terminal end.

This predominance was much more pronounced among the shoots from angled pieces with their terminal ends pointing upwards. On the other hand, when these ends were pointing downwards, the majority of the shoots came from the basal, shallower halves of the angled pieces. Their development and growth were considerably delayed and reduced when compared with the shoots from rhizomes with their terminal ends pointing upwards. This phenomenon was certainly a result of correlative interference from shoots developed nearer the terminal end at the greater depths. Such shoots developed from more rapidly

Table 38. Production of aerial shoots and new rhizomes of *Elymus repens* after burial of 33-cm long rhizome pieces, either in a horizontal position, at different depths, or angled between given depths in a sandy loam soil in the field.[a]

| Burial regime | Dry weight (relative figures) | |
	Aerial shoots	New rhizomes
Horizontal		
Depth 2.5 cm	100	100
Depth 12.5 cm	35	25
Depth 22.5 cm	2	1
Angled, depth 2.5–22.5 cm; terminal ends:		
upwards	84	79
downwards	47	36
Horizontal: mean	45	42
Angled: mean	66	58

[a]Determinations in mid-July after burial of rhizome pieces in the previous autumn or early May. Buried pieces had been cut either including rhizome apices or 4 cm behind the apices. The table presents means of these cutting regimes. These gave, in principle, the same results regarding the production of aerial shoots and new rhizomes. (Figures are based on data from Håkansson, 1968b.)

Table 39. Production of aerial shoots and rhizomes after burial of 11-cm long rhizome fragments cut from 33-cm pieces of *Elymus repens* rhizomes just before burial. Figures express dry weights as percentages of those of the uncut, 33-cm long rhizome pieces dealt with in Table 38.[a]

| Burial regime | Dry weight (percentage of the weight of the 33-cm pieces at burial) | |
	Aerial shoots	New rhizomes
Horizontal		
Depth 2.5 cm	101	108
Depth 12.5 cm	51	48
Depth 22.5 cm	≈ 50	≈ 38
Angled, depth 2.5–10 cm; terminal ends:		
upwards, only terminal parts	76	69
downwards, only basal parts	165	163
Angled: mean	108	98

[a]At horizontal burial, all 11-cm fragments were buried for comparison with the uncut 33-cm pieces. At angled burial, only one of the 11-cm parts was buried. This was the part corresponding to the upper part of an uncut rhizome piece and therefore either a terminal or a basal part.

activated buds, but most of them died without reaching the soil surface.

When rhizome pieces including apices were buried, the most rapid shoot growth took place from the apices. This certainly resulted in a stronger correlative dominance of the shoots from the terminal ends than when shoots developed from pieces without apices. The number of shoots was only slightly lowered by this dominance when the buried pieces were angled, but was considerably reduced at horizontal burial.

The biomasses produced per primary aerial shoot differed between treatments (see Håkansson, 1968b). This can be explained by the fact that the primary shoots differed in time of emergence and vigour, resulting in different growth rates and tillering.

Table 39 describes situations where buried rhizomes had been cut into 11-cm long parts. The dry weights of aerial shoots and new rhizomes produced from the 11-cm parts are described as percentages of those from uncut 33-cm rhizome pieces. These

percentages may be commented upon as follows.

At horizontal burial, three times as many 11-cm parts as uncut 33-cm pieces were buried, i.e. the 11-cm parts were represented by the same biomass as the 33-cm pieces. At the shallow depth of 2.5 cm, the cutting led to increased numbers of emerged shoots: an increase of about 40% when the rhizome apices were cut off and by as much as 250% when the apices remained on the 33-cm pieces and on the terminal part after cutting into 11-cm lengths. The strong increase in the number of shoots from the cut pieces was mainly a result of reduced apical dominance. However, at horizontal shallow burial, the biomass production was only slightly influenced when compared with the great increase in shoot numbers. Only a tendency for an increase was noted: by 1% in aerial shoots and by 8% in rhizomes.

When the depth of horizontal burial was increased to 12. 5 and 22.5 cm, cutting of the 33-cm rhizome pieces into 11-cm lengths led to a strongly lowered number of aerial shoots and a reduction in the rhizome production of about 50% (Table 39; cf. Figs 54–56).

When the buried rhizome pieces were *angled*, only one 11-cm part out of the three in the partitioned 33-cm pieces was buried; this was thus either the terminal or the basal part. The upper points of these parts were placed at the same depth as those of the longer pieces, i.e. 2.5 cm below the soil surface, the lower point then at a depth of 10 cm. When the 11-cm terminal parts of the 33-cm pieces were used, their *terminal ends* were pointed *upwards*. The number of primary aerial shoots from these parts became 20–40% lower than from the 33-cm pieces and, according to Table 39, the rhizome weight was reduced by about 30%.

When the 11-cm basal parts of the 33-cm pieces were buried, their *basal ends* were pointed *upwards*. A much greater number of primary aerial shoots then emerged from the 11-cm parts than from the corresponding 33-cm pieces. This was certainly caused by the absence of correlative dominance of shoots developing from great depths, as for the shoots from the terminal regions of the longer pieces. These shoots emerged to a very slight extent, as they faced two disadvantages; they developed from greater depths and they were restricted by 'intra-plant' competition from the shoots from buds in shallower, and in that sense more favourable, positions. The rhizome production from the shorter basal parts was much higher than from longer pieces with their terminal ends pointed downwards. It was on average about 60% higher. (It was about 110% higher when the 11-cm basal parts were compared with 33-cm pieces with apices and 30% higher when compared with longer pieces without apices; see Håkansson, 1968b.)

The values on the bottom line in Table 39 are means representing all the angled rhizome fragments, i.e. both those with their terminal ends pointing upwards and those with these ends pointing downwards. These values show the interesting result that *when the upper ends were placed at the same depth of 2.5 cm, 11-cm long rhizome parts then, on average, produced approximately*

the same amounts of aerial shoots and new rhizomes, by weight, as did the 33-cm long angled pieces. This implies that the regrowth from rhizome branches buried at different depths by tillage is more dependent on the level of the shallowest parts of the branches than on their mean depth. For angled rhizomes, the influence of their length depends on the position of their terminal end, with activated buds developing correlatively dominating shoots. Regeneration is favoured when activated buds are in the shallow region of an angled rhizome. When the terminal end has become placed in the deeper region, correlative dominance of shoots from this region may suppress growth from shallower buds. This may suppress the total growth from a shorter rhizome fragment less than that from a longer one, whose correlatively dominating shoots are initiated at depths from where they do not emerge or emerge late with weak shoots (compare values on the middle lines in Tables 38 and 39).

Elymus repens is one of the relatively shallow-growing creeping perennials. Particularly in soil that has not been cultivated for a long time, the great majority of the rhizomes usually occur in layers from near the soil surface to a depth of 5–10 cm. In newly ploughed soil, considerable numbers of rhizomes are found down to the ploughing depth, but new rhizomes nevertheless predominantly develop in shallower soil layers. Information often referred to says that the rhizomes go proportionally deeper in loose than in more compact soil. This is true, but most of the new rhizomes grow in the shallower layers even in loose soils.

The spontaneous distribution of new regenerative structures of *E. repens* in differently deep soil layers was studied in plants developed from rhizomes buried at different depths in a field with sandy loam at Uppsala (Håkansson, 1969c). Rhizome pieces, 16 or 8 cm long, were buried horizontally in early spring. In the subsequent late autumn, the plagiotropic rhizomes and other underground stems were dug out from different soil layers. Table 40 illustrates results representing plants from the 16-cm rhizome pieces. Plants from the 8-cm pieces exhibited a similar picture.

Table 40. Depth orientation of new rhizomes[a] of *Elymus repens* plants developed from rhizome pieces buried at different depths in a field with sandy loam. Development in the first growing season after burial.

Depth of soil layer (cm)	Burial depth (cm)				
	5	10	15	20	5–20
Distribution of new rhizomes (%, by fresh weight)[b]					
0–5	96	80	65	65	85
5–10	4	20	33	25	14
10–15	0	0	2	10	1
15–20	0	0	0	<1	<<1
0–20	100	100	100	100	100
Fresh weight of new rhizomes, total (g)					
0–20	2132	1287	672	162	4253

[a]Including underground bases of aerial shoots.
[b]Weights represent rhizomes of plants developed in pure stands from a total of 20 rhizome pieces, 16 cm long, with a fresh weight of 12 g in total at each depth. (From Håkansson, 1969c.)

Table 40 indicates that the majority of new regenerative structures developed in the upper 5-cm soil layer even after deeper burial, although most pronouncedly after shallow burial. Even at a burial depth of 20 cm, the great majority developed above the 10-cm depth. The depth position of the plagiotropic rhizomes was regulated in two ways, namely through the angle of their growth in length and through the depth of the buds from which they originated. The figures in Table 40 include all regenerative stems, vertical as well as horizontal. However, excluding the vertical stems provides largely the same picture of depth distribution as including them. Moreover, the orthotropic underground bases of aerial shoots are regenerative (Håkansson, 1969c). Including them therefore provides the best description of the quantity and distribution of regenerative structures.

As seen in Table 40, there is not only a somewhat stronger concentration of new regenerative structures from rhizomes buried at shallower depths than from rhizomes buried deeper, but also a much higher production of new rhizomes (bottom line in left-hand column). In the absence of tillage, plants from shallower structures therefore take over more and more in the course of time. This easily explains why *Elymus repens* usually exhibits a shallower rhizome growth in fields that have not been ploughed for a long time than in regularly ploughed fields (cf. Palmer, 1962; Cussans, 1973).

Tillage leading to deep burial suppresses the growth of *E. repens* owing to both reduced regrowth and to increased mortality among the buried structures. The predominant allocation of the new regenerative structures to shallow soil layers even after deep burial is a way for the plant to re-establish optimum depths of its rhizomes. This is, at the same time, an agronomic advantage. In situations where soil operations can be repeated, the structures are easily broken by shallow tillage and, furthermore, transferred to greater depths by ploughing. From many aspects, it also facilitates effective chemical control.

To sum up. Tillage wounds and breaks regenerative structures and alters their depths and angles in the soil. Breakage induces an increased number of buds in the structures to develop shoots. A combination of breakage of the regenerative structures and transfer to greater depths is, when possible, worthwhile in the mechanical control of perennial weeds. When parts of the structures after tillage are still located near the soil surface, buds on these parts enable a strong regrowth even where tillage has led to a considerably increased mean depth of the structures. Where this regrowth can be obstructed later on, mechanically or chemically, combined with suppression in competitive crops, good joint effects can be obtained on many weeds among the creeping perennials.

Response to repeated tillage

Soil tillage repeated during the major part of a growing season at well-timed intervals can kill or considerably weaken perennial weeds by depletion of their stored reserves. Effective depletion can be achieved by repeated tillage in seasons when regrowth is not restricted by physiological dormancy, temperature or drought.

Although intensive soil tillage repeated over long periods is not desirable, experiments with such tillage are of interest for the understanding of effects of repetitions within shorter periods.

According to opinions common in old times, tillage on 'black fallows' should be repeated as soon as, or even before, aerial shoots of perennial weeds reappear. This was doubted by some early scientists, who suggested that a somewhat later time would be optimal, as consumption of food reserves continues for a period after emergence. Concerning *Elymus repens*, Kephart (1923) recommended recultivation when new aerial shoots had leaves with a length of about 1 inch (2.5 cm). Dunham *et al.* (1956) suggested that the leaves should be 2–3 inches (5–7.5 cm). Carder (1961) proposed 2 inches and emphasized the importance of tillage shredding the rhizomes. Fail (1956, 1959) reported 100% mortality in one growing season by rotary cultivation repeated when new shoots had leaves with a length of 3 inches. In experiments with operation intervals of either 2–3 weeks or about 5 weeks, he obtained complete rhizome death after six or four operations, respectively.

Early field experiments concerning effects on perennial weeds of repeated tillage (or cutting the plant structures) to different depths in the soil have been carried out in the USA. The experiments mainly concerned deep-growing perennials, not least *Convolvulus arvensis* (e.g. Sherwood and Fuelleman, 1948; Timmons and Bruns, 1951). On the whole, deeper operations required fewer repetitions then shallower to give similar results. However, strict links to stages of development at the soil operations were lacking in those studies. These were therefore the focus in the two experiments at Uppsala presented below.

In order to analyse in more detail the influences of some main factors involved in this complex, experiments were performed with the two creeping perennials *Elymus repens* and *Sonchus arvensis*. The response of these plants to disturbances simulating repeated tillage was studied (Håkansson, 1969a,e). The regenerative structures in both species, i.e. rhizomes and thickened plagiotropic roots, respectively, develop within rather shallow depths in the soil (Figs 50 and 51). With each species, experiments were carried out in different years in a field with sandy loam at Uppsala. The results are summarized in Fig. 58.

Rhizome branches or regenerative roots of the two species were cut into fragments of various lengths (rhizomes from 8 to 32 cm, roots from 6 to 24 cm) and buried at various depths (min. 2.5 cm, max. 10 cm) in spring, or, in one case, in late autumn. In both cases, emergence began in spring. Simulated tillage was repeated when a quarter to a third of the aerial shoots appearing after a preceding tillage had developed a stipulated minimum number of leaves with visible parts at least 4 cm long (see Fig. 58). The operations were effected by digging up the plants with a minimum of breakage and then immediately spreading them horizontally and re-burying them at the same depth as the first burial. Re-burials were continued as long as new shoots emerged after the previous burial and reached the development stipulated for re-burial within the growing season.

In late winter or spring in the following year, dry weights of living regenerative structures and, when present, new shoots were determined. The plants exhibited essentially similar responses at different degrees of fragmentation of the original rhizomes or roots, at different burial depths and in the two experimental years. Their responses are therefore very well described by averages in Fig. 58.

However, differences not described in Fig. 58 are important. Thus, the degree of fragmentation and depth of burial revealed their importance in other ways. They led to differently long burial intervals and to different numbers of burials carried out at stipulated incidence of leaves on emerged shoots. Table 41 shows that the required number of burials of *E. repens* plants decreased with increased burial depth and decreased length of the originally buried rhizome pieces. When re-burial took place at increasingly advanced stages of shoot development, the number of burials naturally decreased.

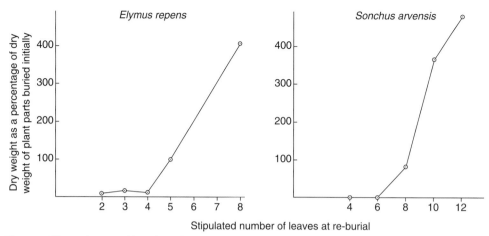

Fig. 58. Effects of repeated burial on *Elymus repens* and *Sonchus arvensis* as means of experiments from 2 years with each species in a field with sandy loam at Uppsala. Graphs are based on data from Håkansson (1969a,e). Burials simulating tillage were repeated as soon as new aerial shoots had developed the stipulated number of leaves. Curves show dry weights of regenerative structures plus (when occurring) new shoots, as determined in late winter or early spring in the year succeeding the 'burial year'. The 'stipulated number of leaves at re-burial' indicates the minimum number of leaves (with visible laminae exceeding 4 cm in length) developed on a quarter to a third of the aerial shoots. The number of burials needed to obtain a certain effect was much greater for *E. repens* than for *S. arvensis*.

Table 41. The response of *Elymus repens* to repeated burial illustrated by the number of burials required in a growing season when re-burial was done at different stipulated stages reached by aerial shoots emerging after a preceding burial. Averages for experiments otherwise dealt with in Fig. 58.

Depth of burial (cm)	Stipulated number of leaves at re-burial[a]	Number of burials Rhizome pieces	
		30–32 cm	8–10 cm
2.5–3	2	8.0	7.0
	3	6.5	5.5
	4	6.0	5.0
	5	4.5	4.0
	8	3.5	3.0
5–10	2	6.3	4.0
	3	5.0	3.7
	4	4.7	3.7
	5	3.7	3.3
	8	3.0	2.7

[a]Figures indicate the minimum numbers of leaves (with visible parts exceeding about 4 cm in length) on a quarter to a third of the aerial shoots.

According to Fig. 58, burials repeated at any of the 2-, 3- or 4-leaf stages resulted in a similarly low degree of survival of *E. repens*.

In one of the two experiments, re-burial at these stages resulted in complete plant death. When re-burials were delayed until the 5-leaf stage, the dry weight of living rhizomes was almost as high as that of the rhizomes originally buried. Repetitions delayed until the 8-leaf stage enabled the plants to increase the biomass of their regenerative structures.

The average numbers of leaves on the aerial shoots were somewhat lower than the numbers stipulated for a quarter to a third of them at re-burial, i.e. the numbers given in Fig. 58 and Table 41. When these numbers were 2, 3, 4, 5 and 8, the great majority of the shoots thus had 1–2, 2–3, 3–4, 4–5 or 6–8 leaves, respectively.

It is interesting to see that the stage when the major proportion of the new aerial shoots had 3–4 leaves was the last stage when burial could be repeated with the same strong effect as burial repeated at earlier stages. This is the stage when the regenerative structures have their minimum dry weight and the plants their lowest tolerance to tillage as a single operation (Figs 49–53).

It should be noted, in addition, that 3–4 leaves on the primary shoots indicates the stage of minimum dry weight of the under-ground parts of plants developed from rhizomes of different lengths and buried at different depths in experiments conducted by Håkansson (1967). However, both increased fragmentation and increased burial depth hasten the exhaustion of the regenerative structures and lower the ability of the plants to regrow after repeated tillage with short intervals. This explains why both increased fragmentation and increased burial depth reduced the number of burials at any treatment with respect to development of leaves stipulated for re-burial (Table 41).

From the results described above it can be concluded that optimum times for repeated tillage in the control of *E. repens* are, *in principle*, when most of the aerial shoots emerged after a preceding tillage have developed 3–4 leaves. Operations repeated at these times will result in the same check of the plants as operations at shorter intervals, but will do so with fewer operations. However, experience obtained from experiments at Uppsala suggests that the exhaustion of

the regenerative structures proceeds somewhat faster when operations are repeated more frequently (Fig. 59).

The number of operations required for complete rhizome death depends on many factors. In Fail's (1956, 1959) experiments, as few as two repetitions of rotary cultivation were sufficient under certain conditions. In those cases, however, each repetition implied additional breakage of the plant structures. In the Swedish experiments with simulated tillage, re-burials were carried out with as little additional breakage of the plants as possible.

Figure 59 is an idealized illustration of how the amount of dry matter in living regenerative structures changes in a growing season when tillage is repeated at different stages of shoot development. In real field situations, the intervals between proper times of operation vary in length because the rate of life activities in the plants varies with factors such as temperature and soil moisture.

Where the option of timing of soil operations is strongly dependent on weather and

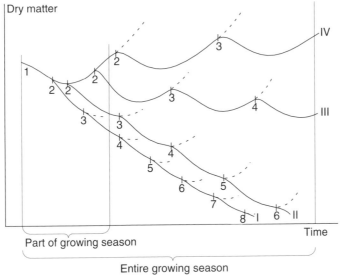

Fig. 59. Dry matter changes in viable regenerative structures (plagiotropic rhizomes plus vertical below-ground bases of aerial shoots) of *Elymus repens* in fields where tillage is repeated when new aerial shoots have developed the following numbers of leaves: 1–2 (curve I), 3–4 (curve II), 5–6 (curve III) or 6–7 (curve IV). Arabic numerals (1, 2, 3, etc.) on the curves denote ordinal numbers of tillage operations. Effective exhaustion of the structures is achieved with a minimum number of soil operations when tillage is repeated at the 3–4-leaf stage, but repetition in earlier stages may lead to a somewhat faster exhaustion. Broken lines intimate dry matter increase expected if the related soil operation is not performed.

soil conditions, repetition of tillage await-ing the theoretically optimal 3–4-leaf stage means risks of unfavourable delays of the operations. It then seems wise to repeat operations earlier, if conditions are suitable.

In situations where tillage is intended to deplete stored reserves, it is reasonable to start with a combination of strong breakage and deep burial. Shallower tillage, mainly aimed at breaking the plant structures, followed by tillage burying them at greater average depths is suitable. The presence or absence of time intervals and desirable inter-vals between deep tillage and final seedbed preparation are discussed on pp. 185–190.

Sonchus arvensis responds to repeated burial in the same way as *Elymus repens* in the early part of the growing season. In late summer, the regenerative structures enter a state of physiological dormancy. In the experiments described in Fig. 58, all structures were invariably killed by burials repeated when the aerial shoots of regrowing plants had developed 4 (3–5) or 6 (5–7) rosette leaves. Burials repeated at more advanced stages of regrowth became gradually less effective.

S. arvensis proved to be more suppres-sed than *E. repens* by repeated burials in the early growing season. The plants were thus completely killed by burial repeated once or twice (i.e. by two or three burials, the first of them when shoots had reached stipulated stages in spring). In a defoliation experiment (Håkansson, 1969e), the plants were com-pletely killed in the early part of the growing season by cutting the shoots at the soil surface three or four times when they had 4 (3–5) or 6 (5–7) leaves.

Because of the very few repetitions required to kill the plants of *S. arvensis* by burial, it is impossible to determine the influences of breakage and burial depth on the number of soil operations required for this effect on the basis of the experiments under discussion. However, breakage stimu-lated shoot activation and depth of burial influenced the times of burials in the same way as for *E. repens*. This, and the fact that more defoliations than burials were needed to kill the plants, suggest that breakage and

burial depth exert, in principle, similar effects on the number of operations in *S. arvensis* and in *E. repens*.

When burial of *S. arvensis* was repeated in the 8-leaf stage (6–9 leaves on the great majority of the shoots), only two repetitions were performed. Following the programme, the second of them was not performed until late summer. Because of the dormancy in late-summer and autumn described earlier, no new shoots then emerged until the next spring. When the plants were buried at stages of 10 leaves or more, only one repetition was performed before the period of dormancy.

The innate, physiological dormancy prevents new shoot development and asso-ciated consumption of stored reserves in late summer and autumn. Intense exhaustion through tillage therefore cannot be brought about in that period, but is evoked in the following spring.

Autumn dormancy is found in the regenerative structures of many perennial weeds. Except in *Sonchus arvensis* (Håkans-son, 1969e; Henson, 1969; Håkansson and Wallgren, 1972a; Fykse, 1974b, 1977) such dormancy is observed in, for instance, *Tussilago farfara* (e.g. Henson, 1969; Kvist and Håkansson, 1985), *Cirsium arvense* (Thind, 1975; Fykse, 1977), *Convolvulus arvensis* (Swan and Chancellor, 1974; Ayerbe *et al.*, 1979) and *Equisetum arvense* (Kvist and Håkansson, 1985). According to experiences at Uppsala, the dormancy is much weaker and less regularly recurring in *Cirsium arvense* than in *Sonchus arvensis*.

Aerial shoots of *Sonchus arvensis* remaining green after the onset of dormancy in late summer or early autumn, continue their assimilation and translocation of assimilates to the regenerative roots. Although tillage does not lead to effective exhaustion of reserve substances during this period, it destroys these shoots, thus reducing further translocation of assimilates to the roots and the amount of food reserves available in the following spring. At the same time, by breaking the roots and increas-ing their mean depth, autumn tillage also causes a weakened growth in the spring. (For more details, see section on stubble and autumn cultivation, pp. 187–189.)

Tussilago farfara and *Equisetum arvense*, which form rhizomes, and *Cirsium arvense* and *Convolvulus arvensis*, which develop plagiotropic thickened roots, are examples of creeping perennials producing regenerative structures at greater depths than do *Elymus repens* and *Sonchus arvensis*. These structures are distributed in the soil profile from rather shallow depths in the topsoil to considerable depths in the subsoil. Maximum depths differ between species and depend greatly on the soil conditions.

When a limited period is available for soil tillage, ploughing should preferably be carried out early in that period and, when possible, should be preceded by tillage that breaks the shallower structures (see section on stubble cultivation, pp. 186–190). Ploughing should cut off the links between the rhizome or root parts in the topsoil and those in the subsoil, so as to prevent translocation of stored substances from the deeper to the shallower structures.

In unbroken structures, buds near the soil surface are normally activated earlier than buds at greater depths. Through correlative dominance, the more resource-consuming development of shoots from buds at deeper positions is reduced. This behaviour, which economizes on plant resources, is counteracted when the links between shallower and deeper parts of the structures are broken. Breaking by tillage in the topsoil not only increases the number of new roots and shoots developed from the topsoil structures of the weeds; it also triggers an increased sprouting from deeper structures. This results in a more wasteful use of stored reserves than shoot development solely from shallow buds.

The largely deep-growing regenerative structures of the perennials discussed above appear to become more or less dormant in late summer and autumn (something that needs more systematic research). It might be suggested that dormancy is not induced in the underground structures if their connections with aerial shoots have been broken when long-day conditions are still prevailing. This needs to be checked experimentally, but if it is true, it is of practical interest, not least regarding the more deep-growing plants.

A Cropping System Approach to Effects of Tillage

Periods when the field is free of growing crops

As tillage usually has many purposes simultaneously, its design must mostly be a compromise with respect to different demands. Even when we look exclusively at effects on weeds, compromises with regard to various weeds occurring together in the field are often required.

For reasons stated earlier, tillage should be performed so as to provide the intended effects with a minimum intensity. Intensive soil tillage should be avoided when good alternatives are available, often through combinations of mechanical and chemical means and rational utilization of competition from crops. Black fallows with tillage repeated throughout the major part of a growing season were previously touched upon mainly to illustrate effects that can be obtained in shorter periods.

Seedbed preparation – effect on weed seeds in the soil

The term 'seedbed preparation' is used here for the final, usually shallow, tillage operations before sowing (cf. pp. 75–78). It should be stressed, however, that the conditions and processes in the final seedbed also depend on previous soil operations. This is true not least regarding the effects on weeds. General principles are summarized in the present section. Viewpoints on field applications are particularly relevant to Scandinavia, but modifications applying to many other areas can be easily conceived.

Tillage on the whole stimulates the germination of weed seeds in the soil, but the degree to which germination rate is increased varies with species and season. It thus affects weed seeds selectively. This was experienced in early experiments (e.g. Brenchley and Warington, 1936, 1945). Both the selectivity pattern and the general rate of germination and emergence vary with the immediate conditions in the soil, not least with regard to moisture and temperature. However, influences of previous conditions, particularly previous temperatures, strongly determine the disposition of the seeds to germinate at a given time (e.g. Baskin and Baskin, 1998; Champion *et al.*, 1998). Previous influences largely determine differences in the effect of tillage on seeds of different weeds in their soil seed banks. Figure 1 illustrates differences between typical *summer* and *winter annual weeds* in a humid temperate climate, explaining differences in their establishment in spring- and autumn-sown crops.

At depths below the superficial soil layers, weed seeds usually germinate to a very low extent without aeration and other effects of tillage. This applies to small seeds in particular, whereas larger seeds, in species such as *Avena fatua* and *Galium aparine*, often germinate more easily at greater depths in the topsoil. Deep tillage provokes an increased germination among many seeds at all depths in the topsoil, but even then proportionally more seeds germinate in shallow soil layers than at greater depths.

Thus, the total number of seeds stimulated to germinate is usually greater after deep tillage than after shallow. However, it is not necessarily so. If, for instance, the great majority of seeds disposed to germinate happens to occur in the superficial soil, shallow cultivation may increase the germination more than deep tillage transferring superficial soil to greater depths. The distribution of weed seeds in the soil profile at a given time varies with previous events, such as times and extents of seed fall in relation to times and types of tillage.

Ploughing moves many weed seeds from deeper levels to positions near the soil surface, which usually results in an increased emergence of weed seedlings.

To avoid an extensive emergence of weed seedlings in a following crop, such tillage should, when possible, not be carried out shortly before sowing. Instead, tillage stimulating weed seed germination should be carried out sufficiently early that the resulting seedlings can be killed by the last cultivation before sowing. This cultivation should be as light and shallow as possible so as not to translocate seeds from deeper to shallower soil layers. In this way, its stimulation of further weed seed germination should be minimized, but weed seedlings and shoots from perennial weeds should still be destroyed.

The crop should be sown immediately after this soil operation and sowing should be done in ways that facilitate rapid crop emergence. In competitive crops such as cereals, the time of crop emergence relative to the weeds strongly influences the growth of both the crop and the weeds (e.g. Fig. 30 and Tables 28–30). Moreover, a strengthened competition from the crop causes a strengthened weed control effect of applied herbicides (e.g. Table 27) and enlarges weakening effects of tillage (e.g. Tables 34 and 35).

In Scandinavia, spring sowing is frequently preceded by autumn ploughing. If this ploughing is not very late and, at the same time, the spring sowing is not extremely early, seedlings of both winter and summer annual weeds occur at the time of sowing. The spring cultivation can be very shallow, particularly in cases when the soil surface has been made reasonably even already in the autumn. In this way, the stimulation of further weed seed germination is minimized at the same time as the earlier, and therefore more competitive, weed seedlings are largely killed. Seeds and young seedlings of perennial weeds are, in principle, affected by tillage in the same way as those of annuals.

Early ploughing in relation to the time of autumn sowing makes it possible to kill more weed seedlings by the following seedbed preparation than does late ploughing. However, repeated tillage over long periods is not desirable with regard to soil conservation and environmental impacts, e.g. through leaching.

Tillage considering the growth rhythm of
Elymus repens

As *Elymus repens* is one of the most important weeds in humid temperate areas, this species is often particularly considered when soil tillage is performed in these areas. Tillage suitable for the control of *E. repens* is also very often an acceptable compromise in the management of many other weeds.

The following description of the response of *E. repens* to tillage is focused on the performance of plants in their annual cycles of vegetative development and growth. The response of other weeds will be briefly discussed in comparison.

As previously stated, the field operations in the normally shallow final seedbed preparation exert in themselves a comparatively weak effect on *E. repens*. However, when preceded by tillage (mostly in the autumn) that has brought about breakage and deep burial of the rhizomes and when, in addition, new shoots have alrady been growing for a period in early spring, even weak additional injuries to the weed will delay its growth in the crop. When this is a competitive crop, even a minor delay is important.

The curves in Fig. 60 describe, in principle, changes in the dry weight of regenerative structures of *E. repens* when growing in a spring-sown cereal crop and/or when subjected to different types of tillage. Figure 61 presents further principles in the effects of autumn tillage. The curves describe typical situations in Scandinavia and other parts of Northern Europe as evaluated from experiments with material planted in the field (Håkansson, 1967, 1968a,b, 1969a,c, 1974) and from other field experiments and observations (e.g. Korsmo, 1930; Åslander, 1934; Fail, 1956; Permin, 1961a,b, 1973a,b; Palmer and Sagar, 1963; Bylterud, 1965; Cussans, 1968; Gummesson and Svensson, 1973; Aamisepp, 1976; Werner and Rioux, 1977; Landström, 1983; Fogelfors and Boström, 1998; Lundkvist and Fogelfors, 1999).The

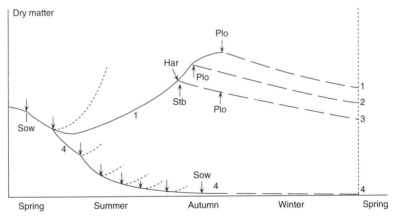

Fig. 60. Dry matter changes in viable regenerative structures (plagiotropic rhizomes plus vertical underground bases of aerial shoots) of *Elymus repens* in various field situations, typical of humid temperate climates such as in Southern and Central Sweden. Curve 1: changes during a period of 1 year, from early spring to the following early spring, as influenced by seedbed preparation connected with sowing of spring cereals (Sow), by competition in the cereal crop, by cereal harvest (Har) and by ploughing (Plo) in late autumn. Curve 2: changes after earlier ploughing. Curve 3: changes after harvest immediately followed by stubble cultivation (Stb) and by late ploughing. Curve 4: no spring crop sown; tillage (marked by arrows) repeated at optimum times on a black fallow followed by autumn sowing (Sow). Black fallows are rarely recommended today.

Broken line to the left: the dry matter increase expected if neither sowing nor soil operations are performed; note that the dry matter minimum in spring is passed later and at a lower level in the presence of a shading crop. Other broken lines at curve 4: increases expected if the soil operation in a given position is not performed.

dry-weight changes shown in Figs 60 and 61 are the net effects of consumption and supply of substances and of mortality.

Figure 60 presents a situation with moderate disturbance of the plants by shallow seedbed preparation immediately followed by sowing of cereals ('Sow') in early spring, carried out as in traditional field practice when preceded by autumn ploughing. When the plants of *E. repens* develop in a crop that produces an early shading, such as barley, they pass the stages of positive net photosynthesis and minimum dry weight of the underground structures later than in the absence of such a crop (compare the solid and broken lines to the left in Fig. 60: cf. Håkansson, 1969d; Skuterud, 1984). This also means that the development of new rhizomes is delayed and restricted through the shading. Parts of the older rhizome branches die, other parts survive, becoming more or less refilled with assimilates.

Cutting aerial shoots of *E. repens* at (cereal) crop harvest ('Har' in Fig. 60) will, in itself, only exert a weak effect on the rhizome growth. Old tall shoots emerged in spring are usually less active, unless harvest is very early, and considerable numbers of leaves of the younger and mostly shorter shoots remain almost intact after harvest at ordinary stubble heights. However, the time and rate of emergence of new aerial shoots in the later periods of the growing season vary between years and fields. In an average situation, there are active green shoots to such an extent that the improved light conditions after harvest immediately, or rather soon, cause a biomass increase in the rhizome system.

Curve 4 in Fig. 60 illustrates the gradual decrease in the dry matter of vital regenerative structures when soil operations have been repeated from spring to the time of sowing a winter crop. It corresponds to curve II in Fig. 59. However, tillage repeated during a shorter part of the year is more interesting than classic 'black fallow tillage'. The part of a growing season that is marked in Fig. 59 may represent spring to early summer – before the establishment of a late crop, sometimes for instance potatoes – or represent the period between harvest and winter.

Tillage between harvest and winter

Tillage after harvest attracts a special interest. In Northern Europe, the response of *Elymus repens* is often of major concern and is discussed below, focusing on *situations where drying is not a reliable method of controlling this weed*. Discussions will first concern *situations when autumn sowing is not planned*.

Curve 1 in Fig. 60 illustrates a typical dry-matter increase in the regenerative structures of *E. repens* from harvest (Har) to ploughing (Plo) in late autumn (Håkansson, 1974). There is no physiological dormancy in the vegetative structures, but low temperatures prevent intense regrowth after late ploughing. The dry-matter decrease during winter is largely caused by death of tissues, not least among older rhizomes, and by the resource-consuming respiration associated with early regrowth. These processes vary with winter conditions.

Curve 2 in Fig. 60 illustrates the effect when photosynthesis is halted by earlier ploughing. As there is no physiological dormancy, the regrowth of shoots can be extensive at higher temperatures. Resource-consuming regrowth and death of tissues can lower the dry weight of regenerative structures before winter.

Curve 3 in Fig. 60 describes the change in dry matter caused by stubble cultivation (Stb) carried out immediately after harvest, long before late autumn ploughing. Photosynthesis is immediately halted by the stubble cultivation. Breakage caused by this cultivation in combination with the increased depth caused by the following ploughing results in a weakening of the structures and a delayed emergence of many shoots in the following spring. This, in turn, causes a proportionally increased competitive suppression in cereals or similarly competitive crops sown in the spring (cf. Table 34).

The effect on *E. repens* caused by stubble cultivation before late ploughing has been evaluated on the basis of aerial shoot weights in the following summer in spring cereals – i.e. weights of shoots representing treatments with stubble cultivation plus late autumn ploughing as a percentage of the corresponding weights at treatments with late autumn ploughing being the sole measure. On average, the inclusion of stubble cultivation reduces the shoot weight of *E. repens* by about 50% in fields of Southern Sweden up to Uppsala. Judging from the results presented in Table 34, this reduction is a joint effect of mechanical weakening and crop competition.

In the coastal parts of Southern Sweden and in Denmark (Permin, 1961b), annual reductions exceeding 50% may be obtained by effective stubble cultivation plus ploughing when compared with ploughing as the sole soil operation in the autumn. In a series of experiments in Southern Sweden, *accumulated effects* of stubble cultivation to depths of 5 or 10 cm *over 6 years* led to a reduction in *E. repens* of 90 or 98%, respectively (Boström and Fogelfors, 1999). In six earlier field experiments in Southern Sweden (Gummesson, 1990), stubble cultivation

after harvest of cereals repeated over 12 years (and of various quality as in field practice) resulted in an average accumulated reduction in *E. repens* of 93%. In a series of field experiments further north in Sweden (about 64° N) where the period from harvest to winter is shorter, the average yearly reduction caused by stubble cultivation after harvest of barley and before autumn ploughing was slightly more than 20%. The average reduction accumulated as a result of this stubble cultivation repeated for 5 years was about 90% when compared with ploughing as the sole tillage in the autumn (Landström, 1983).

Figure 61 sums up the main points of short-term reduction of *E. repens* by stubble cultivation (Stb) carried out immediately after harvest of a cereal crop followed by late autumn ploughing (Plo), in comparison with late ploughing as the sole autumn tillage. Harvest times are 1, 2 or 3. A comparison of

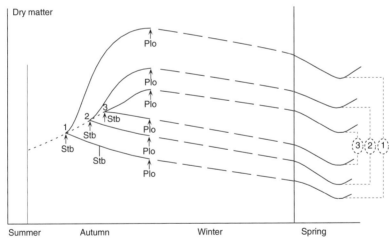

Fig. 61. Dry matter changes in viable regenerative structures of *Elymus repens* in various situations after harvest of a cereal crop (cf. Fig. 60). Harvest carried out at different times (1, 2 or 3) in relation to the following winter. The dry matter may have increased from time 1 to time 3. The three upper curves indicate that, after harvest, photosynthesis results in a dry matter increase if no disturbance (e.g. stubble cultivation) occurs until late ploughing (Plo). The increase caused by photosynthesis is greater after early harvest than after late. The three lower curves show different dry matter reductions expected as a result of stubble cultivation (Stb) immediately following harvest. This cultivation results in a greater reduction after early harvest than after late (if repeated, when required, after early harvest). Therefore, the difference in dry weights experienced in the following spring between treatments *with* and treatments *without* stubble cultivation is greater after early harvest than after late. (Regarding the difference: 1 > 2 > 3, both absolutely and relatively.) Analogously, there is both a risk of a greater increase and a potential for a greater reduction by tillage after harvest in places or in years with a long mild autumn than in situations with a cold autumn and/or an early winter.

corresponding upper and lower curves after harvest times 1, 2 and 3, respectively, shows that *stubble cultivation after an earlier harvest results in a stronger reduction in the regenerative structures, both absolutely and relatively, than cultivation after a later harvest.*

Thus, allowing *E. repens* to grow undisturbed until late ploughing is more disadvantageous after early than after late harvest, but there is, in other words, a greater potential for reducing the abundance of this plant by autumn tillage (or chemical control) after early harvest than after late. Analogously, a long mild autumn means a greater potential for effective control than a short autumn with early winter frosts.

The curves in Fig. 61 describe different situations with regard to the dry weight of regenerative structures of *E. repens* occurring in the spring after different autumn treatments. Regarding the growth from the weed in a crop established in that spring, the relative differences exhibited in the spring will either increase, remain unchanged or decrease, depending on the competitive conditions in that crop. Strong competition will thus suppress plants from originally weaker rhizomes (lower curves) proportionally more than plants from originally stronger rhizomes (upper curves).

Changes in the dry matter of regenerative structures (plagiotropic rhizomes plus vertical underground bases of aerial shoots) of *E. repens* with undisturbed growth in late summer and autumn were measured in planted material (Håkansson, 1967) and in material from ordinarily infested fields (Håkansson, 1974). There was a great variation with location and time of harvest, but on the whole a strong dry-matter increase from the time of cereal harvest to late autumn. It thus appears that there can be an increase of 100% or more after harvest in early August in Southern Sweden. This implies that management measures turning such an increase after harvest into a decrease are very important in keeping *E. repens* at low levels.

If ploughing is the only soil operation that can be carried out in the autumn, it should be done as early as possible (Fig. 60), unless an effective systemic foliage herbicide is applied in the stubble field. After such an application, the field should not be ploughed too early.

A combination of breakage and burial of the broken rhizomes of *E. repens* to greater depths is worth aiming for (Figs 55–57, Table 34). Even cultivation breaking shallow structures immediately before ploughing is justified when the period between harvest and winter is short. This is evident from experiments in Northern Sweden (Landström, 1983). Breaking by autumn cultivation after ploughing is much less effective, which is logical and also supported by experiments in Denmark (Permin, 1961b) and Sweden (Boström and Fogelfors, 1999).

As previously stressed, the average depth of the regenerative structures of *E. repens* can be increased by ploughing because the new rhizomes are produced at shallow depths. Ploughing with techniques inverting the topsoil layer completely (two-step ploughing) is effective on all shallow-growing perennials. Drastic effects of such techniques on *E. repens* were illustrated by Gummesson and Svensson (1973) in field experiments where different techniques of autumn ploughing were compared. Two-step ploughing reduced the shoot production of *E. repens* in the following year by 75–85% of that after ordinary ploughing, when this was done without a skim coulter or trashboard. However, two-step ploughing is expensive. Using a skim coulter or a trashboard in front of the mouldboard and/or ploughing with broad furrows have provided appropriate burial of *E. repens* and other weeds with shallow rhizomes. When preceded by tillage with breaking implements, ploughing with more conventional techniques has proven to be reasonably effective.

Tillage effects on *perennials with structures that become dormant in the later part of the growing season and/or have more deep-growing structures* were discussed on p. 183. Plants with either of these traits are less easily suppressed by autumn tillage than is *Elymus repens*. In Danish field experiments, effects on *Cirsium arvense* and *Sonchus arvensis* were studied beside those on *E. repens* (Permin, 1961b). Skim ploughing, harrowing or deep ploughing were carried out shortly after cereal harvest by mid-August and followed by ploughing in mid-November. These operations reduced the weed growth in the next year's crop when compared with the growth after ploughing in mid-November performed as the only soil operation in the autumn. The average percentage reduction, as measured in spring cereals in the subsequent year, was as follows:

	Skim plough	Harrow	Plough
Elymus repens	65	55	46
Sonchus arvensis	50	40	34
Cirsium arvense	33	24	29

In experiments reported by Boström and Fogelfors (1999), stubble cultivation after harvest (usually harrowing twice at intervals) before late ploughing (in November or early December) had uncertain effects on *Sonchus arvensis* and *Cirsium arvense*, whereas *Elymus repens* was strongly suppressed. In another series of field experiments (Boström *et al.*, 2000), stubble cultivation preceding late ploughing consistently reduced the regrowth of *Cirsium arvense*, although not strongly, whereas the effects on *Sonchus arvensis* varied inconsistently. In conventional field experiments, however, the effects on these weeds are hard to measure accurately due to patchy plant distribution.

In a series of field experiments in Southern Sweden (Gummesson, 1990), stubble cultivation following cereal harvest plus late ploughing repeated over 12 years reduced the shoot weight of *Cirsium arvense* by about 30% when compared with the shoot weight at ploughing as the sole autumn tillage. However, the abundance of *C. arvense* and most other weeds increased or decreased inconsistently over the 12 years even at stubble cultivation. At the same time, chemical control repeated annually caused a strong decrease in most dicotyledonous weeds, including *C. arvense*. The decrease in *C. arvense* caused by herbicides was obviously counteracted to some extent by stubble

cultivation, probably because breakage of the roots obstructed translocation, thus reducing the systemic effect of the herbicides.

As stated in Chapter 5 (see Tables 3 and 5), *stationary perennials* and *perennials with above-ground creeping stems (stolons)* as well as *rhizomatous perennials with low tolerance to tillage* are mostly restrained so much even by light tillage that they become weeds of minor importance in competitive annual crops. Autumn tillage designed to suppress perennials with stronger rhizomes on the whole suppresses these weeds sufficiently.

It could be expected that stubble cultivation followed by late autumn ploughing would gradually reduce the abundance of winter annual weeds in a cropping system in comparison with ploughing as the sole autumn tillage. This, however, has not been clearly confirmed in short-term field experiments, nor in experiments lasting over 12 years for studies of the accumulated effects of stubble cultivation performed annually (Gummesson, 1986, 1990). In five trials with similar annual treatments for 7 years, Boström *et al.* (2000) measured a slight reduction of the plant density of dicotyledonous annual weeds in response to stubble cultivation. The reduction was only about 10% when compared with densities after late ploughing as the only autumn tillage. The percentage reduction was almost the same in treatments with and without broad-spectrum herbicide control of weeds. Surprisingly, no distinct differences were found between summer and winter annual weeds.

Tillage between harvest and autumn sowing

In many areas, the period of time available for tillage is between harvest and sowing of a winter crop. In this period, tillage is sometimes excluded even where it is otherwise a regular element in the tillage programme. When practised, ploughing must often be done shortly after harvest. Regarding weed control, it is desirable that the time from the first tillage (irrespective of method) to the last before sowing is not too short. The first tillage provokes an increased seed germination and activation of vegetative buds of weeds. More seedlings and young shoots of weeds can be killed if the last tillage is carried out a few weeks after the first than if it has to be done much earlier. The final seedbed preparation, even though preferably shallow, should kill or weaken emerging and emerged weed seedlings and young shoots of perennial weeds. Any operation weakening perennial weeds before autumn sowing will reduce their growth in the crop, if this is competitive.

Even when it has to be carried out immediately before ploughing, shallow cultivation is worthwhile in order to break regenerative organs of *Elymus repens* and many other perennial weeds. This should be kept in mind when the time between harvest and autumn sowing is short. After early harvest, an interval between breaking and ploughing that allows for resource-consuming regrowth before ploughing is possible. At the higher temperatures usually following early harvest, rapid regrowth may result in a rapid consumption of food reserves. In many areas, however, this is prevented by drought. Leys for fodder are often followed by winter cereals in parts of Scandinavia. In certain areas, it has been common practice to break the ley in the height of summer. When this is done, very strong effects on *E. repens* have been experienced if breakage, ploughing and some additional shallow cultivation(s) have been carried out at intervals before the autumn sowing. Drought sometimes reduces the effects.

Harrowing and hoeing in growing crops

In growing crops, tillage may be performed either as the only measure of weed management or in combination with chemical control. In earlier times, mechanical measures such as harrowing or hoeing were used besides hand pulling as the only means of direct weed control in a growing crop. In attempts to limit the use of herbicides, mechanical control has attracted a renewed

interest and, in recent years, become the subject of much development work.

Soil operations injure the crop plants to a certain extent. On soils that have developed a surface crust or hardsetting of the surface layer, harrowing and hoeing may have a good immediate effect on the crop plants, which compensates somewhat for the injuries. Moreover, if the crop plants are not too severely injured and the surviving weed plants scarce or harmless, the weakened competition from weeds will overcompensate for the injuries to the crop. The final effect of harrowing and hoeing is largely regulated by the subsequent crop–weed competition.

The most striking effect of shallow tillage on growing plants is usually caused by their covering with soil, although uprooting may also be effective under dry conditions. Weed seedlings are, on the whole, more easily killed and/or more strongly suppressed the smaller they are and the smaller amounts of food reserves they contain. This means that small plants from small seeds are usually more severely hit than are larger plants developed from large seeds in corresponding stages of development. Seedlings are particularly sensitive in stages near the compensation point when their energy reserves are at the minimum. However, plants that are in early cotyledonous stages and have not yet reached their compensation point are often effectively killed or suppressed, because they become better covered with soil than larger plants. Having passed their compensation point, the plants become increasingly tolerant to tillage.

The discussions above mainly concern seedlings. Plants with vigorous vegetative structures representing the more important creeping perennials are less effectively suppressed by *a single* hoeing or harrowing. When there are vigorous rhizomes of *Elymus repens* or roots of *Cirsium arvense*, the soil is hard to treat effectively, not only by harrowing but also by inter- and intra-row hoeing in a crop stand. Rhizomes or regenerative roots of creeping perennials in the upper soil layer obstruct an accurate performance of hoeing and harrowing in a growing crop. In Gummesson's (1990) 12-year experiments previously dealt with, harrowing in cereal crops had an insignificant influence on these weeds and might even have favoured *C. arvense* by disturbing the crop stand, which also led to slightly decreased grain yields.

The weeds, naturally, have greater difficulties recovering after harrowing and hoeing in dry than in rainy weather. Soil cultivation followed by rain also triggers an increased emergence of additional weed plants. Light mineral soils, free of stones, offer the best possibility to achieve good effects on weeds without harm to the crop plants. On clays and loose organic soils, it is harder to achieve satisfactory effects. On clays, the weed plants are usually less effectively covered with soil and it is more difficult to avoid injuries to the crop plants. On organic soils, harrowing and hoeing induce a more extensive additional emergence of weeds than on mineral soils.

Weed harrowing in young crop stands can be carried out in many crops, e.g. cereals, oilseed rape, peas, field beans, flax and beets. In certain crops, careful shallow harrowing may be performed between sowing and emergence. It must then be done when the crop sprouts are still very short, but will, in spite of the early time, kill or suppress many weed seedlings, both among those from seeds germinated immediately after sowing and among plants that have recovered after seedbed preparation. Harrowing should thus be avoided in the period shortly before crop emergence and also some time after, until the crop plants have passed their compensation point.

Cereal plants usually pass their compensation point when they have one well developed leaf and the second is emerging or has just emerged (Fig. 4). In spring-sown cereals, harrowing can be done once or twice when the plants have two well developed leaves and the third is emerging. In autumn-sown crops, suitable times of harrowing vary strongly with climate and soil type. Under Swedish conditions, harrowing in autumn-sown cereals is mostly performed in spring when the cereal plants have just started a rapid growth. If autumn harrowing is possible, which depends on crop development and soil conditions, it is recommended mainly in order to kill or suppress the earliest weed plants, which will otherwise become particularly competitive.

On loose soils, rolling must always be done in association with the sowing so as to prevent the harrow from reaching too great a depth. Proper depths are around 1.5 cm, 2 cm at a maximum.

A description of modern weed harrows is published by Gunnarsson (1997). Formerly, light harrows with chisel tines were used (e.g. Korsmo, 1930, 1954). In recent years, much effort has been made to develop harrows suitable for selective work in different crops. Nowadays, long-finger harrows of various types are often used.

The effects of weed harrowing vary greatly. In a number of classic field experiments, weed reduction from 20 to 80% was a normal variation in spring cereals, depending on weed species, plant size, soil type and weather situation. The effect on the grain yield also varied greatly, depending on the original abundance and percentage reduction of the weeds and the degree of injuries to the crop. The yield often increased by 5–10%, and sometimes much more, but was reduced when crop injuries were not compensated for by sufficient reduction of the weed growth (Bolin, 1924; Drottij, 1929; Petersen, 1944; Sundelin and Gustafsson, 1946).

Habel (1954, 1957) and Koch (1964) studied the effect of shallow harrowing on seedlings from seeds of various sizes and in different stages of development. In brief, they found that the control effect increased with decreasing seed and plant size. They also observed that harrowing caused a somewhat increased additional germination and emergence of weeds. Different species responded differently, which was also found by Chancellor (1964, 1965). The selective effect of harrowing will vary considerably even at minor differences in the time of harrowing. Rasmussen (1991a) has developed a model predicting the outcome of weed harrowing as derived from the weed-killing effect and the damage to the crop.

Recent reports on experiments with modern equipment have, in principle, presented results and stressed problems similar to those described above. However, it seems sometimes possible to obtain slightly fewer injuries to the crops by modern equipment than by older (Rydberg, 1985, 1993, 1994, 1995; Wikander, 1988; Melander *et al.*, 1995; Rasmussen and Svenningsen, 1995; Mikkelsen, 1997; Rasmussen, 1998).

Weed hoeing between crop rows is traditionally carried out in row crops, i.e. crops grown with relatively large row spacings. Typical row crops in Northern Europe are sugar beet, potatoes and many vegetables with row spacings mostly between 40 and 70 cm. Maize is largely planted at row distances around 80 cm. In row crops, hoeing and herbicide application are nowadays often used in combinations that vary with crop, soil, climate and cropping system. In Sweden, oilseed *Brassica* crops were earlier grown as row crops, but after the introduction of suitable herbicides, row spacings around 12 cm are mostly practised nowadays, which facilitates the joint effect of competition and herbicide. The normal row spacing in small-grain cereals in humid temperate areas such as Scandinavia is around 12 cm, but even in such areas the spacing has sometimes been 15–20 cm, or more. With the aim of replacing the use of herbicides by inter-row hoeing, an increase in the row spacing to 20–30 cm has been tried in recent years, mainly in organic farming. In many arid areas, the row spacing in small-grain cereals is traditionally even larger.

In humid areas, the smaller row spacings normally practised in cereals result in higher grain yields and stronger competitive effects on weeds than do larger spacings (see Chapter 7). Within the range 10–20 cm, a row spacing increase of 1 cm results in a yield decrease of 0.3–0.8% (Strand, 1968; Bengtsson, 1972; Håkansson, 1975b, 1984b) and in an up to ten times as high biomass increase in annual weeds (e.g. Table 26; cf. Håkansson, 1975b, 1984b). Similar responses are found in crops such as oilseed rape (Ohlsson, 1975, 1976). The yield decrease caused by increased row spacing made to facilitate hoeing is more or less compensated for by positive effects of hoeing due to weed reduction and, sometimes, also to soil loosening.

As an average from 12 field experiments in spring wheat in Sweden (Heimer and Olrog, 1997), an increase of the row spacing from 12.5 to 25 cm resulted in a weed weight increase of 27% and a grain yield decrease of 3% when no weed control was accomplished. Inter-row hoeing at the 25-cm row spacing, either carried out once early or twice, led to a weed weight reduction of about 50% and a grain yield increase of 11% when compared with the situation at the 12.5-cm row spacing without hoeing or other weed control. Late hoeing, only once, at the 25-cm row spacing resulted in a weed decrease of 25% and a grain yield increase of 3%. According to I. Håkansson (personal communication), inter-row hoeing with modern equipment in a series of experiments in cereals has led to a stronger reduction in weeds and a higher yield increase at a row spacing of 12.5 cm than at spacings of 17 or 25 cm.

Hoeing effects on weeds and crops vary greatly with weed flora, crop character, row spacing, equipment, soil, weather conditions, etc. There is an extensive literature on this topic, not least on the effects of modern equipment for hoeing. The following publications provide further information: Knouwenhoven (1981, 1997), Mattsson *et al.* (1990), Rasmussen and Pedersen (1990), Heimer and Olrog (1997), Johansson (1997) and Terbøl and Petersen (1999).

Intra-row hoeing was formerly only done by hand in combination with hand pulling and, in certain crops, with thinning. In recent decades, together with the development of precision drilling, various mechanical and/or chemical means have been developed for the control of weeds within the rows.

Hoeing does not only influence weed populations by affecting growing plants. Like other tillage, it also stimulates further germination among weed seeds in the soil. Particularly in moist weather, this may result in an extensive emergence of new weed plants, counteracting the intended effect on weeds. In early-closing crop stands, these new weed plants as well as weed plants regrowing after sublethal injuries by the hoeing may be strongly suppressed by competition. In less competitive row crops, a great proportion of these plants will grow vigorously and produce considerable amounts of seeds in the absence of additional control measures.

Hoeing largely affects weed plants through covering with soil, but also through breakage and uprooting. The relative effects of these factors depend on plant size, depth and speed of hoeing, hoeing equipment, moisture content, type of soil, etc. The greater the speed and depth of the hoeing and the smaller the plants, the greater is the covering effect, both absolutely and in relation to the effects of breaking and uprooting. In a young crop stand, hoeing must be rather shallow and slow so as not to cover the crop plants with too much soil. At the same time, however, the weeds are smaller and therefore often sufficiently sensitive even to shallow low-speed hoeing. In an older stand, the hoeing can be deeper and done at a greater speed, and also with techniques moving soil into the row (ridging), in that way also covering smaller weed plants within the rows.

Reduced Tillage

Reasons for minimizing soil tillage were discussed in the introduction to this chapter. Reduction of the tillage in relation to the intensity in common practice in temperate humid areas has usually been based on the prerequisite that desirable weed control can be achieved by herbicides. This applies to zero tillage in particular. In temperate humid areas, moderate reduction of tillage with, for instance, permanent or periodical exclusion of ploughing are practised more often than zero tillage. However, even in combination with reduced tillage, the use of 'reduced' herbicide doses, has produced acceptable results.

It was stated in the foregoing that ordinary soil tillage is the most powerful regulator of the weed flora on arable land; the regulation of weeds is, in fact, often the most important objective of tillage. The selective effect of soil tillage on the plants is naturally modified by the time of the year and the type and intensity of the tillage. The occurrence of weeds with diverse life forms in different types of crop, as described in Chapter 5, represent cropping systems with regularly recurring 'conventional' tillage.

Elymus repens and *Cirsium arvense* are thus important perennial weeds in annual crops at conventional tillage because of their relative tolerance to tillage. Extensive experience indicates that they become much more problematic at reduced tillage unless special measures are introduced. In long-term field experiments in Scandinavia, the replacement of deep ploughing by shallower tillage without the inclusion of chemical control led to a strong increase in *E. repens* (e.g. T. Rydberg, 1992; Børresen and Njøs, 1994; I. Håkansson *et al.*, 1998).

Reduced tillage leads to increased *numbers* and changed *proportions* of weed species, some of which become increasingly problematic if not controlled by herbicides or in other ways (e.g. Cardina *et al.*, 1991; Clements *et al.*, 1994). Experiments indicate that grass species in particular, both annuals and perennials, increase (e.g. Bachthaler, 1974; Cussans, 1976; Thorup and Pinnerup, 1981; Nielsen and Pinnerup, 1982; Andersen, 1987; Légère *et al.*, 1990). In a series of 38 field experiments in Norway, Skuterud *et al.* (1996) found that winter annuals and perennials increased more than summer annuals, and monocotyledons (grasses) more than dicotyledons, when mouldboard ploughing was excluded. Among the perennials, *Elymus repens* and *Cirsium arvense* increased considerably, as did *Taraxacum officinale*, which is even more favoured by exclusion of ploughing than are the first mentioned species.

It is both expected and experienced that perennial plants sensitive to mechanical disturbance will increase in abundance at reduced tillage, unless prevented in different ways. Many perennials of minor importance in cereals at conventional tillage will become harmful (e.g. Buhler *et al.*, 1994). Broad experience in Scandinavia indicates that many stationary and other perennials that are suppressed by tillage but keep up as 'ley weeds' can produce vigorous growth even in competitive annual crops when not weakened by tillage. Examples are *Rumex crispus*, *Taraxacum officinale*, *Artemisia vulgaris* and *Agrostis stolonifera*. In 4–5-year experiments in England, stationary perennials such as *R. crispus* and *T. officinale* and the stoloniferous clover *Trifolium repens* increased, together with the more typical cereal weeds among the creeping perennials. These were in the first place *Elymus repens*, but also *Agrostis gigantea*, *Cirsium arvense* and *Sonchus arvensis* (Pollard and Cussans, 1976, 1981; Pollard *et al.*, 1982; Froud-Williams *et al.*, 1983a).

Some changes in weed populations experienced at reduced tillage may be expected on the basis of information about the weed occurrence in cereals in earlier times, when tillage was less effective than today.

Thus, not only is ploughing more effective today, by inverting the soil better and operating deeper, but so also is the more shallow cultivation. Equipment enabling increased weed suppression by shallow cultivation in seedbed preparation and in growing crops is available today. Furthermore, tine and disc harrows for stubble cultivation, etc. can operate deeper and break perennial weed structures in the upper soil layers more effectively than can older implements.

Improved sowing techniques have resulted in a more rapid and more even emergence of crops than earlier. The *joint* effects of more effective tillage and increased competition in crops such as cereals have exerted an increasingly selective pressure on the weeds. Together with changed competitive conditions caused by increased fertilization, this has resulted in a reduced number of species being important arable weeds (pp. 36–39; literature review

by Fogelfors, 1979). In recent decades, regular use of herbicides has gradually changed not only the general weed abundance, but also the proportions of species, biotypes and life forms of plants present as weeds on arable land.

Although fallows regularly alternated with cereals in permanent arable fields in earlier times, these fallows often favoured rather than disfavoured perennial weeds. Tillage on fallows was not only less intense but mostly also less effective because it did not start early in the growing season. Grazing, common on the fallows, allowed many perennial plants to build up vigorous regenerative structures. Therefore, stationary perennials such as *Taraxacum officinale*, *Rumex crispus*, *Rumex longifolius*, *Rumex obtusifolius*, *Bunias orientalis*, *Artemisia vulgaris*, *Silene vulgaris* and *Vicia cracca* were much more abundant cereal weeds than they are today. Among the stoloniferous perennials, *Ranunculus repens* was considerably more abundant in the cereal crops. The rhizomatous *Achillea millefolium* is seldom seen in cereal fields today, but was more often in older times.

Reduced tillage means very disparate things. Zero tillage (no tillage, no-till) with direct drilling is an extreme. The extensive literature on this subject reports on experiments with various forms of reduced tillage in very different situations.

On the basis of the knowledge obtained, we may first analyse a hypothetical situation where *deep ploughing is replaced by shallower cultivation*, nothing else being altered in the cropping system. This cultivation is assumed to destroy most young seedlings of both annuals and perennials occurring on the tillage occasions.

Absence of ploughing leads to a gradually increased concentration of weed seeds to shallow soil layers where biological activities – involving germination, predation and decay of seeds – are more intense than in deeper layers. The average longevity of the weed seeds will therefore be shorter than at regular ploughing, when the seeds are fairly evenly distributed down to a greater depth in the soil. However, seeds of diverse weeds will respond differently, leading to changes in their mortality and germination and the establishment of seedlings. Thus, not only the general amount of weed seeds producing seedlings will change when there is no ploughing, but also the proportions between species.

During the first years after abandoning ploughing, more weed seeds survive at greater depths than in shallow soil layers, providing only a few seeds are added to the surface soil as a result of weed control (cf. Roberts and Feast, 1972; Semb Tørresen, 1998). If the addition of new weed seeds to the soil is largely prevented by effective weed control, the weed seed bank could be expected to decrease more slowly in the first years without ploughing than at continued ploughing (e.g. Froud-Williams *et al.*, 1983b; Semb Tørresen, 1998). However, different experiments show contradictory results. Differences experienced depend on factors such as the depth distribution of the seeds at the start of the experiments, the degree to which the formation of new seeds is prevented by chemical control, the plant traits, and the soil and climate conditions (e.g. Cardina *et al.*, 1991; Martin and Felton, 1993; Regnier, 1995; Mulugeta and Stoltenberg, 1997; O'Donovan *et al.*, 1997; Vanhala and Pitkänen, 1998).

Légère and Samson (1999) discuss plant traits decisive for the long-term responses of weeds to changes in tillage and other measures and conditions in a cropping system. In addition to the plant life cycle and susceptibility to herbicides, factors such as size, dispersal, germination requirements and longevity of the seeds are stated to be important.

Zero tillage with direct drilling turns out to be impossible without chemical weed control in ordinary agricultural cropping systems of today. It was illustrated in Wisconsin experiments that the weed seed banks in zero-tillage systems are concentrated closer to the soil surface than in systems with shallow soil cultivation (Yenish *et al.*, 1992; Mulugeta and Stoltenberg, 1997). Germination, predation and decay of the seeds then become even more intense than at shallow cultivation

without ploughing. If plants from the germinated seeds and from perennating vegetative structures are not prevented from multiplying, a rapid increase in weeds of various types is both expected and experienced. This is illustrated by the following examples.

In a number of experiments in cereal fields in Norway lasting for a period of years, Semb Tørresen *et al.* (1999) showed that many annual weeds and the perennial *E. repens* (particularly important in the area) increased much more at 'no-tillage' than when tillage of various kinds was performed. After a period of years, the amount of weeds was thus highest at no-tillage and was gradually lower at, in order, spring harrowing, autumn harrowing, spring ploughing and autumn ploughing. Grain yields differed moderately. In experiments reported by Mulugeta and Stoltenberg (1997), the weed infestation increased as a whole at no-tillage without herbicides. However, some of the weeds did not increase even in the absence of chemical control, which illustrates that strong selective effects can always be expected. At zero tillage, the increase is thus more or less counteracted by increased death of seeds and seedlings. This death may be caused by drought, predators or phytophages and other factors selectively affecting the weeds. Brust (1994) and Cromar *et al.* (1999) describe selective predation by invertebrates living at, or near, the soil surface. Cromar *et al.* discuss efforts in *biological control* by using predators with specific feeding habits.

Direct drilling connected with zero tillage requires special treatments of crop and weed residues and shoots of growing weeds. Crop residues have to be removed or burnt, or chopped and spread uniformly on the soil surface in order to facilitate the establishment of a new crop. Herbicides are often applied between harvest and the subsequent drilling in order to kill growing weeds and living remainders of crop plants. Different plant residues may cause strongly differing physical and chemical conditions in the soil with selective effects on germination and seedling growth of weeds (e.g. Blum *et al.*, 1997). The amount and type of vegetation residues above and below the soil surface strongly influence the conditions for germination and plant establishment, not least in dry climates. Herbivores and other organisms attacking seeds or seedlings of weeds and crops are, at the same time, influenced very strongly (e.g. Liebl and Worsham, 1983; Liebl et al., 1992; Johnson et al., 1993; Cromar et al., 1999).

11

Chemical Weed Control as an Element in the Cropping System

Introductory Comments

Chemical weed control and impacts of herbicides in the environment are thoroughly treated in many textbooks. In this book, positive and negative effects are dealt with in broad terms. Chemical weed control is thus chiefly discussed as an element in agricultural cropping systems.

Modern herbicides, which consist of synthetic organic compounds, were introduced for weed control in the 1940s. From around 1950, their use increased dramatically. The new herbicides rapidly resulted in revolutionary changes in weed control strategies in industrialized countries. Chemical control sometimes reduced or changed the use of various mechanical measures and enabled or triggered other changes in the cropping systems.

An increasing number of herbicides with different physiological effects and selectivity were introduced over a few decades from the 1950s onwards. Herbicides with different selectivity allowed the control of progressively more weed species in an increasing number of crops. These herbicides, together with less selective types for use on areas free of crops, enabled a more flexible choice of crops and crop sequences without risking immediate weed problems. In the first half of the 20th century, crop rotations in Sweden almost always included perennial grass–clover leys, mostly with a

duration of 2 years or more. As mentioned previously, changes to one-sided cereal cropping became increasingly common after 1950 and, broadly, it was chemical weed control that made these changes possible.

The possibility of having a free choice of crop sequences is, of course, convenient for many reasons. However, the ability to control weeds sufficiently by herbicides in almost any crop sequence easily encourages practices that may be inappropriate in the long run. One-sided crop sequences in particular have, for example, often resulted in unwanted influences on the soil and/or increased attacks by plant parasites or pests.

Weed control by modern herbicides is effective beyond compare in a historical perspective. Strong immediate effects can be achieved by comparatively cheap and convenient actions. However, this has often caused an overuse of herbicides. Suitable alternative weed management measures and measures for optimizing the use of herbicides were widely neglected in the early era of 'modern chemical control'. Both the potential for, and the occurrence of, negative side-effects were also largely undervalued. However, conditions have improved. The sale and use of herbicides have thus become increasingly subject to Government regulations in the 'Western World'. Restrictions enforced in this way imply that side effects should be considered as carefully as the intended effects.

However, whereas intended effects can be strictly defined and therefore fairly easily measured and evaluated, this does not apply to unintended effects. Unforeseen side effects may thus escape detection for long periods of time until they become obvious and interpreted. Extended knowledge can thus never fully eliminate negative impacts, many of which may appear 'unexpectedly' only after some time of accumulation.

Attempts to optimize the use of herbicides should include environmental problems as well as concerns of immediate economic importance. Both intended and unintended effects of herbicides are therefore discussed in the following.

Intended Effects of Herbicides

Our most important herbicides are strongly selective in their effects on plants, which is the prerequisite for their use in crops. 'Non-selective' herbicides, i.e. compounds similarly toxic to wide groups of plants and solely used for vegetation control outside arable land, are not discussed here. Some of our selective herbicides are effective on wide groups of weeds, but have selectivity patterns enabling their use in specified crops. A common way of gaining a broad weed control in a crop is by the use of mixtures of herbicides.

Herbicides are mostly applied by spraying, either for uptake through the plant foliage, *foliar herbicides*, or for uptake via the soil, *soil herbicides*. Most herbicides are marketed in formulations that also contain other substances, facilitating their handling or increasing their effect on the target plants. The effect of foliar herbicides may be improved by substances increasing their retention, absorption, etc. The formulations are mostly sprayed as solutions or suspensions or, to some extent, as emulsions in the carrying liquid, normally water.

As herbicides are predominantly applied by spraying, the following focuses on this mode of application.

Some herbicides are used for both foliar and soil application, the mode of application depending on soil, climate and crop–weed combination. Large proportions of a herbicide sprayed for foliar uptake usually fall onto the ground, some herbicides then being active also via the soil. This applies to many strongly water-soluble herbicides, if they are at the same time only slightly, or moderately, adsorbed to soil colloids or bound in other ways in the soil. These herbicides may permeate the soil and be taken up by germinating seedlings, roots, etc. if the soil is sufficiently moist.

Uptake via the soil may lead to effects and selectivity patterns differing from those of uptake through the foliage. The joint effects of uptake by the two routes therefore differ with soil, climate and weather conditions, i.e. conditions influencing the proportions and amounts of a herbicide taken up through the foliage and via the soil.

Appropriate use of herbicides is always a complex question. This, of course, is particularly true when evaluated in relation to the use of other measures that could be considered in designing a cropping system. Although sometimes touching wider aspects, the following discussions mainly concern the use of defined herbicide types in defined cropping systems.

The importance of a uniform herbicide distribution

It may be intuitively suggested that a uniform application of a herbicide onto an area is superior to an uneven distribution. This is confirmed in a great number of experiments and can be simply illustrated on the basis of the form of the curve that describes a typical dose–response relationship as shown in the upper diagram of Fig. 62. (The curve often has a maximum at some low herbicide dose, indicating that plant growth is stimulated here, but this is not necessary to consider in the following discussions.)

In Fig. 62, d represents the uniform dose lowering the weed biomass to A_W, expressed as a percentage of the biomass of untreated weeds. Furthermore, d_1 and d_2 (the mean of which is d) stand for uniform doses in

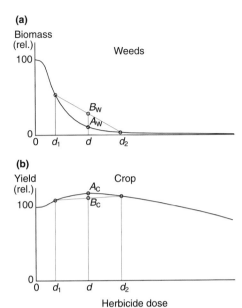

(a)

Biomass (rel.)

Weeds

100

B_W

A_W

0

0 d_1 d d_2

(b)

Yield (rel.)

Crop

A_C

100

B_C

0

0 d_1 d d_2

Herbicide dose

Fig. 62. (a) Typical shape of a curve describing the response (by weight) of weed plants to the dose of a herbicide. A_W denotes the biomass per unit area of weeds surviving dose d of a herbicide at uniform distribution. B_W denotes the average of biomasses after uniform herbicide application at doses d_1 and d_2, the mean of which is also d. (b) Yield of a crop (type: a competitive cereal crop) as influenced by a herbicide applied at different doses to a stand with a weed density considered comparatively high. From the curves in the two diagrams it can be understood that the weeds are much more sensitive to the herbicide than the crop plants are: A_C and B_C in (b) represent situations where the weed biomasses, according to (a), are A_W and B_W, respectively.

alternating equal-sized field plots in the form of rectangles or squares. Their mean weed biomass exceeds A_W, but may be lower than B_W, because edge effects in the plots reduce the deviation from A_W. However, with an increasing size of the alternating plots, which results in decreasing edge effects, their mean biomass approaches B_W.

The curve in the lower diagram of Fig. 62 describes the crop yield as a function of the dose of a uniformly applied herbicide. It is assumed to have a maximum, A_C, at dose d, which may be considered the optimum dose. The crop may be of a type developing competitive stands, e.g. a cereal crop.

The change in its yield with an increased herbicide dose is a result of the conflicting influences of a reduced weed competition and a more or less increased negative effect of the herbicide on the crop. Doses d_1 and d_2 are the same as those in the upper diagram. B_C is the mean of the yields at the uniform doses d_1 and d_2 in alternating large plots (cf. B_W for weeds).

The diagrams in Fig. 62 describe a situation where the herbicide dose that exerts a reasonable effect on the weeds is also tolerated fairly well by the crop. For uneven herbicide application at a mean dose of d, an increased phytotoxic effect in areas with higher doses reduces the yield somewhat. Thus, when the herbicide is applied at doses d_1 and d_2 in alternating plots, it results in a mean yield that is lower than A_C, i.e. approaches B_C when the size of the plots increases, at the same time as the mean weed biomass approaches B_W.

Gradual improvements of sprayers and spraying techniques have facilitated an increasingly even application of herbicides. In spite of that, there are still many adjustments to be made to improve herbicide application and enable reduced herbicide doses.

Joint effects of herbicides and competition in a crop stand

Plants that are weakened by some disturbance are less able to recover in competition with other plants than in the absence of competition. It was shown previously that perennial weeds that had been weakened by mechanical damage were less able to regrow in competition from a crop than without that competition (pp. 114–115; Tables 34, 35). Thus, competition did not only cause a general reduction in their regrowth, but also a *relatively* greater reduction among more strongly weakened plants than among those that had been less weakened. Analogous effects appear when weeds are weakened by herbicides, which has been shown in many experiments with chemical weed control in competitive crops (e.g. Pfeiffer and Holmes, 1961; Suomela and Paatela, 1962; Hammerton, 1970; Håkansson, 1975b; Christensen, 1994). Comparing herbicide effects in cereal species and varieties, Christensen (1994) found that specified weed control effects were achieved at lower doses in crop varieties that were more competitive than in those that were less competitive.

The results of the field experiment in spring barley shown in Table 27 thus indicate an increased herbicide effect with increased competition from the crop. All weeds were annuals. In plots representing ordinary stands of spring barley, the biomass of weeds surviving herbicide treatment was 2–4% of the weed biomass in untreated plots. In plots with large row spacings, this percentage was much higher. In unsown plots, where the weeds grew without competition from a crop, the biomass level was as high as 61% of that in untreated plots.

While a 'normal' herbicide dose was sublethal to many weed plants in the absence of crop competition, the great majority of them were killed when subjected to ordinary competition from the cereal crop.

Experience differs regarding the joint effects of seed rate and foliage-applied herbicides in cereals. Even though an increased seed rate enhances the strength of competition, this does not necessarily lead to an increased herbicide effect. This is understood from certain results shown by Håkansson (1979) and from recalculation of results presented by other authors (Erviö, 1983; Andersson, 1984; Hagsand, 1984; Salonen, 1992a). The probable reason why an increased seed rate sometimes fails to raise herbicide efficacy is that the increased crop leaf cover lowers the amount of herbicide reaching the weeds. A more rapidly developed leaf cover of more competitive cereal cultivars may be the reason why the herbicide efficacy was not greater in these cultivars than in less competitive ones in experiments presented by Lemerle *et al.* (1996) and Brain *et al.* (1999).

When a herbicide is sprayed onto the soil in a pre-emergence application, an increased seed rate (crop plant density) is likely to strengthen the joint effect

of herbicide and competition more consistently.

Efforts to increase crop competition with the aim of lowering herbicide doses without losing intended effects are only worthwhile in comparatively competitive crops, such as cereals and oilseed rape. Techniques facilitating an even plant distribution and a simultaneous emergence of the plants of these crops are preferable to a strongly increased seed rate when foliar herbicides are used.

When the 'herbicide effect' on weeds is measured in a crop stand, it is the joint effect of the herbicide and competition that is measured. Now, both plant competition and herbicides work selectively. When variation is found both in the overall effect and in the selectivity of a given herbicide, this is therefore due partly to variation in the competitive conditions in the crop–weed stand. Factors such as the density and plant traits, the relative time of emergence and age of the crop and weed plants on the occasion of herbicide application are important.

The effect of a specified herbicide application is strongly influenced by weather and soil factors. This is certainly not only due to direct influences of these factors, but also, to a considerable extent, to indirect effects through influences on the competitive conditions. The application technique, the type of formulation, the herbicide concentration in the spray liquid, etc. modify the effect of a herbicide in interaction with competition. Models developed with the aim of predicting the joint effects of competition and herbicide application on crop–weed stands (e.g. Christensen, 1994; Brain *et al.*, 1999; Kim *et al.*, 2002) meet with difficulties caused by the complexity of defined and undefined interactions.

A reduced herbicide dose – a proportionally less reduced effect

In efforts to minimize the use of synthetic chemicals in crop production, a great number of field experiments have been carried out to study the effect of herbicides at doses lower than recommended 'full' doses in specified situations. Results from Sweden and Finland (Fogelfors, 1990; Salonen, 1992a,b; Boström and Fogelfors, 2002a,b) indicate that, in competitive cereal crops, acceptable effects can *often* be obtained at much lower doses. This can be understood from the shape of typical dose–response curves (Figs 62 and 63).

Chemical weed control is performed with herbicide doses in ranges where the dose–response curve is convex (Fig. 63). The curves W_{SS} and W_{MS} in Fig. 63 represent strongly and moderately sensitive weeds, respectively. The 'full' dose (1.0) may give an effect of 97–98% by weight (survival 2–3%). When the dose is 'halved' (0.5), the survival may be 5–20% by weight, depending on the sensitivity of the plants. Considering consequences in the current crop alone, a survival of 20%, by weight, of moderately sensitive weeds (W_{MS}) is often acceptable. This effect can be obtained at lower doses on more sensitive weeds (W_{SS}).

Even when experiments show a satisfactory average effect of a low dose, there is a greater risk of a too weak effect under environmental conditions obstructing the herbicide efficacy. Unforeseen 'bad' weather more easily spoils the intended effect at a lower dose than at a higher.

Provided that the weed density is moderate and the cereal stand uniform and competitive, the yield reduction caused by weeds surviving to 20% may be acceptable (cf. Fig. 62, lower diagram). Thus, if the weeds neither lower the quality of the

harvest nor obstruct the harvesting procedure, a reduced control may very well be justified considering the present crop. In subsequent crops, however, seeds or other propagules from surviving weeds may cause greater problems. Predictable weed problems in future crops therefore should be considered in the degree of control attempted in the present crop.

Recurrent use of herbicides with similar selection patterns causes changes in the proportions of different weeds. Certain weeds are often resistant or less sensitive to all available herbicides suited for use in a given cropping system. They increase and often become problematic if changes in the choice of crops or other changes in the cropping system are not undertaken.

Thus, in cropping systems relying on chemical weed control, many weeds have become problematic for long periods owing to lack of herbicides useful for their control (e.g. Fryer and Chancellor, 1970; Aamisepp and Wallgren, 1979; Hume, 1987; Fogelfors, 1990; Ball, 1992; Boström and Fogelfors, 2002a,b). Grass weeds, both annuals and perennials, were among the weeds to increase in the first decades following 1950, when cropping systems became increasingly reliant on herbicides. However, herbicides controlling these weeds then gradually appeared (e.g. Fryer and Chancellor, 1970; Aamisepp and Wallgren, 1979; Mittnacht *et al.*, 1979; Rola, 1979; Haas and Streibig, 1982).

The proportion of less sensitive weed species will increase gradually when herbicides with similar selection patterns are used over long periods, particularly when they are used at lower doses. Selection within weed species leading to enhanced frequencies of herbicide-resistant types has attracted strong attention in recent decades and has been increasingly documented (see further, pp. 210–211).

The selectivity of a given herbicide may sometimes be altered considerably by changes in the formulation of the commercial product or by adding chemicals to the spray liquid, in this way changing the physical properties of the liquid. If, at the same time, such changes increase the general

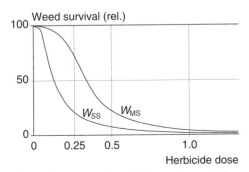

Fig. 63. Response of weeds (by weight) to the dose of a herbicide: W_{SS} and W_{MS} denote weeds that are strongly and moderately sensitive, respectively.

effect of the herbicide, they also permit, or require, lowered doses.

Cereals and oilseed rape, which are competitive because of their early stand closure, are crops where absence of weed control can sometimes be considered. Judging from dose–response relationships (Fig. 64), a lowering of the total amount of herbicides applied over a period of years should normally be done with a lower dose each year when these crops are grown, not by alternation of years without herbicide application and years with application of a 'full dose'.

This principle is supported by experimental results recently presented by Boström and Fogelfors (2002b) and presented graphically in Fig. 65. The results are based on experiments in fields with spring cereals lasting for 10 or 11 years. They indicate that a herbicide application every year at only 25% of the recommended 'full' dose exerted a greater average effect on annual weeds than application of a full dose every second year. A still better effect was achieved with a half dose. Treatment with a full dose every second year allowed a strong

seed set in years without herbicide application. The weed seed production during the experiment period was judged to be lower even when 25% of the full dose was applied every year. Varying responses of diverse weeds may be caused not only by different sensitivity to the herbicides, but also by different competitiveness, different seed longevity, different seasonality of seed maturation and germination, etc.

When the same herbicide is used at low doses over many years, herbicide resistance of kinds caused by many genes has been proven to develop (e.g. Gardner et al., 1998; see further, p. 210). Such resistance develops slowly 'creeping resistance'. Alternation of herbicides with different modes of action can reduce, or delay, that resistance. When, in addition, the alternated herbicides exert different selectivity, interspecific selection is also counteracted.

The response of *perennial weeds* producing vigorous, competitive plants from vegetative organs but weak or relatively few seedlings may differ from the typical response of annuals. This could be expected

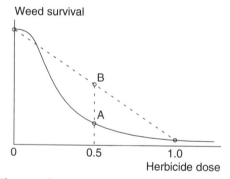

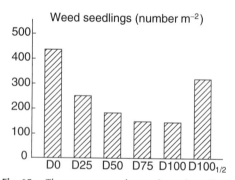

Fig. 64. The response of weeds (by weight) to a herbicide applied at different doses in a competitive crop, as described by the biomass of surviving plants per unit area. The positions of points A and B support the idea that lowering the total amount of herbicides applied over a period of years should, in principle, be effected by lowered doses each year rather than by alternation of years with a 'full' herbicide dose and years without herbicide application, even in competitive crops. A: Situation after application of a dose that is half (0.5) of the recommended full dose (1.0). B: Mean of situations without herbicide application (0) and with application of a full dose (1.0).

Fig. 65. The occurrence of annual weed seedlings following a period of 10 or 11 years with different postemergence applications of herbicides. Means of ten experiments in Southern and Central Sweden. Bars – based on data from Boström and Fogelfors (2002b) – denote seedling densities determined in spring cereals in the year following the last herbicide application. Seedlings counted about 1 week after the flush of weed emergence in the spring had ceased. D0, no herbicide; D25, D50, D75 and D100 denote 25, 50, 75 and 100% of 'full dose' applied each year; $D100_{1/2}$, full dose applied every second year alternating with years without herbicide application. (For further details, see text.)

on the condition that the regenerative structures of these perennials have been strongly weakened in the years with herbicide treatment. In dense crops such as cereals, they will then reproduce proportionally less strongly than typical annual weeds in a year when control is omitted. Effective control every second year alternating with years without control may therefore result in acceptable long-term effects. Data on *Cirsium arvense* and *Stachys palustris* presented by Boström and Fogelfors (2002b) support this suggestion.

Herbicide application techniques

This discussion focuses on application by spraying, when the herbicide is diluted as a solution, a suspension or an emulsion in water, mostly together with other compounds occurring in the commercial formulation or added to the spray liquid. Through nozzles on the spraying equipment, this liquid is atomized into droplets. The spraying technique has been gradually improved on the basis of research and comprehensive development work. The influences of droplet size, spraying pressure, direction of the droplet, etc. have been considered; at the same time, the evenness of the application has been improved. Optimum spraying technique differs with weather, type of crop–weed combination, physical character of the herbicide, the spray liquid, etc., as is illustrated by examples in the following.

The size of the droplets strongly influences both the overall effect and the selectivity of a herbicide. The optimum droplet size depends on many factors: temperature, air humidity, wind speed and others. Increasing temperature and decreasing air humidity raise the rate of evaporation from the droplets. Because larger droplets dry out more slowly than smaller ones, the optimum size of the droplets increases with increasing temperature and decreasing air humidity. A herbicide that reaches the shoots as dry dust or as droplets drying up rapidly produces a weaker effect than a herbicide applied with droplets that dry up more slowly. The

droplet size, the surface tension of the spray liquid and the physical character of the leaf surfaces interactively influence the herbicide quantity that adheres to the plant leaves and penetrates their surfaces. In this way, these factors strongly modify the herbicide effect. The optimum size of the spray droplets thus varies with the surface tension of the liquid and the character of the leaf surfaces.

There is still a great potential for increasing the efficacy of herbicides by improved application techniques, e.g. techniques for better regulation of the size and the size variation of the droplets and for a more uniform distribution of the spray liquid. The reduction of the number of droplets with a suboptimum size in the spray is an important goal. Suboptimally small droplets not only lower the intended effect of the spray, but are also extremely apt to drift (e.g. Arvidsson, 1997). Extensive efforts are in progress to improve the application technique with the aim of minimizing the risks of pollution outside the target area.

Compounds mixed with herbicides in commercial formulations or added to the spray liquid

A herbicide is mostly handled as an ingredient in a commercial formulation where one or several other compounds are intermixed to make it manageable or to improve its effect. With the latter purpose, additives may also be admixed in the spray liquid. An increased efficacy means that intended effects can be attained by reduced herbicide doses.

Compounds included in commercial formulations or added to spray liquids are of particular importance in modifying the effect of foliar herbicides. They are mostly used to increase the retention of a herbicide by leaf surfaces and the subsequent penetration of the herbicide through the surface layers. They thus increase the amount of herbicide reaching the living tissues of the plants. In some cases, compounds are added in order to improve the systemic effect of a herbicide by facilitating its long-distance translocation within the vascular bundles of the plants.

At the same time as the added compounds strengthen the intended herbicide effect, they mostly broaden the effect and may often also change the selectivity pattern. Hence, when combining herbicides with other compounds to increase the effect on weeds in a crop, the optimum dose may be lowered at the same time as the risk of unintended effects must be considered.

A special way of influencing the selectivity of a herbicide formulation is the addition of 'safeners'. These are substances lowering the effect on the crop without reducing the effect on more important target weeds. They are mainly used in the control of grass weeds such as *Avena fatua* and *Alopecurus myosuroides*.

Systemic and contact herbicides

Herbicides differ with respect to their movements within a plant after having penetrated the surface layers. Some are easily translocated to organs far from the sites of penetration. Others largely stay in the living cells close to those sites due to difficulties in moving through parenchymatous tissues. They therefore do not reach the conducting vascular tissues in which rapid long-distance translocation of substances mainly takes place. Foliar herbicides reaching vascular tissues are predominantly translocated in the phloem. Typical *systemic herbicides* are readily translocated, spreading to distant organs in the plant, whereas *contact herbicides* predominantly stay in tissues near the sites where they penetrated the plant surface.

There are no clear distinctions between systemic and contact herbicides. A typical contact herbicide is carfentrazon-ethyl. An extreme contact herbicide is dinoseb, one of the first 'modern' herbicides (now largely banned in weed control). Phenoxy compounds, glyphosate and sulphonylureas are classified as systemic herbicides.

Typical contact herbicides readily kill the cells of the plant tissues immediately below the surface where they entered the plant, in that way destroying the protection of the plant against drying. They therefore rapidly affect sensitive plants, the leaves of which quickly appear burnt. A rapid inhibition of the life activities of weeds is particularly valuable in breaking strong competition from fast growing *annual weeds*. When the aerial shoots of an annual plant are destroyed entirely by a contact herbicide, the plant is killed even though the herbicide has not reached its underground structures. *Perennial plants* can regenerate from structures that are not reached by the herbicide, but can be completely killed by a systemic herbicide translocated in lethal amounts into their regenerative organs.

The degree of long-distance translocation of a systemic herbicide in a plant depends on many conditions, both within and around the plant. Factors facilitating photosynthesis and translocation of photosynthates on the whole facilitate the translocation of foliage herbicides, as these largely move together with the assimilates in the phloem. However, movements also occur to some extent in the xylem. The growth stage of the plant is important. On the whole, a perennial plant is particularly susceptible to a foliar systemic herbicide in a period shortly after it has passed its compensation point. In that period, rapidly increasing assimilate streams from the leaves to other organs facilitate the distribution and effect of the herbicide. The foliage is then also larger and captures more of the herbicide spray than in earlier stages, at the same time as the development of new regenerative structures is still weak. It is known that glyphosate under favourable environmental conditions exerts a strong effect on *Elymus repens* during a long period following the initiation of new rhizomes in the plant.

The correct time to apply a herbicide to a crop is usually a compromise considering the development of both crop and weeds, the latter often emerging during an extended period of time. An early prevention of the growth of weeds is desirable to minimize their competitive effect on the crop.

However, a very early herbicide application often exerts a more harmful phytotoxic effect on the crop than a later application. Analogously, seedlings of weeds (of both annual and perennial species) are usually more easily killed by herbicides within a period rather soon after their emergence. After early application, however, more weed plants often emerge, but in competitive crops, these plants may produce weak growth. Late herbicide application may result in a weaker control effect, at the same time as the weeds have had more time to influence the crop by competition.

In a field where plants from vegetative structures of perennial weeds emerge in addition to weed seedlings, herbicide application can seldom be delayed until the optimum time of controlling them. If the perennials represent species that are very sensitive to the herbicide used, early application directed against weed seedlings may exert a satisfactory effect even on plants emerging from vegetative organs in a competitive crop – providing the herbicide is applied recurrently over a period of years. This applies to the control of *Cirsium arvense* in fields where the predominant annual crops are cereals treated with herbicides such as phenoxy compounds.

Patches of perennial weeds are sometimes treated separately at optimum times. A special case is the control of perennials such as *Elymus repens*, at present by glyphosate after crop harvest.

Pre-emergence soil application of herbicides

While foliage application of herbicides can be decided on the basis of visible weed plants, pre-emergence application has to be justified on the basis of expected weed problems. It may be done before sowing/planting or between sowing/planting and crop emergence. Non-selective, or weakly selective, herbicides can be used in the period between harvest and the next sowing/planting, provided they are rapidly decomposed in the soil without residuals causing harm to the subsequent crop.

Soil-applied herbicides mostly exert their strongest effect when taken up by the plants immediately after germination or the initiation of growth from vegetative structures.

There are two main reasons for incorporation of soil-applied herbicides into the soil by shallow tillage: (i) the herbicide can reach more weeds and reach them better right from their early stages of development. Drought can much more easily lower the effect when the herbicide remains at the soil surface than when it has been incorporated into the soil. In addition, some herbicides evaporate more rapidly on the soil surface, or are decomposed there by ultraviolet radiation. (ii) Tillage makes more seeds in the weed seed bank germinate rapidly and triggers more vegetative buds of perennial weeds to sprout, thus increasing the number of weed seeds and vegetative structures being eliminated by the herbicide. As a direct effect, tillage destroys or weakens plants that have already reached advanced stages in which they are less effectively controlled by the soil-applied herbicides.

Many pre-emergence herbicides are strongly adsorbed by humus or clay colloids. Those that are strongly adsorbed by humus colloids are ineffective at reasonable doses in organic soils. In these soils, either foliage-applied herbicides may be used or soil-applied ones that are only weakly adsorbed. Many herbicides that are very slightly dissociated into charged ions, e.g. triazines and ureas, are adsorbed actively by humus colloids. Strongly dissociated herbicides, whose active parts are negative ions, are only weakly adsorbed to humus colloids, because the net charge of the colloids is negative. When the active part of the herbicide is a positive ion, as in diquat, this is very strongly attached to soil colloids and even to coarser soil particles. Such a herbicide can never be used for soil application. Glyphosate is another example of compounds that are strongly bound to soil particles and therefore never used for soil application.

Certain contents of water in the soil are needed so as to allow sufficiently rapid movements of a soil-applied herbicide for it to exert an acceptable effect. A prerequisite for this effect is, at the the same time, that germination and growth of plants is not strongly restricted by drought. The strength and selectivity of the effect strongly depends on weather, soil properties and seasonality of germination and sprouting of the weeds.

The persistence of herbicides

To produce its intended effect, a herbicide must not be degraded too rapidly after its application. In other words, the persistence of the herbicide should be long enough for its effect to be exerted. It is desirable that the herbicide is then rapidly decomposed without leaving harmful residuals. Herbicides taken up by crop plants should also be broken down rapidly without toxic residuals. Their breakdown within plants is brought about by enzymes. In cases when weed tissues are rapidly destroyed by a herbicide, its degradation is largely, or partly, catalysed by enzymes from soil microbes. Most herbicides intentionally applied to the soil or having reached the ground after foliar application are predominantly decomposed by means of microbial enzymes.

A long persistence of a herbicide in the soil is undesirable for many reasons. The growing of sensitive crops is prevented for a considerable time. Slow decomposition is also undesirable because it prolongs the time when leaching (and/or evaporation) of the herbicide is risked. It thus increases the risk of accumulation at depths in the soil, or at other sites, where its decomposition proceeds even more slowly than in the topsoil of cultivated fields. There, microbial decomposition of herbicides is, on the whole, more intense than elsewhere.

Measurements of herbicide effects

Whenever possible, measurements of herbicide effects on plants or other organisms should be made in dose–response experiments. Effects of different herbicides can be satisfactorily compared only in such experiments. Modifications of the effect of a herbicide caused by the environment, the stage of plant development, etc. can be studied reliably only or mainly by comparing effects measured as functions of the dose.

Typical dose–response relations are described by curves of the type shown in Fig. 62 (upper diagram) and Figs 63 and 64. Various regression equations have been used to fit such curves to experimental data. The dose and/or the response values are usually transformed in different ways to facilitate calculations or graphic presentation. The use of logarithms, square roots, etc. of the dose is normal. Diagrams with log dose scales often make the skewed S-shaped curve symmetrical and require less space for convenient descriptions representing wide dose ranges. Probit analyses developed and described by Finney (1952, 1979) have been frequently used. Transformations of both the dose and the response scales enable linear regressions to be used there. The straight regression lines can then be re-transformed to sigmoid curves. Streibig (1984, 1988, 1992) has discussed statistical methods and developed equations for dose–response relationships.

The response of an organism to a chemical in a dose–response experiment is measured most reliably at doses around the dose reducing the biomass or the number of surviving organisms by 50%. Higher and lower doses are important for determinations of the slopes of the dose–response curves.

Dose–response curves may be calculated for many purposes – e.g. for comparing effects of different herbicides on different plants, effects of given herbicides in different formulations, effects on plants in different environments or in different stages of development. Curves that exhibit the same slope are in other words 'parallel', i.e. they do not intersect. Such curves describe effects that are conveniently characterized solely by the doses representing a reduction to 50%, by biomass (ED_{50}) or by number (LD_{50}) of living individuals of the organism studied.

Dose–response curves describing effects of different chemicals are sometimes not parallel. If they represent treatments carried out under the same environmental conditions, non-parallelism indicates different modes of action of the chemicals compared. Non-parallelism may require many kinds of experiments to allow reliable interpretations. It is too often caused by experimental weaknesses producing non-comparable measurements. In cases of non-parallelism, attention should always be paid to that risk. The slopes of curves obtained from different experiments often differ because of differences in the experimental conditions. When, for instance, the conditions are more heterogeneous in one experiment than in another, this causes a greater variation in the measured values and, as a consequence, less steeply sloping curves.

It has been observed that low doses of a herbicide sometimes have slightly positive influences on the growth of plants, at least *in early growth*. Dose–response curves then exhibit a *maximum* at some low dose level. This has often been experienced with herbicides of very different types in demonstration experiments for the author's students. (Various attempts to explain the phenomenon have been presented, but none of them seems to be generally adopted.) In order to reduce calculation problems in cases of positive low-dose effects, ED_{50} values can be calculated excluding responses at very low doses. This may be justified also because the low-dose values are mostly uncertain; whether they represent positive or negative influences.

The maximum of the crop yield curve shown in Fig. 62 (lower diagram) and discussed on p. 199 should not be confounded with maxima of the type discussed above. The yield maximum described in Fig. 62 falls at dose levels normal in weed control. However, these levels are low as regards the phytotoxic effect on the crop. Certain herbicides might here even produce a slightly positive effect on the crop at normal dose levels, but there are uncertainties and unclear interpretations about this so far. Negative effects are, on the contrary, often experienced, but when the weed abundance is not too low, negative effects should be overshadowed by the positive effect of decreased influences of the weeds.

Unintended Effects of Herbicides

General

Side effects of herbicides are being increasingly considered in screening tests of potential herbicides before they can be introduced as new compounds for weed control. The urgency of such tests has become increasingly clear from experiences of side effects of previously established herbicides. As stressed above, however, chemical compounds may produce important negative effects of types that are still unknown and therefore unexpected. Multi-factorial tests and observations are thus always a necessity in order to discover such effects. Methods of measuring and evaluating toxic and other unwanted side effects of agrochemicals are treated in specialist literature. A comprehensive, gradually updated literature dealing with the fate of herbicides in soils, plants and other organisms, and associated environmental risks is available at present.

Influences of herbicides on plants and other organisms besides the target plants should, of course, always be minimized. This has relevance not only to crops and other plants, but to higher animals and humans; toxic effects on microbes and other small organisms in soil and water should be avoided as their presence in the ecosystems is extremely important. Where microbes and invertebrates in the soil have been injured, it seems that their populations in most cases recover relatively rapidly, but even when so, unfavourable intraspecific and interspecific selection may have occurred without being understood or detected so far.

Acute toxicity to higher animals, including humans, is more easily detected. The use of herbicides with such acute toxicity is therefore largely prohibited today or limited to strictly specified situations. Classic examples are two dinitrophenols, DNOC and dinoseb, which disturb general respiration processes and are therefore harmful to many kinds of organisms even when moderate doses are used in weed control.

Chronic toxicity of herbicides, resulting in slowly developing symptoms, is mostly less easily detected. The individual factors underlying, for instance, a teratogenic effect are often hard to detect among many possible and often interacting factors. When the main factor causing such an effect is a specific chemical, its role behind the effect is therefore easily overlooked for long periods. Observations over longer periods are hence required to detect and evaluate such toxicity, causing, for instance, teratogenic, carcinogenic or mutagenic effects.

Toxicity

A chemical that has proved toxic to a great variety of organisms should either be prohibited as a marketed herbicide or be cautiously handled and only used in well defined situations. If such a chemical has a long persistence, i.e. is very slowly decomposed in soil, water or living organisms, it should never be used as a pesticide (cf. discussion on hydrophilic and lipophilic substances, p. 209).

It goes without saying that compounds solely acting on physiological mechanisms of higher plants are preferable to compounds disturbing vital processes common to wide groups of organisms. For example, inhibition of photosynthesis is the main mode of action of many herbicides, e.g. triazines and ureas. These therefore appear to exert only weak effects on organisms other than plants. However, possible influences of herbicides besides their main effect on plants must always be watched for, e.g. carcinogenic and teratogenic influences on animals; and even physical problems with compounds accumulating in places where they are broken down slowly.

Persistence – movements

Herbicides such as triazines have a long persistence in the soil and may therefore prevent the growing of sensitive crops for more than 1 year. On the other hand, they move comparatively slowly in the soil owing to adsorption to soil colloids, which counteracts strong leaching to ground water and open waters. According to many measurements, however, frequent use of these herbicides leads to their extensive leaching and also accumulation in the groundwater in low-lying terrain.

Degradation of herbicides may occur in three ways: photochemically, chemically or enzymatically. Most herbicides are decomposed predominantly by means of enzymes. In the soil, these enzymes are mainly produced by microorganisms. As biological activity is limited at greater depths in the soil, even compounds that decompose in a few weeks in the topsoil may accumulate over years after leaching to greater depths. In groundwater and wells they will then persist for years. Compounds that are weakly adsorbed, or not strongly bound to soil particles in other ways, leach more easily.

Diquat and paraquat are examples of compounds that are less disposed to leaching although they are very water soluble. The reason is that their active parts, being positive ions, are very strongly bound to particles in soils of any type. Fortunately, they usually exhibit a short persistence.

In humid areas with predominantly downward water movements in the soils, there is always a fairly pronounced risk of leaching of chemicals to the groundwater. The tendency of a chemical to leaching may depend more on its disposition to becoming bound to soil particles than to its degree of water solubility, unless it is insoluble due, for instance, to a strongly lipophilic character. The reason for this is that the quantities of water in the soil are usually sufficient for dissolving even weakly soluble (hydrophilic) herbicides, the amounts of which are normally very small in relation to the quantity of water percolating through the soil in humid areas.

In plant or animal tissues, different herbicides are broken down at different rates by enzymes. Of those that break down slowly in crop plant tissues, residuals may remain in harvested products.

Compounds displaying a long persistence in soil and a low tendency to adsorption or to other bonds in the soil may be very harmful in the landscape as they are easily leached into the groundwater or spread by surface water on the soil. Picloram is a classic example of a compound uniting these properties. Because it is, in addition, toxic to very different plants and to certain other organisms, it can easily cause harm in lower-lying areas of broken ground, in open waters, etc. Comprehensive studies have shown that the compound is not only very persistent in soils but also in plants, including harvested products (Ebbersten, 1972). Its use as a herbicide was therefore very soon prohibited in many countries. Chlorsulfuron is a 'low-dose' herbicide belonging to the first generation of sulphonylureas. It has caused problems because it moves easily in most soils and has a persistence long enough to cause harmful effects on sensitive crops grown in the year following its application.

Strongly volatile compounds may cause problems. In the form of gas, these herbicides may reach sensitive plants and other organisms at sites near the target areas. Examples of volatile compounds are found among esters and carbamates. To reduce adverse effects and loss of intended effects due to evaporation, volatile soil-applied herbicides should be incorporated into the soil by shallow cultivation immediately after their application.

Harm of pesticide pollution

A very adverse combination of properties of a pesticide is a long persistence and a lipoid character. Some of the early synthetic pesticides are extremely persistent lipoids. A well-known example is DDT, which is lipoid and also persistent over several decades. Lipoid compounds dissolve in fats and other lipids; they therefore attach to such substances. If they are at the same time highly persistent in living tissues, they may accumulate to injurious concentrations in animal food chains even when their acute

toxicity to higher animals is weak, as measured by LD_{50} values.

Most herbicides are hydrophilic, although sometimes with a very low solubility in water. Esters of phenoxy acids, e.g. 2,4-D esters, are lipophilic, whereas corresponding salts are hydrophilic. However, after having entered plant or animal tissues, these esters are rapidly split into hydrophilic components. It seems that none of our herbicides persist as unbroken lipoid compounds over long periods, either in living organisms or in the abiotic parts of the soil.

Herbicide drift due to dispersal of gases of volatile substances in the air was mentioned above, as well as the drift of droplets from sprays. Drift of small droplets is in general the most important way of herbicide dispersal by air, causing polluting deposits outside target areas, particularly at sites in the near surroundings. Beside the amount of the herbicide deposited, the properties of the herbicide and the character of the site determine the influences of the pollution on plants and other organisms. Not only in groundwater, but also in open waters, herbicides persist for longer periods than in the upper layers of cultivated soils. One way of reducing direct pollution of open waters is to let strips of fields remain untreated with herbicides along ditches or streams and other open waters. Untreated field margins are also highly justified as refuges for weed plants, thereby providing 'useful' insects, birds, etc. with feed.

An extensive literature based on research focuses on unwanted drifts of herbicides and ways of minimizing this drift. The following examples may be given: Elliot and Wilson (1983) and Miller (1993). Comprehensive experiments were recently conducted by Arvidsson (1997), who critically discusses measures and interpretations of measurements, while at the same time presenting an extensive literature survey.

When discussing herbicide residuals in the environment, possible harm caused by degradation products should not be overlooked. Many intermediary products of decomposition are biologically active. It is important to know their mode of action in plants and other organisms, as well as their persistence and possible accumulation at various sites. Although present knowledge indicates that at least the phytotoxic effects of degradation products are weaker than those of the unbroken herbicides, more knowledge about this is desirable.

Effects of chemicals added to the herbicides when commercial products are formulated are of interest not only with regard to improvements of the intended herbicide effects but also to possible side effects. Conceivable biological activities of these chemicals and their decomposition products should be watched.

Problems in handling herbicides

Problems in handling commercial products for chemical control of weeds and other organisms are being increasingly considered. Improvements have been made regarding their formulation and packing so as to minimize poisoning of people handling them and to reduce the contamination of soil and water. An example of improvements is the increasing use of tablets and granules, which are more convenient and less risky to handle than liquids or dusts.

Sprayers should not be filled or cleaned in places where wells, ditches, other open waters or the ground water can be easily contaminated. It has been shown that contamination of groundwater with herbicides and other pesticides to considerable extents originates from sites where sprayers have been repeatedly filled and cleaned. Filling and cleaning of sprayers often take place in farmyard areas where the original surface soil has been replaced by layers of gravel or sand. Spilled pesticides are easily washed down through such layers.

Filling and cleaning spraying equipment may, for instance, be done on watertight concrete surfaces connected to a cistern for rinsing water. At convenient times, this water can be spread onto the field. Instead of leaching from concentrated spillage, the pesticide then becomes diluted in the field topsoil, where it can break down normally without strong leaching. Where watertight

manure pits with liquid slurry tanks are accessible and suitable for the purpose, filling and cleaning can preferably be done there and the pesticides can be spread with the manure. It goes without saying that these views imply normal amounts of spillage, whereas dumping of large amounts must never occur.

Special arrangements for filling and cleaning spraying equipment are 'biobeds' proposed and tested by Torstensson and Castillo (1997). Biobeds should have room for the entire equipment used at spraying. These beds are about 60-cm-deep layers of fertile topsoil mixed with peat and straw. Spilled pesticides are largely captured and rapidly broken down there through microbial activities, but the bed material must be changed at least every 5–8 years (Torstensson, 2000).

Resistance in response to recurrent use of similar herbicides

Repeated use of the same herbicide, or of herbicides with the same type of physiological action and the same selectivity pattern, easily leads to unfavourable *interspecific selection* among the weed species present. *Intraspecific selection* leading to increased proportions of genotypes resistant to the herbicide(s) in question is a particularly difficult problem that has to be taken very seriously. Pronounced effects of intraspecific selection towards resistance were first observed in fields with monocropping of maize that had been repeatedly treated with triazines. Since those observations, an increasing number of weed species with enhanced resistance to specified herbicides have been described (e.g. LeBaron and Gressel, 1982; Powles and Holtum, 1994; De Prado *et al.*, 1997).

Gardner *et al.* (1998) have analysed the selection for different types of resistance caused by recurrent use of the same herbicide. It seems that successive application of high doses may rather rapidly select for resistance based on one gene. Repeated use of low (near lethal) herbicide doses may, instead, cause a more slowly extending resistance in a weed population, a 'creeping resistance', which is based on the interaction of many genes. Variation of herbicides over time and/or temporary insertion of increased doses will delay the development of creeping resistance.

Although much progress has been made concerning our knowledge of the selection for herbicide resistance, many questions remain to be answered. However, based on the present knowledge it can be said that all reasonable possibilities of alternating herbicides with different modes of action are desirable in attempts to maintain low herbicide doses in cropping systems relying on chemical weed control. Access to herbicides affecting different physiological mechanisms in the plant is therefore important. Besides the concern at general environmental risks, this aspect should be considered when introduction and withdrawal of commercial herbicides are contemplated.

Herbicide and Tillage Problems in Efforts to Develop Sustainable Cropping Systems

Due to the increasing awareness of negative side effects of chemicals used in crop protection, a great number of compounds used as herbicides have been gradually removed from the market. In combination with improved formulations of commercial products and a more careful handling of them, this has contributed towards reducing the risks of negative effects of chemical weed control. However, at the same time as certain herbicides have been discontinued, new ones have been added.

A great number of commercial products containing the same herbicides are both unneeded and unwanted. As stressed above, however, it is desirable to have access to a variety of 'low-risk' herbicides with different physiological activities so as to minimize intraspecific selection for herbicide resistance among weeds. This applies to herbicides with similar intraspecific selectivity patterns as well as to herbicides that have

different selectivity on the species level. Access to both types of herbicides provides flexibility in choosing crops and cropping systems considering the environment and the use of resources.

It has been repeatedly emphasized that intensive soil tillage should be avoided as far as possible. Therefore, *efforts to decrease the use of herbicides and, at the same time, reduce soil tillage are conflicting*, a problem that is discussed in the following.

A rational and responsible use of herbicides must satisfy many demands. Only compounds with a 'benign ecotoxicological profile', i.e. without evident or reasonably suspected environmental harms, should be used. Doses should be minimized on the basis of good knowledge of factors affecting the effects at different doses. At the same time, chemical control should be compared with alternative measures with respect to labour and other immediate costs, energy consumption and environmental risks. The comparisons certainly turn out very differently in different parts of the world.

Manufacturing and handling of herbicides before their field application should not be overlooked when determining the energy consumption, pollution, etc. associated with their use. Herbicide use should be energy-efficient in itself, but may be particularly justified when it replaces weed control based on intensive soil tillage, as this is not only strongly energy-consuming, but also adverse to the soil in various ways. Intensive tillage can be disastrous not only on uneven or hilly ground, but also on shallowly sloping ground in warm climates with high precipitation. Here, even moderate tillage too often triggers rapid gully erosion and/or soil degradation of other types.

Particularly after early harvest, it is necessary to stop the growth of weeds such as *Elymus repens* as soon as possible (Figs 60 and 61). In Sweden as in many other countries, the question then is whether this should be done with a herbicide (glyphosate at present) or by tillage, i.e. by stubble cultivation or early ploughing. Stenberg *et al.* (1999) have shown that tillage in late summer or early autumn causes a much greater nitrate leaching than does tillage (ploughing) delayed to late autumn or spring. No important environmental harm seems to be associated with the use of herbicides such as glyphosate. Considering weed effects, consumption of mineral oil, loss of plant-available nitrogen, estimated environmental harm by the leaching, etc., the choice is usually in favour of chemical control as the first measure following harvest. As well adapted glyphosate application exerts a considerably stronger effect on *E. repens* than stubble cultivation, it can be excluded more often than stubble cultivation in a period of years. However, glyphosate is at present the only very suitable herbicide for *E. repens* control. Its recurrent use may therefore be on the way towards selecting for resistance. Alternation with new herbicides may be the solution to such a problem.

Soil compaction caused by field trafficking with heavy machines is an increasing problem for almost any type of soil. Subsoil compaction is particularly serious because it is largely irreversible and therefore increases over time (e.g. Håkansson, I. and Reeder, 1994). This compaction is an effect of high ground pressure from the wheels of many kinds of machinery. Deep, irreversible compaction of soils is mainly brought about by the wheels with high axle loads. Machinery producing high ground pressures is used not only for soil tillage, but also for other field activities, e.g. when heavy equipment is used for the application of pesticides.

Frequent passes with machinery that compacts the subsoil is strongly in conflict with attempts to maintain a sustainable crop production. It is thus an urgent task to develop arable field equipment exerting a low ground pressure. Although the possibilities are often limited, any potential of such development should be attempted. *Field operation logistics deserve increased consideration.* Although the total cover of wheel tracks in a field is normally much lower at spraying than at ploughing and other intensive tillage, chemical control activities should not be overlooked in this context.

Cropping systems where ploughing is excluded cause more weed seeds to remain near the soil surface than systems with ploughing recurring regularly (cf. pp. 193–196). These seeds have a shorter average longevity than seeds distributed over greater depth ranges, because of a greater rate of germination, a greater exposure to predation and microbial attacks, a greater degree of damage from drought, etc.

In the absence of ploughing, the relatively extensive germination and rapid establishment of weed plants from shallow depths enforce effective control of these plants in young stages. Having established early from shallow depths, these weeds otherwise can become very vigorous, competitive and reproductive. They must therefore be restrained to avoid weed population increases. Periods with repeated shallow tillage may both speed up germination and kill the seedlings, in that way reducing weed populations, including the weed seed banks. However, uninterruptedly recurring tillage over periods of years is seldom practicable or advisable. In attempts to minimize even shallow soil tillage, chemical weed control has become the solution. This is justified if the compounds and their mode of use can stand thorough examination concerning environmental impacts.

Perennial crops with dense stands, such as grass–clover leys, prevent the establishment and reproduction of annual weeds by competition more effectively than most annual crops. However, *in cropping systems where soil tillage is regularly undertaken before sowing/planting and where weeds are sufficiently controlled in the annual crops, populations of both annual weeds and most perennials can be more rapidly reduced in years with annual crops than in years with perennial leys.* The reason is that the weed seed banks are then more rapidly diminished in years with tillage than in the absence of tillage during ley years.

In a crop sequence without chemical weed control, annual weeds are suppressed better in years with well-managed grass–clover leys than in years with cereals and other annual crops, if mechanical weed control is not performed there. This is a fact strikingly experienced in organic farming.

In annual crops such as cereals, increased competition can be achieved by techniques facilitating a rapid closure of the stands (see Chapter 7). In perennial grass–clover leys for cutting, good closure is attained by adapted management. Normal mowing in these leys restrains certain tall-growing weeds – e.g. *Cirsium arvense* and *Sonchus arvensis* – more than the cultivated ley plants. Extensive experience concurs in the opinion that the rhizomatous grass *Elymus repens* endures mowing in the leys better than *C. arvense* and *S. arvensis*. These species normally decline with increasing age of a ley, whereas *E. repens* persists with an unchanged abundance in an average ley. At the same time, however, *E. repens* is less well adapted to mowing than tufted grasses such as *Agrostis capillaris* and *Deschampsia cespitosa* (Tables 5 and 8).

Provided tillage recurs regularly in years with annual crops, stationary perennial weeds are, on the whole, suppressed more successfully than annual weeds in the absence of chemical control. In the absence of both ploughing and chemical control, however, some stationary perennials, e.g. *Taraxacum officinale* and *Rumex* spp., are not always sufficiently suppressed by shallow tillage alone. This applies in particular where sequences of annual crops alternate with perennial leys, allowing the stationary perennials to produce vigorous weed growth (Tables 3, 5 and 8). However, the breakage of a ley mostly requires either ploughing or treatment with broad-spectrum herbicides to allow the preparation of a good seedbed for the following crop.

Without both chemical control and ploughing, most perennial weeds with vigorous rhizomes or creeping roots are insufficiently controlled by shallow tillage. In climatic areas where *Elymus repens* is an invasive weed, 'rescuing' chemical control of this species 'at need' is justified so as to avoid intensive tillage.

The introduction of crops grown for purposes other than traditional production of food and forage will increase the possibilities to design cropping systems where weeds can be reasonably suppressed with a low intensity of mechanical and/or chemical weed control.

A decreased weed control level from, say, 95–98% to 80–85% is usually possible in competitive crops such as cereals and, as illustrated above, this enables lowered herbicide doses and/or reduces the need for intensive soil tillage. However, weeds lowering the quality of harvested products and weeds causing technical problems at harvesting, etc. have to be controlled more strongly.

The use of 'low-dose' herbicides is not necessarily sufficient to meeting requests for reducing the use of herbicides. Thus, a 'low-dose' herbicide is not unconditionally the same thing as a 'low-risk' herbicide. However, a compound acting effectively at a dose of 5–10 g ha^{-1} has some indisputable advantages over a compound that has to be applied with a dose of 1–2 kg ha^{-1}; it is easier to handle, to store and to transport without risks of adverse contacts and pollution. This makes it preferable if it, at the same time, has no adverse environmental impacts in comparison with an alternative 'high-dose' herbicide with a similar intended effect.

12

Special Management Measures

Repeated use of the same type of measure in crop production, together with repeated growing of the same crop or type of crop, sooner or later results in adverse 'monotony' or 'overuse' effects. This mostly causes unmanageable increases in certain organisms – e.g. certain weeds, herbivorous insects, parasitic fungi, etc. In many areas, cropping systems involving intensive soil tillage rapidly cause disastrous soil erosion; in other areas, such systems cause a more slowly progressive deterioration of the soil.

Where repeated use of the same herbicide dominates in direct weed control, selection in the weed populations usually leads to problems. Some weeds may increase and cause problems, often even when they represent very few species. Such a problem usually arises and becomes apparent within a few years. However, there are risks of more dangerous or creeping effects caused by selection within the weed species, which result in increased frequencies of genotypes resistant to the herbicide used (cf. p. 210). Alternation of herbicides with different modes of action is of great interest; and even more so are extensive changes in the entire cropping system by variation of crops and/or changes in the crop sequences, allowing for more all-round combinations of weed control measures.

In Chapter 10, which deals with tillage, methods and consequences of reduced tillage were discussed. Discussions there focus on principles for the guidance of rational and 'free' combinations of measures suitable in different situations. Chapter 11 takes up opportunities and consequences of reducing the amounts of herbicides, for instance by lowering herbicide doses. From previous statements in this book, it can be understood that systematic combinations of different measures can replace recurrent use of monotonous, intensive measures in weed control. Various factors, some of them causing only slight effects themselves, can cause very strong interactive effects.

Various Methods of Mechanical or Physical Weed Control

A number of special weed control measures that have not been explicitly mentioned previously in this book are briefly presented below. They represent alternatives or special measures mostly used in combination with other measures, not least in horticultural crops. Liebman *et al.* (2001) present informative overviews and literature reviews concerning such measures.

Brush weeding and the way brush weeders work are discussed by Melander (1997) and Fogelberg and Dock Gustavsson (1999) in papers where they present results of recent brush weeding experiments. Brush weeders are used to control weeds in seedling stages. They affect plants selectively by varying degrees of uprooting and/or soil covering. The selectivity depends on plant anatomy, plant size (species, stage of development), soil type and weather. The use and further development of brush weeding is of interest primarily for intra-row control of weeds in, for instance, carrots, the young plants of which withstand the treatment better than many weed seedlings.

Soil cultivation in darkness has attracted attention as it triggers fewer weed seeds to germinate than cultivation in daylight. This fact has been studied and described by many authors in recent years (e.g. Ascard, 1994; Jensen, 1995; Gallagher and Cardina, 1998). It is often desirable that the final seedbed preparation immediately associated with sowing induces as few weed seeds as possible to germinate in the following crop. However, in situations where direct broad-spectrum weed control is to be performed in the crop, it may be better to bring about a maximal reduction of the weed seed bank by stimulating weed seed germination as much as possible, even during the final seedbed preparation. The difference in germination brought about by cultivation in light and darkness varies greatly between weed species and time of the year.

Flaming is a form of thermal weeding that has been studied from various aspects. The hot flames of, for instance, propane and butane gas disrupt cell membranes and cause dehydration of the affected plant tissues (e.g. Ellwanger *et al.*, 1973). Effects of various techniques of flaming have been described by many subsequent authors (e.g. Ascard, 1995). The selective effect on young annual plants, i.e. weeds of different species and crop plants, largely depends on differences in the protection of their terminal buds. Differences in the vertical position of sensitive structures can be utilized as a basis for selectivity between crop and weed plants with different growth habits or at different stages of development. Flaming is an example of a weeding method with a comparatively high energy requirement and high CO_2 production. Melander (1998) has studied interactions between soil cultivation in darkness, flaming and brush weeding for weed control in vegetables; Holmøy and Netland (1994) have compared flaming, hoeing and band spraying in cabbage.

Referring to Liebman *et al.* (2001), another few examples are given here. *Near-row tools* are constructed so as to reach weeds close to crop rows without injuring the crop plants too much. The tools work according to different principles and are of interest in many crops, not least in those grown at small row spacings. *Ridge tillage* is described as 'an important soil conservation and weed management system, particularly in regions where maize, sorghum, and soybean are the predominant crops'. The crops are sown on ridges after soil and residues of the previous crop have been transferred from the tops of the ridges into the inter-row depressions. When the new crop plants have reached a certain size, soil and decomposed residue are hilled back onto the ridges. Different types of tools for improved mechanical control of weeds between the rows of crops have been developed. *Electric discharge weeders* are designed to control weed plants that are taller than the crop and therefore can be reached without touching the crop plants. The utility of electricity in this and similar ways is, so far, questionable.

Liebman *et al.* (2001) discuss the use of energy behind different weed control measures. Referring to analyses presented by different authors, they have found some unexpected results. With respect to energy consumption, chemical control is sometimes not as favourable in comparison with mechanical control as is usually imagined. One reason for this may be that energy for raw material, processing, packaging, storage, transportation, etc. is often underrated.

Cover Crops and Mulches

Soil erosion and leaching from soils have attracted increasing attention during the last few decades even in areas where these problems have long been considered to be of minor concern. In tropical regions with high precipitation in particular, gully erosion was found to be a major problem as soon as larger areas were broken for crop production and the soil was periodically laid bare. In more arid regions, wind erosion was experienced as a huge problem soon after large-scale cultivation of land in open areas had begun.

Soil erosion has been counteracted by having crop stands covering the fields with as short interruptions as possible, by leaving vegetation residues *on* and *in* the surface soil, etc. Leaching of plant nutrients, nitrate in particular, is a major problem in humid areas where fertilizer levels are high. Where the period of time from the harvest of one 'main crop' to the establishment of the next is long, a 'cover crop' is often established with the purpose of preventing erosion or leaching and, at the same time, suppressing weeds as far as possible. Cover crops chiefly established to prevent leaching of plant nutrients are often named catch crops. A cover crop serving some of the above-mentioned purposes may at the same time be a *green manure crop*. When this is the case, it is mainly a crop intended to add nitrogen to the soil and it generally consists of leguminous plants in pure stand or in some mixture with grass. Minerals taken up from the soil by the cover crop are made available to subsequent crops when the residues of the cover crop decompose. Many cover crops, like leys, contribute to increasing the humus content in the soil.

The crop plants in a *green fallow* can serve all the purposes mentioned above, particularly if the green fallow lasts for a whole growing season. Sometimes perennial plants are established and the green fallow may then last for more than one year, for instance in a *'set-aside' field*. With regard to its effects on weeds, a green fallow has much in common with leys.

Cover crops should exert strong competition against weeds. When grown in dense stands, they can, in fact, be very competitive, but the suppressive effect is selective, and certain weeds can withstand the suppression rather well. The weed populations present, the crop species and the time and method of establishing the cover crop are critical in determining which weeds can grow and reproduce in the crop. When the cover crop represents perennial plants with the ability to regrow rapidly after cutting, many tall-growing weeds can be restrained

satisfactorily by the joint effects of competition and cutting.

Cover crops should protect from both erosion and leaching and at the same time compete successfully with weeds. Crop and weed species and management measures differ between agricultural areas. Weeds can sometimes be effectively controlled by herbicides. Breakage of a cover crop may be carried out by mechanical measures alone or be preceded by application of a herbicide such as glyphosate.

Mulches are plant residues used as ground covers. Liebman *et al.* (2001) present a broad overview of cover crops and mulches, which serve partly similar purposes. Mulches may originate from plants previously growing in the field or may be material transferred from other areas. There are a range of intermediary forms, from above-ground mulches to plant residues that are more or less incorporated in the soil.

A mulch layer changes the light, temperature and moisture conditions in the superficial soil more or less strongly depending on its thickness and structure. The thicker and the more compact the layer, the lower the light transmission, the diurnal temperature fluctuation and the moisture variation. In broad terms, large-seeded plants are influenced to a lesser degree by a mulch cover than small-seeded plants. The germination of small weed seeds is, on average, more dependent on the temperature level and its fluctuations and on light and other environmental conditions than larger seeds of most crops and many weeds. In addition, seedlings from small seeds have greater difficulties in penetrating a mulch layer and becoming established plants than seedlings from larger seeds.

Many agricultural crops have seeds sufficiently large for seedling emergence and plant establishment, even when a rather thick or dense mulch layer has to be penetrated. There is then a selective effect of the mulch layer in favour of the crop. Mulch is sometimes placed in inter-row spaces to avoid crop penetration difficulties. After the crop plants have established, part of the mulch may be moved into ridges around the crop plants, suppressing small weed seedlings within the rows. As is understood from the above, small- and large-seeded weeds may be differently influenced by a mulch layer. Mulches also influence weed seed germination selectively.

Differences in the incidence of seed predators, herbivores and parasites with selective influences on plants may be associated with different mulches. When mulches decompose, plant nutrients are released, in various amounts and at varying rates depending on the type of mulch. Phytotoxic substances may also be released. Liebman *et al.* (2001) illustrate the principles of selective effects exerted by toxic substances diffusing only into the superficial soil layer, where germinating weed seeds are poisoned, whereas crop seeds placed at greater depths germinate in a chemically safer environment. If small weed seeds germinate at greater depths, they largely exhaust their food reserves without succeeding in establishing plants.

Harvesting – Timing and Methods

Weeds are differently influenced by crop harvesting operations. The strength and selectivity of a harvesting operation depend on the season and technique used. The population dynamics and dispersion of many weeds can be significantly affected. Repeated growing of the same crop or sequence of crops harvested at similar times and by similar methods may select for weeds that easily persist at this harvesting. If weeds that are favoured by such selection are not easily controlled together with other weeds in the field, or are troublesome for other reasons, changes in the cropping system may be justified or necessary. Variations in the timing and methods of harvesting, facilitated by diversified crop sequences, increase the possibilities of avoiding this undesirable selection. Diversified crop sequences generally also permit variations in other measures and conditions, thus facilitating rational systems of weed management.

Combine harvesting

Harvesting of crops such as small-grain cereals and oilseed rape is dealt with here. The timing and the stubble height are decisive both for the general amount of seeds left in the field and for the selective effect of a harvesting operation on weeds. At early harvesting, smaller numbers of weed seeds have been shed and/or are dropped to the ground at harvest. Later harvesting leads to a stronger selection between weeds with early and late maturing seeds, respectively. High (tall) stubble leads to more weed seeds being left in the field than low (short) stubble. A change in the stubble height changes the selective effect on tall and low-growing weeds. The issues discussed above have been studied in several countries by many researchers (e.g. Aamisepp *et al.*, 1967; Fogelfors, 1981).

Because combine harvesting is usually performed considerably later and with a higher stubble than previous harvesting methods using a binder or scythe, more weed seeds have matured and, in certain species, have also fallen to the ground before the time of harvesting. In relation to the number of weed seeds produced in the field, more of them usually drop to the ground at combine harvesting than at harvesting with the older methods. However, industrialized modern agriculture has more effective weed control than previous systems, so the absolute numbers of weed seeds added to the soil seed banks are usually comparatively small even at late harvesting and high stubble. Chemical weed control has strongly masked potential problems with weed seeds dropping to the ground in association with combine harvesting. Attempts to reduce the amounts of herbicides in crop production should therefore be made in connection with the development of measures to suppress weeds in other ways and/or with the development of harvesting techniques that reduce the dropping to the ground of weed seeds.

Influences of harvesting time and stubble height were studied in a field experiment run for 6 years in two-row barley in Southern

Sweden (Hansson *et al.*, 2001). In plots that were not treated with herbicides, three times of harvesting were compared, which were related to the degree of barley ripeness as follows: (i) at 'milk-ripeness', which is the normal time for harvesting green cereal; (ii) 'yellow-ripeness'; and (iii) 'binder-ripeness'. Stubble heights were 10 and 30 cm. In control plots, weeds were effectively suppressed by herbicides each year and the crop was harvested at full ripeness (iv), i.e. at the normal time for combine harvesting.

The weed density increased from year to year in all plots that were not treated with herbicides. The increase was stronger at later harvesting than at earlier, and stronger at high stubble than at low. The weed flora consisted of annual weeds, which increased at different rates, depending on their time of maturation and seed shedding. In plots harvested early (i) and with a low stubble, the weed density had increased only slightly after 6 years without herbicide treatments when compared with the herbicide-treated control. Early harvest with low stubble to produce green cereal may be a possible way of reducing the amount of herbicides needed for controlling weeds in the cropping system.

The direct mechanical influences of combine harvesting on regenerative organs of perennial weeds of importance in Scandinavian fields are fairly slight. Tall, older shoots of species such as *Elymus repens* and many other perennial weeds produce minor amounts of assimilates and the leaves of younger, more active shoots largely escape cutting with high stubble. Other perennials are also less strongly influenced. Their older shoots are often less active or are moribund and many are short or prostrate. Their younger shoots and the young plants of the perennial species developed from seeds are on the whole short and therefore only slightly affected.

Crop harvesting, however, indirectly influences perennial as well as annual weeds by changing their environment. The time of harvesting is one important factor, as it determines how long the weeds are subject to shading and other influences from the crop, although it should be remembered

that the degree of shading decreases during the period of maturation in most crops. In addition, the time of harvesting has important indirect effects on weeds in cropping systems where the performance of tillage or other measures influencing weed plants and weed seeds depend on this time.

Harvesting of root crops

When root crops, such as beets, turnips, or other crops grown for their underground storage structures, e.g. potatoes, are harvested, the soil is directly mixed, often as strongly as with actual soil tillage. There is a considerable potential for including these harvesting operations in systems for improved mechanical weed control through a moderate increase in energy consumption and negative soil impacts. No general rules can be given for this; the weed flora, the climatic and soil conditions and the available equipment are determining factors.

General knowledge of the traits and trait-dependent responses of the major weeds forms a sound basis for the design of improved operating systems. These may, for instance, focus on breakage and burial of vegetative parts of perennial weeds and/or on triggering germination of weed seeds. Potential measures may be based on principles discussed in Chapter 10. The time of the year when the harvesting is performed is decisive, as is the length of the period available or suitable for the operations. Very often, this period should be short and a new crop established rapidly to minimize soil erosion. The operations from harvesting to the establishment of the new crop may be regarded as a 'management unit' in the cropping system.

Breeding for Increased Competitive Ability of Crops

In the last few decades, breeding for enhanced competitiveness of crops versus weeds has attracted an increasing interest. The aim is to strengthen the competitive power of a crop stand not only by agronomic techniques, but also by growing genotypes bred for a high competitive ability. Competition from crops is thus intended to become increasingly important as an integral part of weed management systems.

Efforts to strengthen crop competition are worthwhile mainly in crops grown in stands that normally close fairly early, such as cereals. In, for instance, those root crops and vegetables that must be grown in open stands to give products of a required standard, increased weed suppression by crop competition can only reduce the need for other weed control measures to a minor extent.

A crop variety with a stronger competitiveness may allow lower doses of a herbicide owing to the joint effect of the herbicide and competition. When the crop exerts a stronger competitive effect as a result of more rapid ground cover, a reduction of the amount of herbicide reaching the weeds at foliar application may lower the intended effect of competition.

The potential for increasing the competitiveness of crops has been studied in small-grain cereals in particular, and considerable varietal differences in weed suppression in various cereals have been described (e.g. Dennis and Mortensen, 1980; Challaiah et al., 1986; Wicks et al., 1986; Andersson, 1992; Christensen, 1995; Didon, 2002). At the same time, the ranking order with regard to competitive outcomes was often found to vary with the environment (year and field).

Comparing barley varieties, Christensen (1995) and Didon (2002) found no correlation between yield capacity and competitive ability. They therefore suggest that breeding for increased competitive ability in combination with high yield could be possible. Christensen found that ranking of the varieties with regard to their loss in grain yield caused by weeds corresponded well to their ranking with regard to the dry weight of the weeds with which they competed. Christensen's and Didon's experiments indicated relationships between the competitive ability of the varieties vis-à-vis weeds and the degree to which they shaded

the weeds in decisive growth periods. Even in crops such as soybean, differences between varieties regarding their competitiveness should not be neglected (e.g. Burnside, 1972; Burnside and Moomaw, 1984).

As mentioned in Chapter 7, experimental results hint at the possibility of suppressing weeds by the use of selectively acting substances produced in cultivated plants or other organisms, mainly soil microbes (e.g. Olofsdotter and Mallik, 2001). Interesting results have been obtained in rice in particular. On the basis of various reports and observations, Didon (personal communication) is devoting increasing attention to the allelopathic effects of barley.

Biological Control

'Biological control' of weeds can be understood in different ways. The concept has sometimes been used in a very wide sense, including cultural measures with varying main aims, although sometimes deliberately modified with regard to the suppression of weeds. However, the concept of biological control, or biocontrol, of weeds is nowadays generally used in a more narrow sense. It thus stands for an intentional use of organisms that can attack specified weeds and considerably reduce the biomasses and population densities of these weeds. Impacts of large, grazing animals are usually not included within this concept. Efforts at developing weed biocontrol methods have focused on smaller organisms that are natural enemies of the target weeds, e.g. herbivorous insects and other invertebrates, pathogenic fungi and plant-suppressive bacteria.

There is an extensive literature on biological weed control, presenting research and development work and discussing the potential for and actual use of such control. Some informative publications may be mentioned: Julien and Griffiths (1998), Mortensen (1998), Isaacson and Charudattan (1999), Müller-Schärer and Vogelgsang (2000) and Weissmann (2002).

The following main strategies in biological control of weeds may be distinguished:

- *Classical*, or *inoculative*, biocontrol is a strategy used for controlling a plant that has been introduced and naturalized in a new area. The control agent is an organism introduced from the native range of this plant. Before the initial release of the presumptive control agent, comprehensive tests of its behaviour must be made. The tests should guarantee that the agent, mostly an insect, will not attack plants other than the target weed when this weed, after successful suppression, is no longer an available host. The released control agent must be able to reproduce and disperse on its own. When it is a flying insect, it should be able to rapidly seek out its target plant individuals. In principle, the control agent should not need to be released more than once to remain as an effective control agent for long periods of years.
- *Inundative* biocontrol implies the control of a native or an introduced plant by means of an organism already present in the area of the target plant. This organism, e.g. a fungus, is increased by mass rearing and spread over the whole area where it is intended to be active. It will not disperse outside this area, nor persist as an effective control agent for several years. The organism, named a *bioherbicide*, normally has to be spread annually and may be suitable for suppression of annual weeds in the first instance.
- *Augmentation* strategies are based on collection and redistribution of organisms that are native to the areas of the target plants. The strategies resemble those of bioherbicidal control, but contrary to bioherbicides, the agents persist for several years. Their application may therefore be repeated less frequently and with smaller quantities than the application of bioherbicides.
- *System management* approaches in weed biocontrol mean that an already established weed–pathogen combination or weed–insect association in an area is manipulated by measures resulting in increased injuries to the target weeds. The potential for weed control by this method is attracting increased attention in recent years.

There are some well-known examples of successful biocontrol of weeds by the classical strategy. The most spectacular example is the control of *Opuntia* spp. (prickly-pear cactus) in Australia by means of a moth, *Cactoblastis cactorum*, from South America. The cactus plants were originally introduced to Australia for ornamental purposes, but they rapidly spread and formed dense stands over wide pasture areas. After careful laboratory tests, some 3 million eggs of the moth were released in the mid-1920s. The larvae of the moth then rapidly reduced the populations of prickly pears to minor patches here and there. A 'moth–cactus balance' was established and is still keeping the cactus at low density levels. Many other examples of varyingly successful biocontrol of plants by the classical method could be given. Control agents are insects or fungi. Typical target plants are species introduced into new areas where they have become major problems on low-yield grazed land. Biological control of aquatic plants using classical methods have been developed or are under development.

The potential for classical biocontrol of weeds is limited to situations where one or a few very closely related species are the target plants. If these species occur as 'desirable' plants outside the areas where they are unwanted weeds, the classical strategy cannot be used. The control agent must have a very narrow host range. On arable land, where several unrelated species normally occur as major weeds, biocontrol of the type discussed here has a limited interest. Furthermore, almost all major arable weeds are closely related to important cultivated plants and to many wild plants. There are thus strong risks of uncontrollable attacks on such plants by the released biocontrol agents.

Other strategies mentioned above might have a greater potential for biological control of weeds on arable land. They present lower risks of uncontrolled spread and unwanted injuries to plants outside the target areas. Research on biocontrol of weeds on arable land is therefore focused on such strategies. It seems, however, that no real success has been achieved up to now. For instance, experiments with rust fungi as bioherbicides against *Cirsium arvense* have been conducted by many researchers over a number of years but so far without reliable effects.

Ongoing experiments on the system management strategy are presented by Leiss (2001). They concern the control of *Senecio vulgaris* with the rust fungus *Puccinia lagenophorae* Cooke. Plants of *S. vulgaris* occurring on ruderal habitats were shown to survive winter to a great extent, whereas those growing on arable land rarely did. Plants from both habitats proved to be more strongly infected by the rust fungus in later stages of development than in earlier. The degree of infection was also influenced by environmental factors. Leiss suggests that stimulation of epidemics of rust on *S. vulgaris* on arable land could be achieved by wintering of ruderal plants of *S. vulgaris* as inoculum sources combined with manipulation of environmental conditions in favour of rust infection.

Weissmann (2002) and co-workers have focused on bacteria as weed suppressing agents. Knowing that rhizosphere soil microorganisms include plant-suppressive bacteria, they isolated such bacteria and applied them to plants by foliar spraying. Selective effects have been found, suggesting a potential for regulated suppression of certain weeds by means of bacteria.

The effects of biocontrol agents will certainly be influenced by *plant competition*. Thus, even when plants of a given category are only slightly more affected by a suppressing factor than plants of another category, competition will magnify the relative suppression of the former plants when the two categories compete in mixed stands. Tables 27 and 37 illustrate that a herbicide or a virus effect is strengthened by plant competition. Possibilities of using interactive effects of plant competition and biocontrol agents are worth systematic studies.

In this context, interactions between *mycorrhizal fungi* and weed and crop plants deserve attention. There is a rapidly increasing knowledge of different types of such interactions and their direct or indirect influences on the life in soil ecosystems (e.g. review by Jordan *et al.*, 2000). These interactions seem to be important for the growth and competitive position of many plants, exerting both positive and negative influences. So far, it can only be speculated that mycorrhizal interactions can prove useful as elements in integrated weed control.

13

Important Points for Understanding the Occurrence and Rational Management of Weeds

General Views

Any physical or chemical factor in the environment may influence the establishment and development of a plant species in an area, although certain factors will have such decisive influences that others appear unimportant within normal ranges of variation. The effects of different factors are mostly interdependent, which makes it difficult, and sometimes impossible, to correctly evaluate the influences of individual factors. This easily leads to confusion or misinterpretations. The influence of a defined physical or chemical factor is modified by the competition between the individuals in a plant community. Accordingly, the outcome of competition between the individuals also depends on these factors, often in ways that are hard to interpret correctly.

Once a plant has invaded an arable field as a weed, its further fate in that field, i.e. its ability to reproduce and spread and its long-term persistence, depends on joint influences of a great number of factors, such as climate, basic soil properties and cultural measures. The cultural measures may influence the weed directly and/or exert indirect effects by modifying, among others, edaphic and competitive conditions.

A broad knowledge of the ability of diverse plants to grow and reproduce as weeds in different situations is needed for planning rational management. The following may be the minimum knowledge of a weed or potential weed.

- Its morphology and morphological plasticity and its reproduction and annual rhythm of development and growth.
- Its way(s) of surviving winters or other seasons when plant growth is impeded.
- Its response to various factors or conditions and its general environmental demands, including its competitive ability under varying conditions.
- Its response to various cultural measures (including control), particularly with respect to the response of its individuals in different stages of development.
- Its relations to parasites and herbivores under varying conditions.

It is important to know the genetic constitution of weeds. Knowledge of their intraspecific genetic diversity is a basis for judgement of their potential for selective adaptation to agricultural impacts. The disposition of a weed for rapid development of resistance to herbicides largely depends on its genetic character and variation and on the longevity of its seeds or other propagules in the soil.

Illustrations: Case Studies in Sweden

Central questions under the items listed above are exemplified in the following, on the basis of case studies concerning the performance of some plant species as arable weeds in Sweden. The discussions focus on knowledge of major importance for understanding the responses of these species to varying agricultural conditions and cropping measures. The discussions also, directly or indirectly, illustrate the need for improved knowledge.

Allium vineale – a bulb-forming perennial

Introductory remarks

In certain coastal areas of Scandinavia, mainly in South-eastern Sweden, including the islands of Öland and Gotland, wild *Allium* species (wild garlic, wild onions) have caused problems as weeds in both arable land and grassland for centuries. From the late 1950s on, these problems have gradually decreased owing to radical changes in agricultural practices.

An investigation was started in South-eastern Sweden in 1954 regarding the occurrence and traits of *Allium* species and the problems caused by them (Håkansson, 1963a). Grain of winter cereals, particularly rye, was frequently contaminated with aerial bulbils (small bulbs) of *Allium* plants. Such bulbils cause problems with milling even in very small quantities. *Allium* plants do not create problems in leys for hay but are troublesome when occurring in grassland grazed by cattle in spring and early summer, because they then cause unpleasant flavours in milk and dairy products.

The investigation focused on arable land. Among three *Allium* species appearing there, *A. vineale* was rapidly found to be, beyond comparison, the most abundant weed.

A field survey was carried out in the 'wild-garlic areas' in 1954 and 1955. In these years, *the use of modern herbicides had just begun in the area, but still had no obvious effect on weed populations.*

Wild garlic was generally considered to 'thrive' better on lighter soils than on heavier and in rye better than in wheat. This was based on broad experiences and on information in Swedish literature. The Swedish

name of *A. vineale* is *sandlök* ('sand garlic'), which alludes to its occurrence mainly on lighter soil in drier areas of Sweden. It was also experienced to be problematic in autumn-sown crops and grassland, very rarely in spring-sown crops. According to studies in Great Britain (Richens, 1947), *A. vineale* may be found on any soil type except on fen peat, but as an arable weed there, it occurred predominantly in an area with clay soils, where winter wheat strongly dominated in the crop sequences.

In the Swedish survey in arable fields, bulbil-bearing scapes of *A. vineale* were mainly found in autumn-sown crops and in leys for hay. They were also found considerably more frequently and abundantly on light soils than on heavier and more often in winter rye than in winter wheat. This could easily *have been interpreted as a 'confirmation' of opinions that* A. vineale *under Swedish conditions 'prefers' light mineral soils to heavier and rye to wheat, as well as autumn-sown crops to spring-sown.*

To what extent is this true? To answer this question, the main characteristics of the plant had to be studied: What were the strengths and weaknesses of the species with respect to its ability to establish and reproduce in different crops and cropping systems? Which cultural measures favoured and which disfavoured the plant? *The knowledge obtained is the same as the knowledge needed as a basis for rational management of the plant as an arable weed.*

The study and its results

As the introductory observations in 1954 indicated that *A. vineale* (wild garlic) was the dominating *Allium* weed on arable land, the investigation focused on this plant. Another two *Allium* species – *A. oleraceum* and *A. scorodoprasum* – were often seen at the sites studied, but mainly in the surroundings of the cultivated areas. They were studied to some extent in comparison with *A. vineale*.

General knowledge of the life history and ecology of *A. vineale* was obtained from Kirchner *et al.* (1912–13), Pipal (1914), Korsmo (1930, 1954) and, in particular, from an English study and literature review by Richens (1947). Morphology, reproduction and annual rhythm of growth and development were fairly well characterized through this literature. The following morphological description applies to plant types found in South-eastern Sweden. Experiments mainly concerned their appearance and response to influences of various conditions and measures on arable land (Håkansson, 1963a).

MORPHOLOGY AND REPRODUCTION (Fig. 66). The main types of *A. vineale* in South-eastern Sweden are characterized by compact heads of bulbils, sole or intermixed with flowers in umbels at the top of 30–100-cm long scapes. Bulbils usually strongly dominate by numbers over flowers, which are often missing. In the soil, a scape-forming plant develops a main bulb (a major offset bulb) with a width of 1–2 cm close to the base of the scape inside the sheath of its innermost (youngest) foliage leaf. It mostly also develops one or several minor offset bulbs in the axils of outer leaves. Most individuals in a population usually do not develop a scape but, instead, a terminal

main bulb and often one or more minor offsets. The seed set is very poor and therefore certainly of minor importance for the reproduction and persistence of the plant in a short-term perspective.

ANNUAL RHYTHM OF GROWTH AND DEVELOPMENT (Fig. 67a).   New *bulbs* develop in spring and early summer. At the same time, the parent bulbs gradually die. In *A. vineale*, most of the minor offsets remain dormant for one year or more, thus forming a bank of bulbs in the soil. The new main bulbs and bulbils sprout at various times in the late summer and early spring following their formation. Minor offsets sprout in the same seasons, but usually not until they have survived one year or more of dormancy in the soil.

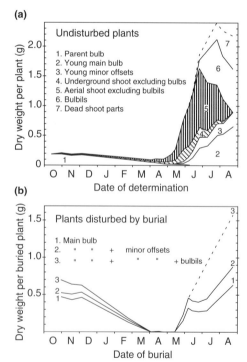

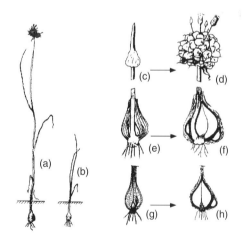

Fig. 66. *Allium vineale.* (a) Scapigerous plant. (b) Plant with a terminal bulb instead of a scape. (c–h) Details at two stages of plant development. (c–f) Scapigerous plant forming one major offset bulb (the main bulb) and some minor offsets in the soil and, on the top of the scape, a head of bulbils mixed with a few flowers. (g, h) Plant forming a terminal bulb and one or a few minor offsets. From Håkansson (1963a).

Fig. 67. (a) Undisturbed growth of *Allium vineale* plants in pure stands developed from bulbs, uniform in size, planted at a depth of 5 cm in a sandy soil on Öland in early October 1956. All plants formed scapes with bulbils. (b) Bulbs produced by plants disturbed by burial (horizontally, at a depth of 10 cm) in different stages of development; weights in the autumn of 1957. Plants at times of burial identical to those in (a).

POPULATION STRUCTURE.   From late June on, scape-forming plants are almost the only plants with visible above-ground structures; non-scapigerous plants complete their development and growth in the early part of the growing season. Field surveys characterizing the general frequency and abundance of the species in arable fields were based on the occurrence of bulbil-bearing scapes in autumn-sown cereals. Bulbils and different underground bulbs were counted in a number of rye fields in the late summers of 1954–1956. Average numbers given per major offset bulb (approximately per scapigerous plant, because of low mortality) satisfactorily describe the structure of a typical *A. vineale* population in rye fields. Units formed in the study years were, per one major offset: 13 terminal bulbs, ten minor offsets and 43 bulbils. Older structures were: 11 minor offsets and < 1 bulbil (in the soil). Seeds were seen very rarely. In one rye field, where the seed set of *A. vineale* seemed to be somewhat more frequent than usual, the number of units formed in the study year was, per one major offset, 11 terminal bulbs, 15 young minor offsets, 45 bulbils, 3.6 flowers and 0.35 seeds. Thus, one seed developed per ten flowers and 200 bulbs (including bulbils).

SENSITIVITY TO TILLAGE.   Experiments with tillage in fields – both fields with natural populations of *A. vineale* and fields with plants from transplanted bulbs – indicated that the species is strongly susceptible to mechanical disturbance in spring. Figure 67b illustrates its low capacity to regenerate after burial for about 1 month from early spring. It was shown that the amount of food reserves in undisturbed plants was at a minimum in that period, coinciding with the dry weight minimum (Fig. 67a). In the same period, the development of scapes and new bulbs was initiated. After burial in this period, very few bulbs developed and they exhibited a reduced viability. Minor offsets formed by plants buried in this period were found to be largely non-dormant or non-viable; these offsets either germinated or died before the next growing season. None of them thus added to the bank of bulbs in the soil.

OCCURRENCE IN AUTUMN-SOWN CEREALS.   The occurrence of *A. vineale* was recorded in fields with autumn-sown cereals where herbicides had not been applied in the study year (and probably not previously). Surveyed fields were rather evenly distributed over the 'wild-garlic areas' in South-eastern Sweden. The occurrence of *A. oleraceum* and *A. scorodoprasum* was recorded at the same time. These species, like *A. vineale*, develop heads of bulbils with comparatively few flowers.

The field surveys were undertaken in 2 consecutive years, during a few weeks before cereal harvest. All three species then had scapes with easily visible heads of maturing bulbils. At each of 397 sites surveyed, three zones were distinguished: (i) untilled headland bordering the tilled field, (ii) the adjacent border strip of the tilled field, about 2 m wide, and (iii) the inner parts of the field. Details are described with the results in Table 42.

Table 42 illustrates that the three species differed in frequency and abundance, not only in the cultivated fields but also in the headlands. This suggests that they are differently adapted to the general environmental conditions in the areas studied. All

Table 42.   Occurrence of *Allium vineale, Allium oleraceum* and *Allium scorodoprasum* in fields with winter cereals and adjacent strips of non-tilled headland at sites evenly distributed over the 'wild-garlic areas' in South-eastern Sweden. (Headland is a roadside or ditch-bank, etc. Field border is a strip, about 2 m broad, of the tilled field along the studied headland.)

	A. vin.	A. ol.	A. scor.
Number of sites with scapes out of 397 sites			
Headland	348	134	27
Field: border	199	33	10
Field: inner	141	8	3
Average abundance: assessed approximate number of scapes m^{-2}			
Headland	1.44	0.32	0.07
Average abundance as a percentage of that in headland			
Headland	100	100	100
Field: border	55	9	5
Field: inner	35	2	1

three species exhibited a lower average frequency and abundance in tilled zones than
in headlands, but this was much more pronounced for *A. oleraceum* and *A. scorodoprasum* than for *A. vineale*. The conclusion
to be drawn from this is that the two former
are less able than *A. vineale* to withstand the
cultural conditions in the arable fields.

However, by comparing all the data,
there are risks of biased results. Different
influences could be confounded due to
possible divergences between the three
species regarding their adaptation to different regions, soils, crops, cropping systems,
etc. represented in the survey area.

*To reduce the risks of such biased
results*, A. oleraceum *and* A. scorodoprasum
were compared with A. vineale *exclusively
at those sites where they were found in the
headlands together with the latter species.*
This applied to 129 sites with *A. oleraceum*
and 27 sites with *A. scorodoprasum* (Table
43). The risk of bias was reduced by the fact
that *A. vineale* was always recorded not only
in the headlands but also within all border
and inner parts of fields where either of the
other two species was seen, and always also
exhibited a higher abundance than these.
Table 43 presents frequency numbers within
the tilled fields as percentages of those in the
headlands. Although the three species might
respond differently to many undefined
environmental factors in the survey area,
it can be concluded with great certainty that
A. oleraceum and *A. scorodoprasum* are less
able than *A. vineale* to stand the cultural
conditions prevailing in the arable fields in a
period of years up to the study year. Now,
what is the reason for this?

The three species exhibited a rather
similar annual life rhythm and developed
main bulbs, minor offsets and aerial bulbils
to similar extents. However, the following
difference between the species was observed
in plant material dug out in different fields
in the spring of the first study year: No
dormant bulbs were found other than among
certain bulbs of *A. vineale*. In late August
of Year 1 (1955), young minor offsets of
the three species were planted in the field.
Among those of *A. vineale*, 16% produced
emerged plants in Year 2 and another 27% in

Table 43.   Average abundance of *Allium* species,
as a percentage of the abundance in the headland. Values corresponding to those in Table 42,
but now only representing sites where the species
placed in pairs in this table occurred together in
headlands.

	In 129 sites		In 27 sites	
	A. vin.	A. ol.	A. vin.	A. scor.
Headland	100	100	100	100
Field: border	52	8	37	5
Field: inner	27	3	24	1

Year 3. In August of Year 3, 54% were still
dormant and 3% were dead. From the bulbs
of the other two species, no emergence was
recorded later than in May of Year 2 and no
bulbs of these species remained dormant in
August of Year 3.

As the seed set is poor in all three
species, none of them will build up seed
banks of appreciable importance for their
short-term persistence. It is therefore suggested that *the dormancy among the minor
offset bulbs in* A. vineale *is the main reason
why this species is more capable than the
others of establishing persistent populations
on arable land.* More comprehensive studies
might reveal dormancy in scattered minor
offsets and/or bulbils in *A. oleraceum* and *A.
scorodoprasum*, but if so, it would certainly
be to a minor extent when compared with
the dormancy of the minor offsets of *A.
vineale*.

Lazenby (1962a) studied great numbers
of bulbils of *A. vineale*. He found a delayed
sprouting to some extent, particularly
among bulbils planted at great depths. In
the Swedish studies of the population
structures of *A. vineale*, in fact, a few
bulbils found in the material dug out from
the soil were suspected to be 'at least 1 year
old'. Based on all observations, however, it
was suggested that the minor offsets of *A.
vineale* are the only bulbs representing the
three *Allium* species that build up important
banks of dormant units.

In a field with sandy soil strongly
infested with *A. vineale*, the longevity of
the minor offsets occurring in the soil was
studied. The field was tilled repeatedly so as

to prevent production of new bulbs from Year 1 (1955) on. The offsets were sieved out from the soil in the late summers from Year 1 to Year 4. The number of viable minor offsets per square metre was as follows.

Year 1:	336
Year 2:	86
Year 3:	36
Year 4:	8

The repeated *tillage probably speeded up the sprouting and death of the offsets.* Thus, in the experiment described above, 54% of the *A. vineale* offsets remained dormant in Year 3 after planting in soil henceforth undisturbed by tillage.

GROWTH IN YEARS WITH AND WITHOUT SPRING TILLAGE (Table 44).   A field with sandy soil, uniformly infested with *A. vineale*, was ploughed and harrowed in the autumn of 1958 (Year 1). Some plots were sown with winter rye in that autumn. Others were sown with barley or planted with potatoes in the spring of Year 2, or were left as unsown fallows with repeated tillage starting either in early spring or in mid-June. Winter rye was sown in the whole experimental field in the autumn of Year 2. Shortly before this rye crop was sown, the occurrence of different bulb types, including bulbils, was determined. In Year 3, the occurrence of scapes

Table 44. Relative values of fresh weights of bulbs (in 'Year 2') and bulbils (in 'Year 3') of *Allium vineale* in an experiment started in the autumn of 'Year 1' in a field strongly infested with the weed.

Conditions in Year 2	Bulbs in autumn, Year 2		Bulbils in Year 3	
	Main bulbs	Minor offsets	Per unit area	Per kg rye grain
No spring tillage				
Winter rye	100	100	100	100
Fallow	257	166	121	135
Spring tillage				
Barley	1.8	25	1.2	0.0
Potatoes	2.0	21	0.5	4.0
Fallow	1.4	25	2.3	7.0

and bulbils was determined in the rye crop. The results are summarized in Table 44.

The table shows a high bulb production in Year 2 where rye had been grown and an even higher production on fallow tilled only after mid-June when compared with the production where barley or potatoes had been grown or the fallow had been tilled from early spring. In all three cases where spring tillage was carried out, very small numbers of bulbs were produced. The numbers of minor offsets were relatively greater than the numbers of the other categories of bulbs after spring tillage. This was a result of the survival of dormant units. Lazenby (1962b) illustrated a stronger effect of soil cultivation carried out before spring sowing than before autumn sowing.

It is interesting to note that, in a fallow tilled only from mid-June, *A. vineale* reproduced better than in rye. This was probably not only due to weak effects of the late tillage, but also to the absence of competition from a crop. *Fallows, often used in these areas up to the 1950s, were mostly not cultivated until mid-June. In this way, the farmers unintentionally favoured* A. vineale *in their cropping systems.*

OCCURRENCE ON HEAVIER AND LIGHTER SOIL AND IN WINTER RYE AND WINTER WHEAT.   As previously stated, *A. vineale* was found more often and more abundantly in winter rye than in winter wheat and more often on lighter soils than on heavier. However, *in a few cases where the two autumn-sown cereals were observed side by side in comparable cropping systems, the weed grew more vigorously in wheat than in rye.* Now, *farmers grew rye more frequently than wheat on light soils and vice versa on heavier soils* (Fig. 69). So, could the 'apparent preference' of *A. vineale* for rye be caused by a preference for light soils? However, another explanation seemed likely.

Results summed up in Fig. 67b and Table 44 show that tillage in spring has a stronger suppressive effect on the weed than tillage in other seasons. It was, further, a well-known fact that *spring crops were grown less frequently in the cropping systems on lighter soils than on heavier* because of the dry springs in the area. It therefore seemed likely that these two circumstances in combination were the reason why *A. vineale* as an arable weed was found more frequently and more abundantly in rye than in wheat and more often on light soils than on heavier. This was supported by further analyses of the field survey data and by a competition experiment.

The soil texture had been characterized in all the 397 fields underlying the data in Table 42. (Only mineral soils occurred.) The sites were divided into four groups with respect to the soil texture. Figure 68, which describes the occurrence of *A. vineale* at the sites representing each soil group, indicates that, within the tilled fields, the plant was found more frequently on lighter soils than on heavier and was, on average, also more abundant there. At the same time, no clear influence of soil texture was found in the untilled headlands. *This suggests that the cultural conditions in the arable fields more than the soil type influenced the occurrence of* A. vineale *in the climatic areas now concerned.* This suggestion is further examined below.

Figures 69 and 70 represent 72 sites where information could be obtained on the proportions of different crops, including fallows, represented in firm crop rotations or sequences of crops grown in the previous 10 years. At these 72 sites (included among the 397 sites represented in the field survey), *A. vineale* always occurred in headlands and surrounding areas. It was therefore judged to be a potential arable weed also at sites where it was not seen in the winter cereals. Thus, its presence or absence as an arable weed was judged to be chiefly an effect of cultural conditions.

Figure 69 indicates that spring-sown crops, as expected, were grown less frequently on lighter soils than on heavier and, in addition, that rye was grown much more than wheat on the former soils and vice versa on the latter. *A. vineale* was recorded in proportionally more fields on lighter soils than on heavier. In both cases its occurrence was judged to depend mainly on the frequency of spring crops, i.e. the frequency of years with spring tillage.

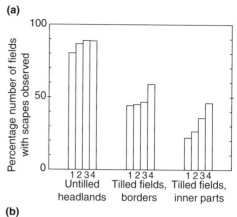

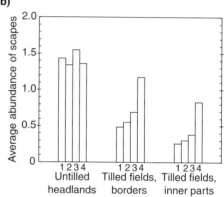

Fig. 68. Assessments of the occurrence of scapigerous plants of *Allium vineale* in winter cereal fields and adjacent untilled headlands on different soils represented at the following numbers of sites.
1.   Clay soils, 36 sites.
2.   Loam soils (clayey fine sand and clayey moraine), 106 sites.
3.   Fine sand and moraine poor in clay, 128 sites.
4.   Coarse sand and gravel, 127 sites.

(a) Number of sites with observed scapes of *A. vineale* as a percentage of the total number of sites investigated. (b) The average abundance, by number of scapes m^{-2}, according to assessments.

The distribution of fields with and without *A. vineale* presented in Fig. 70 supports the expectation that the ability of *A. vineale* to establish persistent populations on arable land largely depended on *the frequency of years with spring tillage. The critical frequency level seems to have been 30–40% under the agricultural conditions prevailing in South-eastern Sweden up to the mid-1950s, i.e. before regular use*

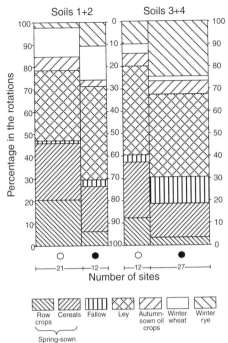

Fig. 70. Presence or absence of *Allium vineale* in arable fields characterized by different percentages of spring-sown crops in crop rotations (or sequences).

  •, *A. vineale* present

  ○, *A. vineale* absent

Description based on the same 72 sites as in Fig. 69, all of them thus in areas where *A. vineale* was seen as a headland plant. Sites representing the islands of Öland and Gotland and the adjacent mainland are distinguished; so are also sites with heavier (1 + 2) and lighter soils (3 + 4). Soils characterized in the legend to Fig. 68.

Fig. 69. Percentage occurrence of various crops in crop rotations (or sequences) at sites with heavier (1 + 2) and lighter soils (3 + 4), respectively. Presence or absence of *Allium vineale* in the winter crops.

  •, *A. vineale* present

  ○, *A. vineale* absent

Description based on 72 sites, all of them in areas where *A. vineale* was seen as a headland plant. Soils characterized in more detail in the legend to Fig. 68.

of herbicides had affected the weed populations.

Wild garlic problems were not recorded in any field where spring tillage recurred in more than 40% of the years in a crop rotation or a 10-year crop sequence. The sites in Fig. 70 are distributed with respect to geographical areas and soil types. No clear difference between regions, or between heavier and lighter soils, can be discerned, which reduces the risk of mistaking influences of the geographical area or the soil texture for an influence of the frequency of spring tillage.

Now, *what about wheat and rye?* A *field experiment* was conducted studying plants of *A. vineale* originating from bulbs of varying sizes planted in early autumn in pure

stands, in winter wheat, in winter rye and in ley for hay. This was on a moraine loam on Öland where both the crops and the weed were expected to do well. The cereals were sown at two seed rates, 'low normal' and 'high normal'. Different amounts of fertilizer were applied (Fig. 71a; Table 45a).

The garlic plants *grew more vigorously in wheat than in rye*, as did other weeds too, probably chiefly because rye grows more rapidly than wheat in early spring. The production of new bulbs, including aerial bulbils, was higher in wheat than in rye, by number and by weight. It was even higher at the higher seed rate in wheat than at the lower in rye. (For details, see Håkansson, 1963a.) This applies to each of the fertilizer levels tested.

A. vineale always grew better in the cereal stands than in the ley stands. This was also the case in an additional field experiment, where the growth of *A. vineale* planted in pure stands, in rye and in ley was compared at different levels of fertilizer on a sandy soil on Öland (Fig. 71b, Table 45b). It can be seen from Table 45a and b that the plant, when growing in competition with the cereals, was not favoured, but rather disadvantaged, by the fertilizers applied. It may therefore be suggested that generous nitrogen fertilizer application disadvantages the plant in cereal crops. Experimental results shown by Lazenby (1961a, 1971b) support this suggestion.

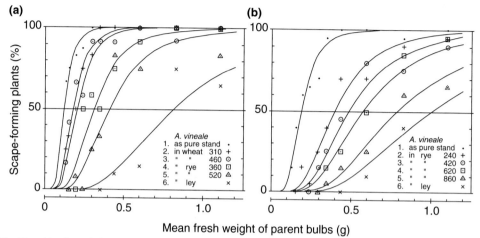

(a)

(b)

Scape-forming plants (%)

Mean fresh weight of parent bulbs (g)

A. vineale
1. as pure stand
2. in wheat 310
3. "   "   460
4. "   rye   360
5. "   "   520
6. "   ley

A. vineale
1. as pure stand
2. in rye   240
3. "   "   420
4. "   "   620
5. "   "   860
6. "   ley

Fig. 71. Number of plants of *Allium vineale* forming scapes in Year 2 (1959) expressed as a percentage of the number of bulbs planted in the autumn of Year 1 (1958), at different fresh weights of the planted bulbs. Field experiments on Öland: (a) at Vickleby and (b) at Sandby, further described in Table 45. Means over the fertilizer treatments presented in that table. Plants grown in pure stands, in winter wheat, in winter rye or in leys. Different cereal plant numbers per m^2, resulting from different seed rates (given by figures with the keys to symbols). Curves fitted using probit analysis.

Table 45. Effect of fertilizer application on the production of cereal grain and new bulbs in 'Year 2' in two field experiments on Öland where bulbs of *Allium vineale* were planted in the autumn of 'Year 1' in pure stands, in winter cereals and in first-year grass–clover leys. Means of cereal plant densities in the same experiments are dealt with in Fig. 71.

Fertilizer application[a]	Cereal grain (kg ha⁻¹)		Dry weight of new bulbs per planted bulb (g) (and relative[b])				Ratio of bulbs to cereal grain[c]	
	Winter wheat	Winter rye	Pure stand	In wheat	In rye	In ley	Wheat grain	Rye grain
(a) Experiment at Vickleby, on a moraine loam with a moderate content of humus								
0	1785	2385	4.16	0.86 (21)	0.61 (15)	0.31 (7)	1.86	1.13
1	2110	3060	3.68	0.94 (26)	0.64 (17)	0.34 (9)	1.72	0.92
2	2600	3290	4.99	0.73 (15)	0.44 (9)	0.33 (7)	1.09	0.59
(b) Experiment at Sandby, on a sandy soil with a moderate content of humus								
0		893	1.54		0.48 (31)	0.30 (19)		2.89
1		1150	2.43		0.49 (20)	0.33 (14)		2.26
2		1354	2.60		0.52 (20)	0.33 (13)		2.06
2a		2753	3.19		0.41 (13)	0.34 (11)		0.79
2b		1248	3.17		0.58 (18)	0.38 (12)		2.47

[a]0, No fertilizer application.
1, 50 kg P + 120 kg K ha⁻¹ in September, 31 kg N on May 2.
2, 100 kg P + 240 kg K ha⁻¹ in September, 62 kg N on May 2.
2a, 100 kg P + 240 kg K ha⁻¹ in September, 62 kg N on April 3.
2b, 100 kg P + 240 kg K ha⁻¹ in September, 62 kg N on May 27.
[b]Figures in parentheses: percentage of dry weight in pure stand.
[c]Weight of bulbs, including aerial bulbils, per plant to weight of cereal grain per plant.

Figure 71 illustrates the percentage number of individuals developing scapes as a function of the weight of the planted bulbs. Determinations in various experiments of this investigation showed that an increased proportion of scapigerous plants in a plant population implies an increased total production of new bulbs – an increased production of underground bulbs, by weight and by number of minor offsets, plus an increased production of bulbils. The curves therefore indicate a more vigorous growth and a higher production of bulbs, including bulbils, in wheat than in rye. They also indicate a more vigorous growth in the cereals than in the leys for hay.

The cereal yields presented in Table 45 were somewhat lower than the average for the 1950s, certainly due to an extremely dry period in May and June (1959) in combination with unusually high summer temperatures. Probably also due to this, the yields differed only slightly with the seed rates (Håkansson, 1963b). The weed growth also differed unusually slightly. The higher rye yield after early application of nitrogen (nitrate) in treatment 2a (Table 45b) was certainly a result of a better effect of the nitrogen applied when the shallow soil was still moist in early April.

In Lazenby's (1961a,b) experiments, the effects of physical soil properties and fertilizer application differed. However, effects of these properties on emergence and early growth of a plant often differ strongly from the effects on the subsequent growth and final plant production.

Besides the very strong influence of spring tillage on the occurrence of *A. vineale* in the cropping systems, there were certainly important effects of other factors representing crops, cultural measures, soils and weather. The considerable differences now shown in the competitive effect of wheat, rye and ley stands on *A. vineale* are examples. If, for instance, leys at the time of this field survey had not been grown to rather similar extents on all farms represented in Fig. 69, the proportion of leys in a crop sequence would probably have proved influential. There were certainly also influences of soil properties other than texture which cannot

be seen in an approach to the question of the type now being presented.

INTERPRETATION OF RESULTS.   Experiences from this study stress the need for considerable care when causalities in ecology are to be evaluated. Statistical analyses showing strong positive correlations of a weed to light soil and to rye, respectively, too easily could have been understood as the plant's preference for rye and for light soil. Thus, a field study such as that presented above could too easily have been interpreted as a confirmation of the general opinion that *A. vineale*, as an arable weed in Scandinavia, strongly prefers light soils to heavier and winter rye to winter wheat.

The occurrence of *A. vineale* recorded in arable crops in South-eastern Sweden in the mid-1950s, *before chemical weed control was regularly used*, may be briefly commented on as follows. *First, the strong suppressive effect of spring tillage on* A. vineale *immediately explains why vigorous plants of this species appeared only sparsely in spring-sown crops. The strong effect of this tillage overshadowed most other influences on the occurrence and long-term persistence of* A. vineale *on arable land in South-eastern Sweden. As a weed on arable land,* A. vineale *was found more frequently and abundantly on light soil and in winter rye than on heavier soil and in winter wheat, respectively. This can be explained mainly by the fact that spring tillage recurred less frequently on lighter soils than on heavier and rye more often than wheat on the former soils, i.e. where spring tillage recurred less frequently.*

EFFECTS OF HERBICIDES.   As stressed above, herbicides had not influenced the weed populations very much up to the time of the field survey in 1954 and 1955. However, the use of herbicides rapidly increased in the 1950s. Effects of herbicides were therefore also studied.

It was immediately experienced that, among the new chemicals used at that time, the systemic herbicides MCPA and 2,4-D could exert appreciable effects on *A. vineale*. Their effects were studied in dose–response experiments. Not surprisingly, stronger effects were obtained with esters than with salts, but otherwise the two types worked similarly.

Effects of the herbicides at different stages of plant development were studied in particular. Maximum effects were obtained in late spring, when new bulbs and scapes were starting a rapid growth, i.e. in the later part of the period when the plants were strongly sensitive to mechanical disturbance, or even somewhat later. Special measurements indicated that this was not only because the larger leaf areas in later stages intercepted more of the herbicide than the smaller areas in earlier stages.

Bulbs developed after sublethal herbicide treatment exhibited a reduced viability; minor offsets, in addition, showed a decreased degree of dormancy. Most of them therefore sprouted or died within a year. This secondary herbicide effect is very important in addition to direct killing of the plants.

OCCURRENCE AND MANAGEMENT OF *A. VINEALE* DISCUSSED ON THE BASIS OF THE INVESTIGATION. MCPA was the main herbicide used in the area in the 1950s. It was normally applied for the control of annual weeds, which is earlier in the spring than the optimum time for controlling *A. vineale*. Although MCPA and most other herbicides introduced after the 1950s largely exerted only moderate effects on *A. vineale*, their accumulated long-term effects became sweeping. Regular chemical control in autumn-sown cereals was probably the main reason why *A. vineale* rapidly declined as an arable weed in South-eastern Sweden in a couple of decades following the 1950s.

Increased fertilizer use may have contributed by disadvantaging *A. vineale* in its competition with cereals. It appears that nitrogen fertilizer has weakened the position of this weed in cereal stands. However, neither the experimental results shown in Table 45 nor the results presented by Lazenby (1961a,b) sufficiently affirm this assumption.

For long periods up to the 1950s, 2-year leys (range 1–3 years) for hay were regular elements in the crop sequences (Fig. 69). After that, increasingly many farmers gave up cattle production and the establishment of leys for fodder. As herbicides were henceforth regularly applied to the cereals, even autumn-sown cereals replacing the leys disadvantaged *A. vineale* as an element in the crop sequences. Where leys were replaced by cereals and other crops in which *A. vineale* was easily controlled, this species declined more rapidly than where leys were retained. In extreme cases, winter rye had sometimes been the only crop in fields with very coarse soils; in these fields, *A. vineale* had become particularly abundant. However, these soils were largely abandoned as cropped areas in the 1960s and 1970s. The combination of changes described here led to a rapid reduction in arable fields infested with *A. vineale*.

A. vineale is certainly still a potential arable weed on any mineral soil in the climatic areas under discussion. In the former 'wild-garlic areas', it is still a frequent plant in open places adjacent to arable fields. It will therefore rapidly infest these fields as soon as changes in the cropping systems allow its re-establishment there. A discontinuation of herbicide use would cause the return of *A. vineale* as an important winter-cereal weed in crop sequences where the frequency of years with spring tillage is low. If, in addition, the seedbeds of spring cereals in the future are made ready in the autumn, *A. vineale* would grow at least as vigorously in the spring cereals as in the autumn-sown, thus becoming a persistent weed even in crop sequences with a high frequency of spring cereals. To what extent the higher fertilizer levels used today would be a counteracting factor has to be shown by further experiments.

Looking at its geographical distribution, it appears that, at least in Scandinavia, the ability of *A. vineale* to become an arable weed is restricted to areas with low precipitation in late spring and early summer, i.e. in the period when new bulbs develop. In Scandinavia, such areas are the coastal parts of South-eastern Sweden, including the islands of Öland and Gotland, and another few minor areas where *A. vineale* was a weed in previous times.

Where precipitation in late spring and early summer is higher, *A. vineale* obviously 'prefers' ground that dries up quickly after rains, such as shallow soil on rocks, slopes with coarse soil, embankments with gravel, etc. Reasons for that might be weak competition from other plants and/or low degrees of infection by fungi. On deeper soil in those areas, both outside and inside arable fields, the moisture conditions favour the development of lush, competitive vegetation and might also favour infections by fungi. During the investigation in South-eastern Sweden, it seemed that injuries by fungi were somewhat more frequent on heavier than on lighter (dryer) soils. If so, however, the injuries are obviously not sufficiently frequent to prevent *A. vineale* from becoming a persistent plant on both heavier and lighter soils in the coastal areas of South-eastern Sweden with their low precipitation in spring and early summer.

Sinapis arvensis – a summer annual

Being a summer annual plant, *Sinapis arvensis* is an arable weed particularly establishing in spring-sown crops. Before modern herbicides were being regularly applied in cereals from the 1950s onwards, it was a very common weed in the more fertile agricultural areas in Southern and Central Sweden. It was often seen with extensive stands of flowering plants in spring cereals, particularly in areas with fertile soils (e.g. Bolin, 1912, 1922; Granström and Almgård, 1955; Hanf, 1982). In its competition with spring cereals, it may be favoured by high soil fertility (e.g. Granström, 1962) and moderate to high pH levels (Ellenberg, 1974).

Up to the 1950s, *S. arvensis* could mostly grow undisturbed by direct control in cereals, whereas it could be controlled by hoeing and hand weeding in row crops, such as sugar beet and potatoes. In these weakly competitive spring crops, however, plants that survived early control measures or

emerged late could grow vigorously and produce great numbers of seeds. To avoid this, these plants were, together with other weed plants, controlled by additional late weeding, largely by hand weeding. In well-managed row crops, *S. arvensis* was mostly effectively controlled, as were also most summer-annual weeds.

It was thus in spring cereals that flowering plants of *S. arvensis* were seen most abundantly. It was these plants that produced most of the *S. arvensis* seeds that were added to the soil seed bank, even where weed harrowing, treatment with iron sulphate or, in the 1930s and 1940s, application of calcium cyanamide were practised in the spring cereals. Even though *S. arvensis* was among those weeds that were best controlled by the chemicals mentioned, it was on the whole less effectively checked. It seems that this weed easily persisted in crop sequences where spring cereals recurred in 2–3 years out of 10. In Southern and Central Sweden, spring cereals alternated with winter cereals, grass–clover leys for hay (2–3 years) and, in many areas, row crops in rotations of 5–8 years.

The occurrence of *S. arvensis* as an arable weed decreased drastically when modern herbicides were regularly applied in cereals from the mid-1950s onwards. Because of its sensitivity to the herbicides used, *S. arvensis* has been largely kept at a low level in arable fields.

In the 1960s and 1970s in particular, livestock production was discontinued on a great number of farms. On these farms, leys for fodder were excluded from the crop sequences. The leys were replaced by annual crops, mainly cereals, to varying extents spring cereals. Without effective control by the new herbicides, the increased proportions of spring cereals in the crop sequences would have favoured *S. arvensis*, as well as many other weeds favoured by spring cereals in the absence of effective control (cf. description of *Avena fatua* below).

Whereas spring cereals had previously strongly favoured the growth and persistence of *S. arvensis*, the effective control by herbicides drastically changed the situation. Spring tillage increased the germination of

seeds in the soil seed bank of this summer-annual weed (Fig. 1, p. 7). Its seed bank
was therefore reduced more strongly in
years with spring tillage followed by crops
where its seed production was effectively
obstructed, than in years without spring
tillage, i.e. in years with leys and autumn-
sown crops (cf. p. 55).

Weak or absent suppression of seed
production of *S. arvensis* in spring cereals
has strongly negative effects where crops
such as oilseed rape alternate with cereals.
At least up to now, *S. arvensis* cannot be
as effectively controlled in rape as in the
cereals. An increased establishment of *S.
arvensis* plants in rape not only increases the
seed bank of the weed, but also contaminates
the harvested rape seed. The similarly
shaped *S. arvensis* seeds cannot be fully
removed and may strongly lower the quality
of the oilseed product even when present in
small quantities.

As reported by many agronomists, the
decline of *S. arvensis* after the introduction
of effective herbicides was particularly
rapid where spring cereals represented
a high proportion in crop sequences in
relation to winter cereals and leys. The main
reason is certainly that the seed bank then
decreased proportionally more rapidly in
years with spring cereals than in years with
winter cereals and leys.

The exclusion of chemical control
in organic farming has led to a comeback
of *S. arvensis* and many other species as
troublesome weeds where they are not counteracted by suitable crop sequences or other
measures (e.g. Rydberg and Milberg, 2000).
There are certainly sufficient numbers of
seeds left in agricultural areas to enable a
rapid comeback of weeds such as *S. arvensis*
even after a long period of suppression to
low levels.

Where plants such as *S. arvensis* with a
potential to become weeds in annual crops
cannot be effectively controlled there, inclusion of perennial leys in the crop sequences
facilitates their suppression. Crops for green
manuring and 'green fallowing' similarly
contribute towards reducing the proportion of 'weed-favouring crops' in a crop
sequence. Weed control effects of harrowing

and hoeing in cereals and other annual crops
were discussed in Chapter 10. Various ways
of improving the competitive effect of crops
such as cereals can contribute to weed
suppression, as stressed in Chapter 7.

Avena fatua – a summer annual in the Nordic countries

Avena fatua is the only species of wild
oats of importance in Sweden today. In
this country, it is a typical *summer-annual*
plant. According to Bolin (1926), *A. strigosa*
also has been a weed in parts of Sweden,
mainly in Western Götaland. It still appears
occasionally in Sweden.

A. fatua was a common cereal weed in
large parts of Sweden in earlier times. It
declined with the revolutionary changes
in the cropping systems that started in
the late 18th century and were gradually
implemented in the 19th century. With
these changes, *A. fatua* became a more and
more uncommon plant in Sweden.

In the early part of the 20th century, *A.
fatua* had almost disappeared from Swedish
arable fields. In flora from the 1940s, it was
described as a rare plant. About 1950, however, it reappeared and rapidly became an
abundant weed in many fields in some of
the main agricultural areas of Sweden. In
two decades following the 1950s, it spread
into most agricultural areas of the country,
becoming a feared weed, foremost in spring
cereals. Reasons for the changes in its
occurrence are discussed below.

Being a summer annual in the Nordic
countries, *A. fatua* grows and sets seed
mainly in spring crops, unless it is effectively controlled. In the permanent arable
fields of earlier times, cereals were usually
grown as the only crops in alternation
with fallows in fixed rotations. The fallows
recurred either every second or every third
year, seldom every fourth. Crops other than
cereals were mainly grown outside the
cereal–fallow rotations. Winter fodder for
the livestock (kept on every farm) was taken
from meadows outside arable land. On permanent arable land, *A. fatua* could establish

and reproduce, especially in spring cereals. This is described by many authors from the 18th century, e.g. by Linnaeus. However, this weed could often produce vigorous growth even in winter cereals due to weak crop competition and on fallows because of inefficient tillage. It could build up its soil seed banks through seeds shed before and during harvest and sown with the cereal seed.

Radically changed cropping systems in the 19th century involved an expansion of the permanent arable land into adjacent meadows and other areas. At the same time, grass–clover leys for winter fodder were established on arable land grown in alternation with annual crops, mainly cereals. Tilled fallows were also frequently included in more or less fixed rotations. In many areas, hoed row crops were also introduced and, to a minor extent, other crops too. The fallows were more and more effectively tilled from the early 19th century on, though not always resulting in effective weed control.

Around 1900, these new cropping systems were established in all important agricultural areas of Sweden. The reproduction of *A. fatua* was decreased by leys for hay, which were almost always included in the arable crop rotations. It was also decreased by fallows, which were more effectively tilled now than previously, and by hoed row crops, which were also grown in many areas. The reproduction of *A. fatua* was much more restricted in autumn-sown cereals than in spring-sown. Spring cereals, where *A. fatua* was at an advantage because of the absence of effective control, mostly occupied less than 25% of the years in the crop sequences. The seed production of *A. fatua* over a period of years was then obviously too low to maintain the soil seed bank. *A. fatua* had thus become a rare plant in Sweden by the early 20th century.

It seems that *A. fatua* needs a higher proportion of favourable crops than *Sinapis arvensis* to maintain its soil seed bank.

When *A. fatua* reappeared as a frequent weed in the 1950s, many opinions circulated about the reasons. A widespread opinion was that seeds of particularly competitive plant types had been imported from abroad. The reappearance was probably hastened by imported seeds, but these did not necessarily represent more competitive plant types than the earlier types in Sweden. Instead, the main reason for the reappearance of *A. fatua* was almost certainly an increased growing of crops in which the plant could reproduce sufficiently to become a persistent weed.

Thus, from the late 1940s, the proportion of cereals, including spring cereals, increased at the expense of leys in some of the most important agricultural areas around Lake Mälar and Lake Hjälmar – and it was in these areas that *A. fatua* first reappeared as an abundant weed in spring cereals. A strong preference of *A. fatua* for spring cereals over autumn-sown wheat and rye was illustrated in a field experiment by Åberg (1959). No herbicides enabling an effective control of this weed in the cereals were available at that time.

From the 1950s on, the growing of cereals, including spring cereals, increased on more and more farms due to discontinued growing of leys for fodder. In spite of that, many important dicotyledonous weeds could be sufficiently controlled by herbicides in the cereals. (As described above, *Sinapis arvensis*, which is extremely sensitive to the herbicides, decreased rapidly.)

However, with the expanded growing of spring cereals in large parts of Sweden, *A. fatua* was seen in increasingly many districts in the country. Field surveys in various parts of Sweden in the mid-1970s and early 1980s support the logical supposition that the increasing spread of *A. fatua* resulted from the increased growing of spring cereals at the expense of leys for fodder (e.g. Gavlestam, 1977; Gohde, 1977; Olofsson, 1981). Table 46 illustrates this. Strong seed control and hand weeding were stipulated by legislation, but produced insufficient effects on the weed while effective herbicides were not available or commonly used.

Herbicides against *A. fatua* in barley and wheat, mainly tri-allate, were used to some extent from the mid-1960s, but did not prevent a continued spread of the weed.

An increasing number of herbicides against *A. fatua* became available in the

Table 46. Occurrence of *Avena fatua* on farms with different acreage of spring cereals as a percentage of the total acreage of arable land. Field surveys in 1976 in the parishes of Torp and Stöde about 300 km north of Stockholm in Sweden.[a] (From Gohde, 1977.)

Percentage acreage of spring cereals	Percentage number of farms with occurrence of *Avena fatua*	
	Torp	Stöde
< 20	0	0
20–40	8	2
40–60	18	19
60–80	15	17
> 80	40	25

[a]The reduced growing of leys with a corresponding increase in the acreage of spring cereals started 5–10 years earlier in Torp than in Stöde.

1970s (Aamisepp, 1991). In the early 1980s, chemical control had largely diminished the occurrence and spread of *A. fatua* in Sweden.

The distribution pattern of *A. fatua* in Sweden has changed from the 1970s to 2000. In the areas where the plant first reappeared as an arable weed, the frequency of fields where it is abundant has decreased considerably. The main reason is almost certainly the use of effective herbicides, although continued strong seed control, hand weeding, etc. may have contributed. Effective herbicides were applied to an increasing extent up to the late 1980s and have then been used 'as needed'. At the same time, *A. fatua* has continued spreading into new parts of the country due to a continuously increasing cultivation of spring cereals in parallel with a decreased acreage of grass–clover leys. The spread of *A. fatua* into new areas is probably still occurring to a lesser extent. Changes comparable to those in Sweden have also been reported from Finland (Erviö and Salonen, 1987).

Elymus repens and Agrostis gigantea – perennial rhizomatous grasses

The rhizomatous grass *Elymus repens* is an arable weed in large parts of the humid temperate regions (cf. Korsmo, 1930; Palmer and Sagar, 1963; Holm *et al.*, 1977). It is an important weed in most agricultural areas in North-western Europe and occurs, or is a potential weed, on almost all arable land in Scandinavia and Finland. It can become troublesome in all crops and on any type of soil. Another rhizomatous grass often seen as an arable weed in humid temperate areas is *Agrostis gigantea*, the appearance of which will be briefly described in the following in comparison with *E. repens*.

Going by car from Southern Sweden up to about 1300 km northwards in 1971, I studied the occurrence of *E. repens* and *A. gigantea* in the border strips of arable fields and their adjacent headlands. I did this at 75 sites chosen at random along the road. *E. repens* was found at all sites, both within the tilled fields and in the headlands. *A. gigantea* was seen at most of the sites in Southern and Central Sweden, mainly in the field margins and the headlands, but usually at lower abundance than *E. repens*. Its occurrence as an arable weed gradually decreased further north, from Uppsala to Umeå.

Because of the ubiquity of *E. repens* in arable fields and adjacent areas in the Nordic countries, attempts to cultivate the plant for fodder have been made, particularly in Finland (e.g. Teräsvuori, 1929). Later attempts have also been made, but *E. repens* has always produced low and varying yields when compared with specially bred ley grasses. Dense patches of *E. repens* in perennial grass–clover leys obviously produce lower amounts of fodder, but the plant seldom reduces the fodder quality.

As described above, cereals replaced leys in the crop sequences on increasingly many farms in Sweden from the 1950s on. The changes involved a rapid increase in weed control by herbicides effective on most weeds except grasses. At the same time, tillage for the particular purpose of controlling perennial weeds was considerably reduced. The changes resulted in an increased abundance of grass weeds. It was then not only *Avena fatua* that caused concern; increased problems with *E. repens* gave rise to much concern too. There were also some problems with another two rhizomatous grasses, *A. gigantea* in the first instance, *Holcus mollis* more locally.

Table 47 describes the occurrence of *E. repens* and *A. gigantea* as weeds in an agricultural area in South-western Sweden. It is based on a field survey carried out in 1972, at the same time as more systematic efforts had been started to control these weeds in Swedish agriculture (using stubble cultivation and/or TCA). The design of the survey is presented with the results in Table 47.

Sites representing 325 fields with cereals or oilseed rape were surveyed. Fields on mineral soil varying from heavy clay to fairly light sand were included. At each site, the following three zones were studied: (i) 'Headland': an untilled strip, e.g. a roadside or ditch-bank along the tilled field, within a predetermined range of 25–50 m in length; (ii) 'Border': the adjoining border strip of the tilled field, about 1.5 m wide; and (iii) 'Inner': area observed when walking in a circle reaching 25–40 m from the border. The occurrence of plants is presented in figures shown and explained in Table 47. *E. repens* is commented on first.

E. repens was recorded on the headlands of as many as 320 of the 325 sites

Table 47. Occurrence of *Elymus repens* and *Agrostis gigantea* in fields with either autumn-sown rye or wheat or spring-sown oats, barley or oilseed rape. Description based on a field survey in 1972 in an area with different mineral soils in Västergötland, South-west Sweden. Assessments made from early July to mid-August in a total of 325 fields from 80 farms with different crop sequences. Farms randomly selected from local registers. (Data rearranged from Leuchovius, 1972.)

Category of field with respect to crop type, occurrence of ley in crop rotation and soil type (No. of sites)	*Elymus repens*			*Agrostis gigantea*		
		Tilled field			Tilled field	
	Headland[a]	Border[a]	Inner	Headland[a]	Border[a]	Inner
Number of sites with occurrence of the species noted						
All categories (325)	320	318	274	239	154	86
Average abundance according to assessments[b]						
All categories (325)	3.7	3.2	2.6	2.3	1.2	0.6
Where both weeds seen[c] (239)	3.6	3.3	2.6	3.0	1.6	0.9
Autumn-sown cereals (116)	3.7	3.2	2.6	2.4	1.7	1.0
Spring-sown cereals (171)	3.7	3.3	2.5	2.3	1.0	0.5
Spring-sown oilseed rape (38)	3.7	3.3	2.1	1.7	0.5	0.1
Ley[d] in crop sequence (165)	3.6	3.2	2.7	2.6	1.4	0.8
No ley in crop sequence (160)	3.9	3.3	2.4	2.0	1.0	0.5
Clay soil (84)	3.9	3.2	2.3	1.3	0.4	0.1
Clayey loam (113)	3.7	3.1	2.6	2.2	1.1	0.6
Sandy soil poor in clay (128)	3.6	3.2	2.6	3.1	1.9	1.0
Average abundance as a percentage of that in headland						
Clay soil (84)	100	82	59	100	31	8
Clayey loam (113)	100	84	70	100	50	27
Sandy soil poor in clay (128)	100	89	72	100	61	32

[a]Headland, roadside or ditch-bank, etc. Border, strip, about 1.5 m broad, of the tilled field along the headland studied.

[b]Assessments using an ordinal scale of 0–5: 0, no shoot seen; 1, occasional shoot or minor group; 2, minor groups here and there; 3–5, a more frequent distribution of groups with up to 10 shoots m^{-2}, up to 100 shoots m^{-2}, or up to much more than 100 shoots m^{-2}, respectively. Assessments were made by the same person throughout, and shoot counts for calibration of assessments were often made. Numbers in this ordinal scale are not equidistant, but their mean values as presented in the table will give an acceptable description of differences observed.

[c]Sites where both *E. repens* and *A. gigantea* were recorded in headland or field border.

[d]Grass–clover leys cut for fodder, duration 2–3 years, undersown in spring cereals. Where leys regularly recurred in a crop sequence or rotation, they were mostly represented in 25–35% of the years.

investigated. In the border strips of the tilled fields, it was found at 318 sites, in the inner parts at 274 sites. The values in the table indicate that the plant had great opportunities to grow in all the annual crops studied, in autumn-sown cereals as well as in the spring-sown crops.

Its average abundance in the annual crops was similar whether the crops were grown in sequences without leys for fodder or in sequences including such leys. Where leys were grown, they were usually grass–clover leys with a duration of 2–3 years, averaging 25–35% of the years in a crop sequence.

The soil texture showed no clear influence on the occurrence of E. repens in the headlands, but the figures for the inner parts of the fields suggest a weak preference for lighter soils over clay soil (see bottom lines in the table).

Even though E. repens was expected to occur abundantly, the average abundances recorded in the crops were surprisingly high. One reason for this is certainly that effective tillage (by stubble cultivation and/or fallows) had been almost completely abandoned and not yet compensated for by suitable chemical control. The increasing nitrogen fertilizer use during the 1950s and 1960s may have contributed. Thus, a high supply of nitrogen obviously favours E. repens (e.g. Thörn, 1967; Ellenberg, 1974; Hoogerkamp, 1975; Grime, 1977; Williams, 1977; Courtney, 1980; Tilman, 1987; Wedin and Tilman, 1996; Hansson and Fogelfors, 1998).

The results shown in Table 47 are in accordance with general experiences and statements that E. repens grows well on very different types of soil. However, its preferences for soil type are disputed. It is sometimes stated to prefer lighter and looser mineral soils to heavy clay (e.g. Korsmo, 1930; Salonen et al., 2001a). In any case, it grows to greater depths in loose than in compact soils. At least in weakly drained fields, it seems to thrive on coarse soil (Mukula et al., 1969). Under dry conditions, it might be less competitive on coarse mineral soils than on soils with a higher capacity for water retention, such as humus and clay

soils. The area represented in Table 47 is characterized by relatively good moisture contents throughout the growing season. Together with high fertilizer levels, favourable moisture conditions may counteract the influence of the soil texture on E. repens.

Table 47 illustrates that A. gigantea occurred less frequently and less abundantly than E. repens. It was found to grow more on lighter mineral soils than on clays. This is in accordance with many statements that A. gigantea occurs more on lighter than on heavier soils (e.g. Bolin, 1922; Korsmo, 1930; Hubbard, 1954; Thurston and Williams, 1968).

A. gigantea did not only occur less frequently and less abundantly than E. repens in the studied area; both its frequency and its average abundance were also proportionally much lower within the tilled fields than in the adjacent headlands (bottom lines in Table 47). The interpretation of this is that A. gigantea was less able than E. repens to withstand the cultural conditions in the tilled fields. The risk of a biased interpretation is reduced by comparing figures based on sites (239 in number) where the two species were seen together.

The vegetative reproduction of E. repens by means of rhizomes is favoured by a moderately cool temperate climate with much light and a long growing season. In Northern Scandinavia and Finland, where E. repens is a very important weed, long days compensate for short summers. The data in Table 16 (p. 49) illustrate that E. repens, which is a C_3 plant, has difficulties in producing rhizomes at tropical temperatures. This particularly applies where the amount of light accessible to the weed resembles that in ordinary cereal stands. However, the production of rhizomes has proved low at high temperatures even at a considerably enhanced intensity of light (Håkansson, 1969d). It is therefore not surprising that important and widespread rhizomatous perennials among the arable weeds in warm climates are C_4 plants (cf. Table 15, p. 47).

According to Tables 5, 12 and 47, E. repens grows vigorously in very different types of crop in humid temperate areas.

Undisturbed plants (Figs 49, 50 and 52) develop primary aerial shoots from early spring. They start to form rhizomes when unshaded shoots have developed 3–4 leaves in full daylight at temperatures prevailing during most parts of the growing seasons in Scandinavia. In shade, more leaves develop before rhizome formation starts (Håkansson, 1969d).

When the rhizome system of an *E. repens* plant is looked upon as a unit, no all-embracing innate dormancy is observed in any period of the year (Håkansson, 1967). Regeneration can therefore start from rhizome buds immediately after tillage in any season when not prevented by drought or low temperatures. Regrowth, consuming food reserves in early stages, can thus, in principle, be triggered by tillage at any time of the growing season (Figs 52 and 59–61).

Activation of buds and continued new plant growth are regulated by interactions within the plant. Dominance exerted by the shoots from early activated buds restricts the activation of other buds. At the same time, the degree of bud activity differs owing to factors other than an all-embracing innate dormancy. Low bud activity in late spring, reported by Johnson and Buchholtz (1962) as 'late-spring dormancy', results from shortage of food reserves (Håkansson, 1967). The extent and velocity of bud activation and shoot growth are, in addition, influenced by many environmental factors, such as nitrogen supply, temperature and moisture (e.g. McIntyre, 1971, 1972; Leakey *et al.*, 1978a,b,c,d).

When plants of *E. repens* grow in shading crop stands, their rhizome production is lowered more than their aerial shoot growth, but may still be considerable (Håkansson, 1969d; Williams, 1970b; Skuterud, 1984). Photosynthesis may continue until late in the growing season, allowing rhizomes to grow and increase their contents of food reserves in late autumn (Håkansson, 1967, 1974).

When the access to light increases at ripening and harvest of a shading crop, intensified photosynthesis may allow a considerable increase in the rhizome dry weight and food reserves before winter. The extent of this increase depends on the length of the period between harvest and winter and the temperature and light conditions in that period. In Southern Sweden, the weight of new rhizomes may be more than doubled after cereal harvest (Håkansson, 1974). Considerable suppression can then be achieved by post-harvest cultivation (Figs 60 and 61) changing an increase in the rhizome dry weight into a decrease. (Effects of drought, see p. 166.)

Because *E. repens* is almost self-sterile (Beddows, 1931), its seed set greatly depends on the location of different clones within an area. The seed production thus varies greatly, but is on the whole considerable. Although the seeds have no, or no lengthy, innate dormancy (e.g. Williams, 1968), a considerable soil seed bank is often built up (Korsmo, 1930; Palmer and Sagar, 1963; Williams, 1970a; Williams and Attwood, 1971; Melander, 1988). However, in fields where the occurrence of rhizomes is not strongly reduced by control, seeds only moderately influence the short-term population dynamics of the species. On the other hand, they play a decisive role when control has strongly limited vegetative regeneration. In the long-term perspective, the seeds are always important by enabling a continuous formation of new biotypes.

In earlier times, cereals were the only crops grown in the permanent arable fields in alternation with weakly tilled fallows. The situation was almost unchanged until the drastic alterations of the 19th century took place. From extensive literature published in the 18th century, it can be understood that *E. repens* had great potential to persist both on fallows and in cereal crops.

As *E. repens* appears to be favoured by high nitrogen levels, it was probably a less competitive weed in older times than in modern, high-input agriculture. Descriptions suggest that *E. repens* was a less harmful weed formerly than today, even though the cereal stands were thinner and matured earlier than today.

In the older cropping systems, *E. repens* was also subjected to *relatively* stronger competition from other weeds, which represented more species and higher total densities than today. Many species that are now rare weeds were obviously competitive at the lower fertilizer levels in earlier times. Both in the stubble fields and in the rather ineffectively tilled fallows, competition from many weeds and suppression by grazing probably contributed towards keeping *E. repens* in check.

Changes in the cropping systems in the 19th century involved the introduction of leys for fodder into the permanent arable fields, where they were grown in alternation with spring- and/or autumn-sown cereals in various proportions and mostly also with fallows. The fallows became gradually more effectively tilled. Row crops, where weeds were checked by hoeing, were also increasingly grown in many areas. In the more important agricultural areas of Southern and Central Sweden, the leys were usually kept for 2 (up to 4) years. Further north, leys usually had a duration of more than 4 years alternating with a few years of spring cereals and, sometimes, a minor acreage of other crops.

The conditions just described were generally established in the early 20th century and remained largely unchanged from about 1910 to the late 1940s. In these cropping systems, *E. repens* could grow vigorously in both cereal stands and leys according to contemporary literature. Ploughing and harrowing and, in row crops, hoeing were the main means of controlling *E. repens*.

The final seedbed preparation for spring cereals was mostly made later than today. *E. repens* was probably somewhat more affected by the later spring tillage than by earlier (cf. Figs 49, 52, 53 and 60). However, late sowing could result in thinner and less competitive cereal stands.

Stubble cultivation was carried out in many areas of Southern Sweden. It was mostly carried out shortly after harvest, by skim plough, by cultivator or by a combination of various operations, and followed by ploughing in late autumn. It was experienced in field practice that early stubble cultivation preceding late ploughing was more suppressive to *E. repens* than was

late ploughing alone. This is now easily explained on biological grounds (Fig. 61) and confirmed in field experiments (e.g. Permin, 1961b; Aamisepp, 1976).

Ley swards were sometimes broken by tillage in the height of summer after one cutting for hay. The sward was often ploughed up directly, but sometimes preceded by some type of shallow breakage. Where the ley sward had been ploughed up early, repeated shallow tillage could follow before a winter cereal crop was sown. With this procedure, *E. repens* and many other weeds could be considerably suppressed (particularly where the ley had been broken before ploughing). However, the ley sward was frequently unbroken until 3–4 weeks before the autumn-sowing. In cases where the ley was succeeded by spring cereals, it was often unbroken until ploughing in late autumn. In these situations, the control of *E. repens* was weak.

A green-fodder crop was often sown in part of a spring cereal field. After the crop had been moved in the summer, the field could be tilled repeatedly. Very good effects on *E. repens* and other weeds were reported. The green-fodder crops were often sown at a comparatively high seed rate and were then more competitive than an ordinary cereal stand.

'Black fallows', i.e. summer fallows tilled repeatedly, were often included in regular crop rotations, largely aimed at controlling weeds such as *E. repens* and *Cirsium arvense*. These fallows were used particularly often in areas where hoed row crops were not regularly grown. Black fallows mostly preceded autumn-sown cereals. If tillage started early, these fallows were known to reduce *E. repens* drastically. Often, or in some areas, tillage was not begun until late spring or early summer, resulting in a weaker, but still reasonable effect on *E. repens* if repeated thoroughly. The effect on weeds became in general weaker and also more selective. (Certain early weeds were even favoured by such a fallow, e.g. *A. vineale*, as seen in Table 44.)

By regular stubble cultivation and appropriate breakage of the leys, *E. repens* seems to have been satisfactorily suppressed without inset of black fallows. However, due to weather conditions, or for other reasons, stubble cultivation was not always carried out, or was not sufficiently performed. In row crops, vigorous rhizomes of *E. repens* made effective hoeing difficult; these crops could then favour rather than suppress this weed. In such cases, and in areas where stubble cultivation was not practised, black fallow was an ordinary element in the crop sequences to check weeds in general and perennial weeds in particular.

The grass–clover leys were, *on average*, probably fairly 'neutral' with respect to their influence on *E. repens* in a cropping system. This judgement is supported by results from the field survey presented in Table 47 and, more directly, by surveys indicating that *E. repens* occurs in similar amounts in young and old leys (Hagsand and Thörn, 1960; Thörn, 1967; Raatikainen and Raatikainen, 1975: see Table 8; Hagsand and Landström, 1984). The influence of an individual ley depends on its botanical composition, the evenness and density of its stand, its mineral nutrition, etc. However, *E. repens* could certainly be more strongly influenced by the time and method of breaking the ley and the subsequent measures up to the establishment of the next crop.

The monocropping of cereals that became increasingly common in Sweden from the 1950s on was enabled by the use of new herbicides. As mentioned before, the strong herbicide effect on many weeds led to a considerable decrease in weed control by tillage. Where stubble cultivation and/or black-fallow tillage had been used, these measures were largely discontinued in Sweden in the late 1950s. This was also due to their rising costs. None of the new herbicides used was effective on grass weeds and many of them therefore increased, not least *E. repens*. The increase of this plant may have been facilitated by the rapidly increasing use of nitrogen fertilizers (see above). Situations favouring the increase of grass weeds characterized large parts of Europe in the 1950s and 1960s (cf. Fryer and Chancellor, 1970). The field survey in 1972, presented in Table 47, was conducted just before measures improving this situation were generally introduced.

From the early 1970s onwards, efforts to control *E. repens* increased, mainly by stubble cultivation and/or by sodium trichloroacetate (TCA). TCA, a soil-applied herbicide, is effective on plants in their initial period of growth from rhizome buds or seeds (Håkansson, 1969b; Håkansson and Johnsson, 1970). It was therefore used together with shallow tillage triggering new growth from rhizome buds. TCA was usually combined with stubble cultivation, mainly before sowing of spring crops. It was sometimes applied in combination with harrowing in very early spring, followed by TCA-resistant crops, such as sugar beet or rape.

The foliar herbicide glyphosate was used in Sweden from 1975. Being more effective than TCA, glyphosate then rapidly replaced TCA in the control of *E. repens*. Glyphosate was mostly used after harvest. Due to a stronger effect, it has largely replaced stubble cultivation in the control of *E. repens* (e.g. Landström, 1983; Aamisepp, 1991; Hallgren and Fischer, 1992). Application of glyphosate, followed by ploughing after some weeks, can often control perennial grasses as effectively as black fallow. This has reduced the need for very intensive soil tillage. With glyphosate used at intervals 'as needed', and/or stubble cultivation practised in certain cases, the abundance of *E. repens* decreased in Swedish arable fields during the 1970s and 1980s.

A. gigantea has also decreased for similar reasons, and obviously more strongly than *E. repens*. It is probably somewhat more sensitive to tillage. The decline of *A. gigantea* may also be a result of the increased use of fertilizer. Contrary to *E. repens*, *A. gigantea* seems to be less competitive on richer soils than on poorer. High nitrogen levels disadvantage *A. gigantea* in competitive crop stands (e.g. Williams, 1977; Courtney, 1980).

Catch crops of grasses and/or clovers undersown in cereals have been increasingly used both in Sweden and elsewhere in recent years, in the first instance in order to reduce leaching of nitrate. These crops should, at the same time, satisfactorily suppress weeds. Experiments suggest that catch crops undersown in cereals do not influence weed growth very much before the main crops (usually cereals) are harvested, but exert their major effect after the harvest of the main crop. According to Swedish experiences, winter annuals are the most obviously affected weeds. The density and the staying power throughout the autumn are the most important properties of a catch crop in relation to its suppressive effect on weeds (e.g. Rasmussen, 1991b; Kvist, 1992; Ohlander et al., 1996).

The suppressive effects of catch crops on *E. repens* seem to vary strongly, but considerable effects can be obtained (Cussans, 1972; Dyke and Barnard, 1976). Where *E. repens* can increase its rhizome biomass substantially after crop harvests (Figs 60 and 61; Håkansson, 1967, 1974), it could be suggested that dense catch crops can considerably reduce this increase. At the same time, catch crops reduce the opportunities for mechanical control of *E. repens* in the autumn. It is still questionable as to what extent competition from catch crops can fully compensate for the lack of such control.

Crops for 'green manuring' or 'green fallows' have been increasingly used in Swedish cropping systems in recent years. When their stands are dense and even, they compete strongly with both annual and perennial weeds. Tall-growing weeds may be suppressed by the joint effects of cutting and competition. Depending on the crop plant type, the possibilities of using chemical weed control differ. *E. repens* and other plants in the vegetation cover can be effectively controlled by glyphosate applied in due time before this cover is broken mechanically.

The strong position of glyphosate in the control of *E. repens* today involves a great risk of selection for resistance to this herbicide. It is of interest to find effective herbicides to be used in alternation with glyphosate. At present, there are no herbicides suitable for regular use as alternatives to glyphosate for the suppression of *E. repens* in different cropping systems.

Cirsium arvense – a perennial dicot with plagiotropic roots

The composite perennial *Cirsium arvense* has been known as a troublesome arable weed from ancient times. However, after having been a frequent and noxious weed from time immemorial, its occurrence in Swedish arable fields declined drastically in a few years when new herbicides, dominated by phenoxy compounds (mainly MCPA), were introduced in Sweden in about 1950. Where chemical weed control was then regularly used, *C. arvense* rapidly became a 'non-problematic weed'. Where, in recent decades, the doses and/or the frequency of herbicide use have been lowered or, in organic farming, completely excluded, the plant is gradually coming back as a serious weed.

A classic study of the life history of *C. arvense* was published in Denmark by Lund and Rostrup (1901). Among later publications on this issue, Detmers (1927), Hayden (1934), Bakker (1960), Sagar and Rawson (1964) and Hamdoun (1967) may be mentioned, as well as literature surveys by Holm et al. (1977) and Donald (1994).

C. arvense perennates and multiplies vegetatively by plagiotropic, thickened roots and, to some extent, by underground stem parts. It is exclusively, or largely, dioecious. Flowers may be 'imperfect' with either stamens or a pistil, but sometimes 'perfect', bearing both stamens and a pistil. In perfect flowers, however, either the pistil or the stamens abort, resulting in functionally male or female plants (Lund and Rostrup, 1901; Detmers, 1927; Bakker, 1960; Hodgson, 1968). Dense groups of shoots mostly represent pure clones of male or female plants. The female plants may produce seeds when they are growing less than 60–90 m from male plants (Hayden, 1934; Bakker, 1960). The 'seeds' are 3–4 mm long achenes with plumose pappi, which largely break off at maturity. The seeds are therefore usually less successfully spread by wind than is often assumed.

The species is genetically variable (e.g. Hodgson, 1964; Palaveeva-Kovachevska,

1966), various types being differently apt to become arable weeds. According to Holm et al. (1977), observations in Canada have shown that smooth-leaved plants propagate vegetatively much better than spiny-leaved ones. In Sweden, smooth-leaved or moderately spiny-leaved plants seem to be much more common on arable land than more spiny types.

Being a long-day C_3 plant, *C. arvense* is a common species and a frequent arable weed, foremost in temperate regions, but also in somewhat warmer areas (Table 12), but it is absent in the hot tropics (Tables 14 and 15). This is, most probably, both because it is a long-day plant and a C_3 plant. Its persistence on cultivated land largely depends on its ability to produce strong underground systems of regenerative organs. Being a C_3 plant, its ability to produce these organs will be impaired at high tropical temperatures (cf. Table 15, p. 47). It prefers areas with a moderate precipitation and seems to have difficulties in persisting where the soil easily dries out owing to climate or ground factors. Its difficulties under dry conditions may at least partly be due to its C_3 photosynthesis. It is stated to be relatively demanding for light and also favoured by a high soil fertility (Korsmo, 1930; Ellenberg, 1974; Holm et al., 1977).

However, within the environmental preferences and limitations mentioned, it can grow vigorously in both annual and perennial crops and on very different types of soil. Outside arable fields, it is seen in many types of open areas, particularly in areas influenced by man, from where it can spread into arable land. It may invade arable fields by creeping roots from adjacent headland, and root fragments are spread wider through tillage and activities on the field and, of course, also by seeds transported in many ways, including wind, even though transportation by wind is not as extensive as is often believed.

Regenerative roots of *C. arvense* are mostly 3–7 mm thick (range 1.5–10 mm). They branch and grow in different directions and at different inclinations, i.e. horizontally, upwards or downwards at varying angles, thus penetrating the soil from shallow levels to depths usually much greater than the ploughing depth. In loose soils, the roots may reach depths of several metres (e.g. Hayden, 1934). In Sweden, the maximum depth mostly seems to be between 0.5 and 1 m, which is in accordance with results from the USA (Hodgson, 1968).

It is difficult to determine the maximum age of the regenerative roots. Most of them probably die within less than 2 years. Their *mean* length of life on arable land may even be less than 1 year.

New aerial shoots originate from buds developing anywhere on the roots or from buds at nodes of basal stem parts. Many observations show that very small root fragments are regenerative. Hamdoun (1967) recorded shoot emergence from a 2.5 cm long root fragment buried at a depth of 50 cm, which may be an extreme situation.

Underground stem bases of aerial shoots are regenerative only when they have at least one node and, according to my own observations, not until the shoots have reached the age when the tissues of their bases are firm.

The primary aerial shoots in a growing season emerge in a period of 10–15 days in the spring. On average, these primary shoots of *C. arvense* emerge later than those of *E. repens*, obviously due to slower growth at low temperatures. According to Swedish investigations, the underground plant parts pass a dry-weight minimum when aerial shoots have developed 6–8 leaves exceeding a length of 4 cm. This is slightly before inflorescence buds are visible and is the stage of minimum regenerative capacity, i.e. minimum capacity to recover after disturbance by tillage (Kvist and Håkansson, 1985; Dock Gustavsson, 1997). The number of leaves coinciding with this minimum differs somewhat with the depth of burial and length (amount of food reserves) of root fragments from which the shoots originate (Dock Gustavsson, 1997). Hodgson (1971) reports a food reserve minimum at the same time as flower buds are just becoming visible.

Both underground structures and aerial shoots of undisturbed plants grow rapidly in a period immediately following the weak stage. Some of the thin young roots then begin to grow in thickness, making them regenerative. By growing in length and by thickening and branching, the roots thus penetrate the soil in various directions. Parts of the older roots increase in dry weight after their dry weight minimum has been passed, but large parts gradually die back (Lund and Rostrup, 1901; Dock Gustavsson, 1997). During the summer, new aerial shoots develop from the new regenerative structures, predominantly from those at shallow depths. This resembles the behaviour of *Sonchus arvensis* as described by Håkansson (1969e). The new roots continue to grow until the supply of assimilates has ceased in the autumn.

The aerial shoots are sensitive to frost. They therefore die in autumn or early winter. Hayden (1934) reported weak emergence of new shoots in late summer. He ascribed this to exhaustion of food reserves. According to Hamdoun (1967), new shoots were less easily produced in the later part of the growing season than earlier. My own preliminary observations in Sweden support this, but as there is no shortage of food reserves at that time, these observations suggest that the regenerative organs enter a state of weak physiological dormancy in late summer. If so, this dormancy is not so strong and so regularly occurring as that in *Sonchus arvensis* (Håkansson, 1969e; Henson, 1969; Håkansson and Wallgren, 1972a; Fykse, 1977).

Dormant seeds build up a soil seed bank. Many newly matured seeds germinate readily, as many as 95% in proper environments (Hayden, 1934). Kolk (1962) found that young seeds germinated well in bright daylight, whereas old seeds germinated better in weak light. When seeds were stored in wet sand during winter, they were stratified more strongly and, therefore, germinated better in the spring than after storage in dry sand. At constant temperatures, maximum germination occurred around 30°C. Bakker (1960) reported extensive germination at similar temperatures, and also at temperatures diurnally fluctuating from 10 to 28°C, which resembles situations that often occur at the soil surface in the spring in temperate areas. Seeds covered with soil germinated to a much lesser extent than those exposed to light at the surface of moist soil. Kollar (1968) found in field studies that seeds that had dropped to the ground at maturity germinated at a rate of 15% on the soil surface, but not at all after burial at a depth as small as 1.5 cm.

According to Bakker (1960), seedlings may have difficulties in emerging even from shallow depths. When seedlings establish, they can develop regenerative thickened roots within 7–9 weeks under favourable growth conditions, but they have difficulties in establishing and surviving in competition with rapidly growing plants. Amor and Harris (1975) found that the seed viability declined strongly in the first 2 years and that the seeds contribute only slightly to the spread and establishment of *C. arvense* on grassland.

The persistence of *C. arvense* as a weed thus largely depends on its vegetative growth and multiplication, both on grassland and in arable fields. However, when the vegetative structures have been strongly suppressed, seeds from the soil seed bank and seeds incoming from the surroundings are of the utmost importance. The seeds are also always important as originators of clones representing new genotypes, which facilitate the adaptation of the species to changed environmental conditions and, thereby, facilitate its long-term persistence, its spread and its establishment in new areas.

Because of its root distribution to great depths in the soil, *C. arvense* is less easily controlled by tillage than are the more shallow-growing *Elymus repens* and *Sonchus arvensis*. This was discussed in detail on p. 182. The regenerative roots and stems of *C. arvense* are easily killed by drought or frost (Hamdoun, 1967) – obviously more easily than the rhizomes of *E. repens*. It is difficult, however, to bring such a large proportion of them to the soil surface that death from drought or frost results in an effective control of the weed, at least in climates that are not very extreme. However, proper tillage followed by competitive crops has sometimes proved to be acceptably suppressive. Up to now, however, no measure has been as effective as chemical control by systemic herbicides such as phenoxy compounds.

In relation to its potential for producing strong regenerative structures, *C. arvense* is not very competitive. Its young seedlings grow slowly, and its regrowth from vegetative structures is slow when compared with the early growth of small-grain cereals. At the same time, plants of *C. arvense* have great requirement for light (cf. Holm *et al.*, 1977). According to Hodgson (1971), *C. arvense* recovers more slowly than lucerne after mowing. This slow recovery may be an important reason why *C. arvense* normally declines from year to year in a grass–clover ley cut for fodder (cf. Tables 5, 8; Dock Gustavsson, 1994a,b). Considerable suppression in the ley together with effects of breakage of the ley by adequate tillage (see the following) have proved to suppress *C. arvense* in a crop sequence; probably almost as effectively as hoeing in a row crop. Another perennial weed with creeping roots, *Sonchus arvensis*, is similarly, or even more strongly, suppressed by a ley in the crop sequence.

On the basis of the properties of *Cirsium arvense* described above, changes in the occurrence of this species in relation to changes in cropping systems are briefly discussed in the following.

For hundreds of years before the revolutionary changes in the cropping systems took place in the 19th century, cereals were almost the only crops grown in the permanent arable fields. As described earlier, spring and/or winter cereals were grown for 1 or 2 (seldom 3) years in alternation with 1 year of fallow. Because competition from the cereal stands was weak and the fallows were weakly tilled, *C. arvense*, like most weeds, could survive both in cereals and on fallows. In the cereals, some manual weeding occurred. In contrast to the rhizomes of *E. repens*, the roots of *C. arvense* are palatable to swine, which were often let out on the fallows ('swine fallows'), where they could eat considerable amounts of thistle roots and many other plant parts. *E. repens* was mainly suppressed by grazing cattle. On the whole, however, these weeds largely survived the cereal–fallow rotations.

The introduction of crop rotations including grass–clover leys for hay in the 19th century was certainly positive with respect to the control of *C. arvense*. Leys with a duration of mostly 2 years or more alternated with cereals and, often, also with a year of fallow and/or some row crop. The row crops were weeded by hoeing and hand-weeding. The fallows were often, but not always, gradually more intensively tilled. Improved seedbed preparation sometimes resulted in more even and, therefore, more competitive cereal stands. However, the treatment of fallows and the establishment of cereal stands were often far from efficient. Leys were grown on almost every farm. According to Tables 5 and 8, leys do not favour *C. arvense*. However, the leys were at those times based on low-quality seed and their stands were weaker and more weed-ridden than today. Because they were not cut until high summer, sometimes even later, they were also less suppressive to *C. arvense* than leys that are cut earlier, as they are nowadays (Dock Gustavsson, 1994a,b). Compared with cereals, however, leys were even then considered to be suppressive to *C. arvense*. This may have been true on the whole, as even the cereal stands were less suppressive than today. Where summer fallows were repeatedly and properly tilled, good control of *C. arvense* may have been achieved, but the fallows were often far from thoroughly treated. Row crops such as beets, turnips and potatoes were regarded as suppressive to all weeds. They were mostly comparatively well hoed and hand-weeded.

In spite of gradual improvements in weed control in the 19th century, *C. arvense* was reported to be a very troublesome weed, particularly in spring cereals, but also in winter cereals. However, in dense, well-wintered stands of autumn-sown cereals, *C. arvense* was considered to be better suppressed than in spring cereals. The reason may be that it is relatively light-demanding and, at the same time, grows rather slowly in early spring. Shoots of *C. arvense* in young cereal stands were sometimes pulled up by hand or cut just below the soil surface with a narrow sharp-edged iron tool, a 'thistle iron'. However, even when many possible control measures were carried out, *C. arvense* could be extremely troublesome, often character-ized as the worst weed on arable land.

As previously stated, the cropping systems described here reached their final structure in the early part of the 20th century and were then on the whole unchanged up to the late 1940s. However, commercial fertil-izers enabled a slow increase in fertilizer use. New implements and machinery allowed a more rapid and better soil tillage and better techniques for sowing and harvesting. Earlier sowing increased the competitive effect of spring cereals on *C. arvense*, which grows more slowly than the cereal plants at low spring temperatures.

The stands of the leys cut for hay were gradually improved and could be broken by better tillage implements than before. In the 1930s, the leys were more often than earlier described as 'clearing crops', suppressive not only to annual weeds, but also to *C. arvense* and *Sonchus arvensis* (Bolin, 1926). By the mid-1930s, leys were frequently cut more than 2 weeks earlier than they had been at around 1900. Populations of both annual and perennial weeds were either lowered or, as *Elymus repens*, kept 'in balance' by the leys. Typical ley weeds, mainly stationary perennials (cf. Tables 5 and 8), were previously more frequent weeds in annual crops than in the 1930s and 1940s. With increasingly effective tillage, they had now become less frequent in well-managed annual crops.

Where, in addition to leys, well-managed fallows were regular elements in the crop sequences, or were put in as rescue measures, *Cirsium arvense* and many other weeds could be kept at reasonable levels (Åslander, 1933, 1934). However, *C. arvense* was not suppressed as effectively as *E. repens* by tillage after harvests, mainly owing to its deep-growing regenerative roots. In the 1930s, *C. arvense* was mostly consid-ered a more troublesome cereal weed than *E. repens* in Southern Sweden. It was, at the same time, more conveniently controlled by hoeing and hand-weeding in row crops.

From the early 1950s onwards, when grass–clover leys were excluded on increasingly

many farms, monoculture of annual crops, mainly cereals, became widespread. The rapidly increased use of herbicides in cere-als that took place at the same time enabled an effective control of many dicotyledonous weeds, not only many annuals, but also some perennials, particularly *C. arvense*. Chemical weed control in cereals, mainly by phenoxy compounds or mixtures containing such compounds, soon became almost routine, even where leys were still grown. This resulted in a rapid decline of *C. arvense* in Swedish arable fields. With this chemical control, *C. arvense* decreased even though at the same time tillage was reduced, resulting in the increase in *E. repens* described above. When used for silage, the first cut in leys is now made more than a month earlier than cutting for hay in the 1930s. This increases the suppression of *C. arvense* in the ley.

C. arvense is sensitive to most herbicide formulations that have been used in Swedish cereal fields. From about 1960 until recent years, *C. arvense* was therefore sel-dom seen in abundance in 'conventionally cropped fields' within the main agricultural areas of Sweden.

As chemical control is now being reduced in many farms, or completely abandoned in organic farming, the situation is changing. In organic farming in particular, *C. arvense* is rapidly returning as a problematic weed.

In the absence of chemical control, monoculture of cereals easily leads to a rapid return of *Cirsium arvense*, unless intensified mechanical control is set in to an extreme extent. The inclusion of perennial leys and/or green manuring or green fallow crops in the crop sequences can certainly improve the situation. The stands of these crops should preferably be cut in early summer and cutting should be repeated at least once more in the growing season. Before the sward is ploughed down, break-age by shallow tillage is preferable. Through combinations of such measures, *C. arvense* might be kept at reasonably low levels, resembling the levels obtainable in cropping systems with well-managed leys in the 1930s.

In the 1930s, however, the spread of *C. arvense* from the surroundings of the arable fields was better counteracted than today. The vegetation on the headlands (roadsides, ditch-banks, etc.) was normally cut in the height of summer. This limited the spread of *C. arvense* by seeds and reduced the extension of its roots into the cropped fields. In recent decades, the headland vegetation is frequently left uncut. Where *C. arvense* is regularly controlled by herbicides in arable fields, a moderate import of its propagules is of minor importance. Where, on the other hand, chemical control is now being reduced or has been discontinued, the prevention of such import is of great significance.

In recent decades, attempts have been made, or are ongoing in some countries, to develop measures for biological control of *C. arvense*. Experiments with fungi have dominated. Up to now, however, they have not been very successful.

References

Aamisepp, A. (1976) Control of *Agropyron repens* in cereal stubble. *Agricultural College of Sweden. 17th Swedish Weed Conference, Reports*, D 17–D 19.

Aamisepp, A. (1991) Utveckling av ogräspreparat under 40 år, 1946–1985. *Sveriges lantbruks-universitet, Institutionen för växtodlingslära. Växtodling* 31, 8–41.

Aamisepp, A., Steckó, V. and Åberg, E. (1967) Ogräsfröspridning vid bindarskörd och skördetröskning. *Lantbrukshögskolans Meddelanden* A81. Uppsala, 32 pp.

Aamisepp, A. and Wallgren, B. (1979) Ogräs i stråsäd. Verkan av kemisk ogräsbekämpning och andra odlingsåtgärder, 1950–1978. *Aktuellt från lantbruksuniversitetet* 280. *Mark/Växter*. Uppsala, 14 pp.

Åberg, E. (1956) Luserncirkulationen vid Ultuna 1937–1955. [English summary: The alfalfa rotation at Ultuna 1937–1955.] *Kungliga Skogs- och Lantbruksakademiens Tidskrift* 95, 209–246.

Åberg, E. (1959) Studier över olika åtgärder mot flyghavre, *Avena fatua* L., i ett växtföljdsförsök. [Summary: Studies of various methods against wild oats, *Avena fatua* L. in a crop rotation experiment.] *Växtodling [Plant Husbandry]* 10, 40–53.

Van Acker, R.C., Lutman, P.J.W. and Froud-Williams, R.J. (1997a) The influence of interspecific interference on the seed production of *Stellaria media* and *Hordeum vulgare* (volunteer barley). *Weed Research* 37, 277–286.

Van Acker, R.C., Lutman, P.J.W. and Froud-Williams, R.J. (1997b) Predicting yield loss due to interference from two weed species using early observations of relative weed leaf area. *Weed Research* 37, 287–299.

Åfors, M. (1994) Weeds and weed management in small-scale cropping systems in northern Zambia. *Crop Production Science* 21. Swedish University of Agricultural Sciences, Uppsala, 190 pp.

Aleman, F. (1989) Threshold periods of weed competition in common bean (*Phaseolus vulgaris* L.). *Crop Production Science* 4. University of Agricultural Sciences, Uppsala, 42 pp.

Alex, J.F. (1982) Canada. In: Holzner, W. and Numata, N. (eds) *Biology and Ecology of Weeds*. Dr W. Junk, The Hague, pp. 309–331.

Alström, S. (1990) Fundamentals of weed manage-ment in hot climate peasant agriculture. *Crop Production Science* 11. Swedish University of Agricultural Sciences, Uppsala, 271 pp.

Altieri, M.A. and Liebman, M. (eds) (1987) *Weed Management in Agroecosystems: Ecological Approaches*. CRC Press, Boca Raton, Florida, 354 pp.

Amor, R.L. and Harris, R.V. (1975) Seedling establishment and vegetative spread of *Cirsium arvense* (L.) Scop. in Victoria, Australia. *Weed Research* 15, 407–411.

Andersen, A. (1987) Ukrudtsfloraen i forsøg med direkte såning og pløjning ved forskjellige kvaelstofniveauer. [Effects of direct drilling and ploughing on weed populations.] *Tidsskrift for Planteavl* 91, 243–254.

Andersson, B. (1984) Seed rates and MCPA doses in spring barley. *Swedish University of Agricultural Sciences. 25th Swedish Weed Conference*, Uppsala, pp. 51–61.

Andersson, B. (1986) Influence of crop density and spacing on weed competition and grain yield in wheat and barley. *Proceedings of the European Weed Research Society Symposium 1986, Economic Weed Control*, pp. 121–128.

Andersson, B. (1987) Utsädesmängder i vårsäd med och utan ogräsbekämpning. Resultat från fältförsök. [Summary: Seed rates in spring sown cereals with and without weed control.] *Swedish University of Agricultural Sciences, Department of Crop Production Science, Report 174.* Uppsala, 52 pp.

Andersson, B. (1992) Sorter i stråsäd – ogräskonkurrens. [Varieties of cereals – competition with weeds.] *Sveriges Lantbruksuniversitet. Ekologiskt Lantbruk* 13, 169–175.

Andersson, L., Milberg, P., Schütz, W. and Steinsmetz, O. (2002) Germination characteristics and emergence time of annual *Bromus* species of differing weediness in Sweden. *Weed Research* 42, 135–147.

Andersson, T. (1997) Crop rotation and weed flora, with special reference to the nutrient and light demand of *Equisetum arvense* L. *Acta Universitatis Agriculturae Sueciae. Agraria* 74, 38 pp.

Andreasen, C. (1990) *Ukrudtsarternes forekomst på danske saedskiftemarker.* [Summary: *The Occurrence of Weed Species in Danish Arable Fields.*] Den Kongelige Veterinaer- og Landbohøjskole. Licentiat/afhandling, Copenhagen, 53 pp. and 11 appendices.

Andreasen, C., Stryhn, H. and Streibig, J.C. (1996) Decline of the flora in Danish arable fields. *Journal of Applied Ecology* 33, 619–626.

Andrés, I.M. (1993) A revised list of the European C₄ plants. *Photosynthetica* 26, 323–331.

Arny, A.C. (1927) Successful eradication of perennial weeds. *Ontario Department of Agriculture, 48th Annual Report of the Agricultural Experiment Station Union* 1926, pp. 58–64.

Arvidsson, T. (1997) Spray drift as influenced by meteorological and technical factors. *Acta Universitatis Agriculturae Sueciae. Agraria 71.* Doctoral thesis, Swedish University of Agricultural Sciences, Uppsala, 144 pp.

Ascard, J. (1994) Soil cultivation in darkness reduced weed emergence. *Acta Horticulturae* 372, 167–177.

Ascard, J. (1995) Thermal weed control by flaming: biological and technical aspects. *Sveriges Lantbruksuniversitet, Institutionen för Lantbruksteknik, Alnarp. Rapport 200,* 61 pp.

Åslander, A. (1930) *Våra viktigaste åkerogräs och deras utrotande.* Bonnier, Stockholm, 137 pp.

Åslander, A. (1933) En undersökning av åkertistelns skottbildning. *Centralanstalten för Försöksväsendet på Jordbruksområdet. Meddelande 429,* 37 pp.

Åslander, A. (1934) Helträdan, dess betydelse och brukande. *Kungliga Landtbruksakademiens Handlingar o. Tidskrift 429.* Stockholm, 35 pp.

Aura, E. (1975) Effects of moisture in the germination and emergence of sugar beet (*Beta vulgaris* L.). *Journal of the Scientific Agricultural Society of Finland* 47, 1–70.

Avholm, K. and Wallgren, B. (1976) Renkavle, *Alopecurus myosuroides* Huds., och åkerven, *Apera spica-venti* L. – uppträdande och biologi. *Lantbrukshögskolan. Institutionen för växtodling. Rapporter och Avhandlingar* 53. Uppsala, 30 pp.

Ayerbe, L., Gomez, M., Albacete, M. and Garcia-Baudin, J.M. (1979) *Convolvulus arvensis* L., Como planta adventicia en los cultivos Españoles 3. Prensencia de reguladores de crecimiento en sus rizomas. *Instituto Nacional de Investigaciones Agrarias Anales. Serie: Proteccion Vegetal* 11, 131–143.

Bachthaler, G. (1974) The development of the weed flora after several year's direct drilling in cereal rotations on different soils. *Proceedings of the 12th British Weed Control Conferrence,* pp. 1063–1071.

Baker, H.G. (1965) Characteristics and modes of origin of weeds. In: Baker, H.G. and Stebbins, G.L. (eds) *The Genetics of Colonizing Species.* Academic Press, New York, pp. 147–172.

Bakker, D. (1960) A comparative life-history study of *Cirsium arvense* (L.) Scop. and *Tussilago farfara* L., the most troublesome weeds in the newly reclaimed polders of the former Zuider Zee. In: Harper, J. (ed.) *Biology of Weeds.* Blackwell Scientific Publications, Oxford, pp. 205–222.

Ball, D.A. (1992) Weed seed bank response to tillage, herbicides, and crop-rotation sequence. *Weed Science* 40, 654–659.

Barralis, G., Chadoeuf, R. and Lonchamp, J.P. (1988) Longevité des semences de mauvaises herbes annuelles dans un sol cultivé. *Weed Research* 28, 407–418.

Baskin, C.C. and Baskin, J.M. (1998) *Seeds – Ecology, Biogeography, and Evolution of Dormancy and Germination.* Academic Press, San Diego, California, 666 pp.

Bazzaz, F.A. (1990) Plant–plant interactions in successional environments. In: Grace, J.B. and Tilman, D. (eds) *Perspectives on Plant Competition.* Academic Press, San Diego, California, pp. 239–263.

Beddows, A.R. (1931) Seed setting and flowering in various grasses. *Bulletin, Welsh Plant Breeding Station, Series H* 12, 5–99.

Bell, A.R. and Nalewaja, J.D. (1967) Wild oats cost more to keep than to control. *North Dakota Farm Research* 25, 7–9.

Bengtsson, A. (1972) Radavstånd och utsädesmängd för vårvete och korn. [Summary: Row Spacing and Seed Rate in Spring Wheat and Barley.] *Lantbrukshögskolan Meddelanden A* 160. Uppsala, 28 pp.

Bengtsson, A. and Ohlsson, I. (1966) Utsädesmängdsförsök med vårsäd. [Summary: Trials with Different Seed Rates of Spring Wheat, Two-rowed barley and Oats.] *Lantbrukshögskolan Meddelanden A* 43. Uppsala, 35 pp.

Benvenuti, S. (1995) Ecologia della 'seed bank': aspetti fisiologici ed implicazioni agronomiche nella gestione della flora avventizia. [Summary: Seed bank ecology: Physiological aspects and agronomical implications for weed control.] *Rivista di Agronomia* 29, 204–219.

Benvenuti, S., Macchia, M. and Miele, S. (2001) Light, temperature and burial depth effects on *Rumex obtusifolius* seed germination and emergence. *Weed Research* 41, 177–186.

Berti, A. and Zanin, G. (1994) Density equivalent: a method for forecasting yield loss caused by mixed weed populations. *Weed Research* 34, 327–332.

Bibbey, R.O. (1935) The influence of environment on the germination of weed seeds. *Scientific Agriculture* 16, 141–150.

Bibbey, R.O. (1948) Physiological studies of weed seed germination. *Plant Physiology* 23, 467–484.

Black, I.D. and Dyson, C.B. (1997) A model of the cost of delay in spraying weeds in cereals. *Weed Research* 37, 139–146.

Blackshaw, R.E. and Harker, K.N. (1998) *Erodium cicutarium* density and duration of interference effects on yield of wheat, oilseed rape, pea and dry bean. *Weed Research* 38, 55–62.

Bleasdale, J.K.A. (1963) Crop spacing and management under weed-free conditions. *Symposium of the British Weed Control Council* 3, 90–101.

Bleasdale, J.K.A. and Nelder, J.A. (1960) Plant population and crop yield. *Nature* 188, 342.

Blum, U., King, L.D., Gerig, T.M., Lehman, M.E. and Worsham, A.D. (1997) Effects of clover and small grain cover crops and tillage techniques on seedling emergence of some dicotyledonous weed species. *American Journal of Alternative Agriculture* 12, 146–161.

Bolin, P. (1912) *Våra vanligaste åkerogräs och deras bekämpande.* Frizes Bokf. AB, Stockholm, 168 pp.

Bolin, P. (1922) De viktigaste ogräsarternas olika frekvens och relativa betydelse som ogräs. *Centralanstalten för Försöksväsendet på Jordbruksområdet* 239. Stockholm, 36 pp. + 38 maps.

Bolin, P. (1924) Redogörelse för försök år 1923 med olika metoder för ogräsens bekämpande. *Centralanstalten för Försöksväsendet på Jordbruksområdet. Meddelande* 275. Stockholm, 85 pp.

Bolin, P. (1926) *Åkerogräsen och deras bekämpande.* Åhlén and Åkerlunds Förlag, Stockholm.

Børresen, T. and Njøs, A. (1994) The effect of ploughing depth and seedbed preparation on crop yields, weed infestation and soil properties from 1940 to 1990 on a loam soil in south eastern Norway. *Soil and Tillage Research* 32, 21–39.

Boström, U. and Fogelfors, H. (1999) Type and time of autumn tillage with and without herbicides at reduced rates in southern Sweden. 2. Weed flora and diversity. *Soil and Tillage Research* 50, 283–293.

Boström, U. and Fogelfors, H. (2002a) Response of weeds and crop yield to herbicide dose decision-support guidelines. *Weed Science* 50, 186–195.

Boström, U. and Fogelfors, H. (2002b) Long-term effects of herbicide application strategies on weeds and yield in spring-sown cereals. *Weed Science* 50, 196–203.

Boström, U., Hansson, M. and Fogelfors, H. (2000) Weeds and yields of spring cereals as influenced by stubble-cultivation and reduced doses of herbicides in five long-term trials. *Journal of Agricultural Science, Cambridge* 134, 237–244.

Brain, P., Wilson, B.J., Wright, K.J., Seavers, G.P. and Caseley, J.C. (1999) Modelling the effect of crop and weed on herbicide efficacy in wheat. *Weed Reseach* 39, 21–35.

Brenchley, W.E. and Warington, K. (1936) The weed seed population of arable soil. III. The re-establishment of weed species after reduction by fallowing. *Journal of Ecology* 24, 479–501.

Brenchley, W.E. and Warington, K. (1945) The influence of periodic fallowing on the prevalence of viable seeds in arable soil. *Annals of Applied Biology* 32, 285–296.

Brown, V.K. and Southwood, T.R.E. (1987) Secondary succession: patterns and strategies. In: Gray, A.J., Crawley, M.J. and Edwards, P.J. (eds) *Colonization, Succession and Stability.* Blackwell Scientific Publications, Oxford, pp. 315–337.

Brust, G.E. (1994) Seed-predators reduce broad-leaf weed growth and competitive ability. *Agriculture and Environment* 48, 27–34.

Buhler, D.D., Stoltenberg, D.E., Becker, R.L. and Gunsolus, J.L. (1994) Perennial weed populations after 14 years of variable tillage and cropping practices. *Weed Science* 42, 205–209.

Burnside, O.C. (1972) Tolerance of soybean cultivars to weed competition and herbicides. *Weed Science* 20, 294–297.

Burnside, O.C. and Moomaw, R.S. (1984) Influence of weed control treatments on soybean cultivars in an oat-soybean rotation. *Agronomy Journal* 76, 887–890.

Burnside, O.C., Wilson, R.G., Weisberg, S. and Hubbard, K.G. (1996) Seed longevity of 41 weed species buried 17 years in eastern and western Nebraska. *Weed Science* 44, 74–86.

Bylterud, A. (1965) Mechanical and chemical control of *Agropyron repens* in Norway. *Weed Research* 5, 169–180.

Bylterud, A. (1983) Gras som ugras. Utbredning og omfang i Norge. *Nordiska Jordbruksforskares Förening (NJF) Seminarium 52: Gräsarter som ogräs*, Kemira försökscentral, Esbo, Finland, pp. 11–18.

Carder, A.C. (1961) Couchgrass control in Alberta. *Alberta Department of Agriculture, Agricultural Extension Service* 149, 11 pp.

Cardina, J., Regnier, E. and Harrison, K. (1991) Long-term tillage effects on seed banks in 3 Ohio soils. *Weed Science* 39, 186–194.

Carlsson, H. and Svensson, B. (1975) Beståndstäthetens inverkan på potatisens avkastning och kvalitet. *Lantbrukshögskolan. Institution för växtodling. Rapporter och Avhandlingar* 35. Uppsala, 40 pp.

Cavers, P.B. and Harper, J.L. (1964) Biological Flora of the British Isles. No. 98. *Rumex obtusifolius* L. and *Rumex crispus* L. *Journal of Ecology* 52, 737–766.

Challaiah, Burnside, O.C., Wicks, G.A. and Johnson, V.A. (1986) Competition between winter wheat (*Triticum aestivum*) cultivars and Downy brome (*Bromus tectorum*). *Weed Science* 34, 689–693.

Champion, G.T. Grundy, A.C., Jones, N.E., Marshall, E.J.P. and Froud-Williams, R.J. (eds) (1998) *Aspects of Applied Biology 51, Weed Seedbanks: Determination, Dynamics and Manipulation*. The Association of Applied Biologists, Wellesbourne, UK, 296 pp.

Chancellor, R.J. (1964) Emergence of weed seedlings in the field and the effects of different frequencies of cultivation. *Proceedings of the 7th British Weed Control Conference*, pp. 599–606.

Chancellor, R.J. (1965) Weed seeds in the soil. *Weed Research Organisation. First Report 1960–1964*, pp. 15–19.

Christensen, S. (1994) Crop weed competition and herbicide performance in cereal species and varieties. *Weed Research* 34, 29–36.

Christensen, S. (1995) Weed suppression ability of spring barley varieties. *Weed Research* 35, 241–247.

Clark, B.E. (1954) Factors affecting the germination of sweet corn in low-temperature laboratory tests. *New York State Agricultural Experiment Station, Cornell University, Geneva, N.Y. Bulletin 769*. 24 pp.

Clements, F.E. (1916) Plant succession. An analysis of the development of vegetation. *Carnegie Institute of Washington, Publication 242*. 512 pp.

Clements, D.R., Weise, S.F. and Swanton, C.J. (1994) Integrated weed management and weed-species diversity. *Weed Science* 42, 205–209.

Collins, R.P. and Jones, M.B. (1985) The influence of climatic factors on the distribution of C_4 species in Europe. *Vegetatio* 64, 121–129.

Connolly, J. (1986) On difficulties with replacement-series methodology in mixture experiments. *Journal of Applied Ecology* 23, 125–137.

Connolly, J. (1987) On the use of response models in mixture experiments. *Oecologia (Berlin)* 72, 95–103.

Courtney, A.D. (1968) Seed dormancy and field emergence in *Polygonum aviculare*. *Journal of Applied Ecology* 5, 675–684.

Courtney, A.D. (1980) A comparative study of management factors likely to influence rhizome production by *Agropyron repens* and *Agrostis gigantea* in perennial ryegrass swards. *Proceedings of the 1980 British Weed Protection Conference – Weeds*, pp. 469–475.

Cousens, R. (1985a) A simple model relating yield loss to weed density. *Annals of Applied Biology* 107, 239–252.

Cousens, R. (1985b) An empirical model relating crop yield to weed and crop density an a statistical comparison with other models. *Journal of Agricultural Science* 105, 513–521.

Cousens, R. and Mortimer, M. (1995) *Dynamics of Weed Populations*. Cambridge University Press, Cambridge, 332 pp.

Cousens, R., Brain, P., O'Donovan, J.T. and O'Sullivan, P.A. (1987) The use of biologically realistic equations to describe the effects of weed density and relative time of emergence on crop yield. *Weed Science* 35, 720–725.

Crawley, M.J. (1986) The structure of plant communities. In: Crawley, M.J. (ed.) *Plant Ecology*, Blackwell Scientific Publications, Oxford, pp. 1–50.

Crocker, W. (1916) Mechanisms of dormancy in seeds. *American Journal of Botany* 3, 99–120.

Cromar, H.E., Murphy, S.D. and Swanton, C.J. (1999) Influence of tillage and crop residue on postdispersal predation of weed seeds. *Weed Science* 47, 184–194.

Currie, J.A. (1973) The seed-soil system. *Seed Ecology. Proceedings of the 19th Easter School in Agricultural Science, University of Nottingham 1972*, pp. 463–480.

Cussans, G.W. (1968) The growth and development of *Agropyron repens* (L.) Beauv. in competition with cereals, field beans and oilseed rape. *Proceedings of the 9th British Weed Control Conference*, 131–136.

Cussans, G.W. (1972) A study of the growth of *Agropyron repens* (L.) Beauv. during and after the growth of spring barley as influenced by the presence of undersown crops. *Proceedings of the 11th Weed Control Conference*, pp. 689–697.

Cussans, G.W. (1973) A study of the growth of *Agropyron repens* (L.) Beauv. in a ryegrass ley. *Weed Research* 13, 283–291.

Cussans, G.W. (1976) The influence of changing husbandry on weeds and weed control in arable crops. *Proceedings of the British Crop Protection Conference – Weeds 1976*, pp. 1001–1008.

Dahlsson, S.-O. (1973) Ojämna bestånd i vårvete och korn. Studier av arrangerad ojämn uppkomst. [Summary: Uneven stands of spring wheat and barley. Studies of arranged uneven emergence.] Dissertation Lantbrukshögskolan. Institutionen för växtodling, Uppsala, 229 pp.

Dennis, B. and Mortensen, G. (1980) Varietal differences in weed suppression in barley. *Nordiska Jordbruksforskares Förening (NJF) Symposium in Noresund, Norway. Utnyttjande av konkurrensen från kulturväxterna i ogräsbekämpningen*, 1–13.

De Prado, R., Jorrín, J. and García-Torres, L. (eds) (1997) *Weed and Crop Resistance to Herbicides*. Kluwer Academic Publishers, Dordrecht, 340 pp.

Detmers, F. (1927) Canada thistle, *Cirsium arvense*. *Ohio Agricultural Experiment Station Bulletin* 414, 45 pp.

Didon, U.M.E. (2002) Growth and development of barley cultivars in relation to weed

competition. *Acta Universitatis Agriculturae Sueciae. Agraria 332*. Doctoral thesis. Swedish University of Agricultural Sciences, Uppsala, 83 pp.

Dock Gustavsson, A.-M. (1989) Growth of annual dicotyledonous weeds. Analyses using Relative Growth Rate (RGR) and Unit Production Ratio (UPR). *Crop Production Science* 5. Swedish University of Agricultural Sciences, Uppsala, 106 pp.

Dock Gustavsson, A.-M. (1994a) Åkertistelns reaktion på avslagning, omgrävning och konkurrens. *Sveriges Lantbruksuniveritet Fakta Mark/Växter* 13. Uppsala, 4 pp.

Dock Gustavsson, A.-M. (1994b) Åkertistelns förekomst och biologi. *Sveriges Lantbruksuniversitet Växtskyddsnotiser* 58(3), 79–84.

Dock Gustavsson, A.-M. (1997) Growth and regenerative capacity of plants of *Cirsium arvense*. *Weed Research* 37, 229–236.

Donald, C.M. (1963) Competition among crop and pasture plants. *Advances in Agronomy* 15, 1–118.

Donald, W.W. (1994) The biology of Canada thistle (*Cirsium arvense*). *Reviews Weed Science* 6, 77–101.

Downton, W.J.S. (1975) The occurrence of C_4 photosynthesis among plants. *Photosynthetica* 9, 96–105.

Drottij, S. (1929) Redogörelse för försök med ogräsharvning åren 1924–1927. *Centralanstalten för Försöksväsendet på Jordbruksområdet. Underavdelningen för Växtodling.* Meddelande 348. Stockholm, 23 pp.

Dunham, R.S., Buchholtz, K.P., Derschied, L.A., Grigsby, G.H., Helgeson, E.A. and Staniforth, D.W. (1956) Quackgrass control. *Minnesota Agricultral Experiment Station, Station Bulletin* 434, 12 pp.

Dyke, G.V. and Barnard, A.J. (1976) Suppression of couchgrass by Italian ryegrass and broad red clover undersown in barley and field beans. *Journal of Agricultural Science*, Cambridge 87, 123–126.

Ebbersten, S. (1972) Pikloram (4-amino-3,5,6-triklorpikolinsyra). Studier av persistens i jord och växter samt metodstudier rörande biologisk bestämning av pikloramrester. Doctoral thesis. Lantbrukshögskolan, Institutionen för Växtodling, S-750 07 Uppsala, 173 pp.

Ellenberg, H. (1974) Zeigerwerte der Gefässpflanzen Mitteleuropas. *Scripta Geobotanica* 9. Goeltze, Göttingen, 97 pp.

Elliot, J.G. and Wilson, B.J. (1983) The influence of weather on the efficiency and safety of pesticide application. The drift of herbicides.

Occasional Publication No. 3. BCPC Publications, Thornton Heath, UK.

Ellwanger, T.C. Jr, Bingham, S.W. and Chapell, W.E. (1973) Physiological effects of ultrahigh temperatures on corn. *Weed Science* 21, 296–299.

Elmore, C.D. and Paul, R.N. (1983) Composite list of C_4 weeds. *Weed Science* 31, 686–692.

Erviö, L.-R. (1972) Growth of weeds in cereal populations. *Journal of the Scientific Agricultural Society of Finland* 44, 19–28.

Erviö, L.-R. (1983) Competition between barley and annual weeds at different sowing densities. *Annales Agriculturae Fenniae* 22, 232–239.

Erviö, L.-R. and Salonen, J. (1987) Changes in the weed population of spring cereals in Finland. *Annales Agriculturae Fenniae* 26, 201–226.

Espeby, L. (1989) Germination of weed seeds and competition in stands of weeds and barley. Influences of mineral nutrients. *Crop Production Science* 6. Uppsala, 172 pp.

Fail, H. (1956) The effect of rotary cultivation on the rhizomatous weeds. *Journal of Agricultural Engineering Research* 1, 68–80.

Fail, H. (1959) Mechanical eradication of couch and twitch. *World Crops* 11, 241–244.

Ferdinandsen, C. (1918) Undersøgelser over danske Ukrudtsformationer paa Mineraljorder. *Tidsskrift for Planteavl* 25, 629–919.

Finney, D.J. (1952) *Probit Analysis.* Cambridge University Press, Cambridge.

Finney, D.J. (1979) Bioassay and the practice of statistical inference. *International Statistical Review* 47, 1–12.

Firbank, L.G. and Watkinson, A.R. (1985) On the analysis of competition within two-species mixtures of plants. *Journal of Applied Ecology* 22, 503–517.

Fiveland, T.J. (1975) Forekomsten av de viktigste frøugras i åkerjorda i Norge 1947–1973. [Summary: The occurrence of important annual weed species on arable land in Norway 1947–1973.] *Meldinger fra Norges Landbrugshøgskole* 54(18). Ås Norbok, Oslo, 24 pp.

Fogelberg, F. and Dock Gustavsson, A.-M. (1999) Mechanical damage to annual weeds and carrots by in-row brush weeding. *Weed Research* 39, 469–479.

Fogelfors, H. (1973) Några ogräsarters utveckling under skilda ljusförhållanden och konkurrensförmåga i kornbestånd. Dissertation Lantbrukshögskolan. Institutionen för Växtodling, Uppsala, 74 pp.

Fogelfors, H. (1979) Floraförändringar i odlingslandskapet. Åkermark. *Sveriges Lantbruksuniversitet. Institutionen för Ekologi och Miljövård. Rapport 5.* Uppsala, 66 pp.

Fogelfors, H. (1981) Ogräsflorans förändring vid uppsamling av agnar, boss och halm vid skördetröskning. *Sveriges Lantbruksuniversitet. Institutionen för Ekologi och Miljövård. Rapport 8.* Uppsala, 27 pp.

Fogelfors, H. (1990) Different doses of herbicide for control of weeds in cereals – final report from the long-term series. *Swedish University of Agricultural Sciences. 31st Swedish Crop Protection Conference. Weeds and Weed Control,* Uppsala, pp. 139–151.

Fogelfors, H. and Boström, U. (1998) Anpassa höstbearbetningen efter ogräsfloran – håll tillbaka både ett- och fleråriga arter! *Fakta Jordbruk 8, SLU Info.* Uppsala, 4 pp.

Frantíc, T. (1994) Interference of *Chenopodium suecicum* J. Murr. and *Amaranthus retroflexus* L. in maize. *Weed Research* 34, 45–53.

Frantzen, J., Paul, N.D. and Müller-Schärer, H. (2001) The system management approach of biological weed control: some theoretical considerations and aspects of application. *BioControl* 46, 139–155.

Froud-Williams, R.J., Drennan, D.S.H. and Chancellor, R.J. (1983a) Influence of cultivation regime on weed floras of arable cropping systems. *Journal of Applied Ecology* 20, 187–197.

Froud-Williams, R.J., Drennan, D.S.H. and Chancellor, R.J. (1983b) Influence of cultivation regime upon buried weed seeds in arable cropping systems. *Journal of Applied Ecology* 20, 199–208.

Fryer, J. and Chancellor, R. (1970) Herbicides and our changing weeds. *Botanical Society of the British Isles Rep.* 11, 105–118.

Fykse, H. (1974a) Untersuchungen über *Sonchus arvensis* L. I. Translokation von C-14-markierten Assimilaten. *Weed Research* 14, 305–312.

Fykse, H. (1974b) Studium av åkerdylle. Utbredning i Noreg, vokster og kvile, dels jamført med nærstående arter. *Statens plantevern, Norway. Forskning og forsøk i landbruket* 25, 389–412.

Fykse, H. (1977) Untersuchungen über *Sonchus arvensis* L., *Cirsium arvense* (L.) Scop. und *Tussilago farfara* L. *Scientific Reports of the Agricultural University of Norway* 56. No. 27. 22 pp.

Fykse, H. (1986) Experiments with *Rumex* species. Growth and regeneration. *Scientific Reports of the Agricultural University of Norway* 65. No. 25, 11 pp.

Fykse, H. (1999) Utvikling av ugrasfloraen i åkeren. *Grønn Forskning 02/99,* 43–53. Planteforsk Apelsvoll Forskingssenter, Kapp, Norway.

Gallagher, R.S. and Cardina, J. (1998) The effect of light environment during tillage on the recruitment of various summer annuals. *Weed Science* 46, 214–216.

Garcia, J.G.L., Macbryde, B., Molina, A.R. and Herrera-Macbryde, O. (1975) *Prevalent Weeds of Central America*. International Plant Protection Center, Oregon State University, Corvallis, Oregon, 162 pp.

Gardner, S.N., Gressel, J. and Mangel, M. (1998) A revolving dose strategy to delay the evolution of both quantitative vs major monogene resistances to pesticides and drugs. *International Journal of Pest Management* 44, 161–180.

Gavlestam, M. (1977) Flyghavreinventering i Eskilstuna kommun 1976. Sveriges lantbruksuniversitetersitet, Institutionen för växtodlingslära. Examensarbete 678. Student work, Uppsala, 11 pp.

Gohde, N.-I. (1977) Flyghavre i Medelpad, Västernorrlands Län. Sveriges lantbruksuniversitetersitet, Institutionen för växtodlingslära. Examensarbete 679. Student work, Uppsala, 10 pp.

Goldberg, D.E. and Landa, K. (1991) Competitive effect and response: Hierarchies and correlated traits in the early stages of competition. *Journal of Ecology* 79, 1013–1030.

Grace, J.B. and Tilman, D. (eds) (1990) *Perspectives on Plant Competition*. Academic Press, San Diego, California, 484 pp.

Granström, B. (1962) Studier över ogräs i vårsådda grödor. [Summary: Studies on weeds in spring sown crops.] *Statens Jordbruksförsök. Meddelande* 130. Uppsala, 188 pp.

Granström, B. (1963) Ogräsbeståndet och stråsädens radavstånd. *Jordbrukstekniska Institutet. Meddelande* 302, 8–18.

Granström, B. (1972) Kemiska eller andra metoder i ogräsbekämpningen? *Lantbrukshögskolan. 13:e svenska ogräskonf. Ogräs och ogräsbekämpning. 1. Föredrag*, B 1–B 5.

Granström, B. and Almgård, G. (1955) Studier över den svenska ogräsfloran. [Summary: Studies of the Swedish weed flora.] *Statens Jordbruksförsök. Meddelande* 56. Uppsala, 23 pp.

Gray, A.J., Crawley, M.J. and Edwards, P.J. (eds) (1987) *Colonization, Succession and Stability*. Blackwell Scientific Publications, Oxford, 482 pp.

Grime, J.P. (1977) Evidence for the existence of three primary strategies in plants and its relevance to ecological and evolutionary theory. *American Naturalist* 111, 463–474.

Grümmer, G. (1961) The role of toxic substances in the interrelationships between higher plants. In: Milthorpe, F.L. (ed.) *Mechanisms in Biological Competition*. Cambridge University Press, Cambridge, 219 pp.

Grümmer, G. (1963) Das Verhalten von Rhizomen der Quecke (*Agropyron repens*) gegen trockene Luft. *Weed Research* 3, 44–51.

Gulliver, R.L. and Heydecker, W. (1973) Establishment of seedlings in a changeable environment. *Seed Ecology Proceedings of the 19th Easter School in Agricultural Science, University of Nottingham* 1972, pp. 433–462.

Gummerson, R.J. (1986) The effect of constant temperatures and osmotic potentials on the germination of sugar beet. *Journal of Experimental Botany* 37, 729–741.

Gummesson, G. (1975) Weeds in cereals. *Agricultural College Sweden, Weeds and Weed Control. 16th Swedish Weed Conference*, K 2–K 19.

Gummesson, G. (1979) Changes in weed species composition in field trials. *Agricultural College Sweden, Weeds and Weed Control. 20th Swedish Weed Conference*, 173–178.

Gummesson, G. (1981) Ogräsproblem i oljeväxtodlingen. *Sveriges Utsädesförenings Tidskrift* 91(4), 199–210.

Gummesson, G. (1982) Kemisk bekämpning av ogräs – alternativ och kostnader. *17th Rikkakasvipäivä Viikissä* 12.1.1982, A 22–A 38.

Gummesson, G. (1986) Chemical and non-chemical control – changes in the weed stand following different control measures. *Swedish University of Agricultural Sciences. 27th Swedish Weed Conference: Reports*, 236–256.

Gummesson, G. (1990) Results from a long-term series of trials – chemical and non-chemical weed control. *Swedish University of Agricultural Sciences. 31st Swedish Crop Protection Conference: Reports*, 152–164.

Gummesson, G. and Svensson, K. (1973) Tvåskiktsplog och plog med förplog vid bekämpning av flyghavre och kvickrot. *Lantbrukshögskolan Meddelande* A 202. Uppsala, 26 pp.

Gunnarsson, A. (1997) *Sammanställning över i marknaden förekommande ogräsharvar maj 1997*. Jordbruksverket. Rådgivningsprogrammet för Ekologiskt Lantbruk. Jönköping, Sweden, 57 pp.

Haas, H. and Streibig, J.C. (1982) Changing patterns of weed distribution as a result of herbicide use and other agronomic factors. In: LeBaron, H.M. and Gressel, J. (eds) *Herbicide Resistance in Plants*. Wiley, New York, pp. 57–79.

Habel, W. (1954) Über die Wirkungsweise der Eggen gegen Samenunkräuter sowie die Empfindlichkeit der Unkrautarten und ihrer Altersstadien gegen den Eggvorgang. Dissertation Arbeit aus dem Institut für Pflanzenschutz der Landwittschaftlichen Hochschule, Universität Stuttgart-Hohenheim, Germany, 58 pp.

Habel, W. (1957) Über die Wirkungsweise der Eggen gegen Samenunkräuter sowie deren Empfindlichleit gegen den Eggvorgang. *Zeitschrift für Acker- und Pflanzenbau* 104, 39–70.

Hagsand, E. (1984) Seed rates and Oxitril doses in six-row barley in Norrland. *Swedish University of Agricultural Sciences. 25th Swedish Weed Conference*, 62–68.

Hagsand, E. and Anier, H. (1978) Tuvtåtel i norra Sverige. *Röbäcksdalen meddelar. Rapport från Norrlands Lantbruksförsöksanstalt Röbäcksdalen 1978: 10. Växtodling.* Umeå, Sweden, 78 pp.

Hagsand, E. and Landström, S. (1984) Ensidig grovfoderodling i norra Sverige. *Sveriges lantbruksuniversitetersitet, Inst. för växtodlingslära. Rapport 143.* Uppsala, 74 pp.

Hagsand, E. and Thörn, K.-G. (1960) Norrländsk vallodling. Resultat av vallinventering i Västernorrlands, Jämtlands, Västerbottens och Norrbottens län 1955–1957. *Skogs- och Lantbruksakademiens Tidskrift 1960.* Suppl. 3. Stockholm, 156 pp.

Håkansson, I. and von Polgár, J. (1979) Effects on seedling emergence of soil slaking and crusting. *8th Conference International Soil Tillage Research Organization, ISTRO*, Hohenheim, West Germany, 115–120.

Håkansson, I. and von Polgár, J. (1984) Experiments on the effects of seedbed characteristics on seedling emergence in a dry weather situation. *Soil and Tillage Research* 4, 115–135.

Håkansson, I. and Reeder, R.C. (1994) Subsoil compaction by vehicles with high axle load – extent, persistence and crop response. *Soil and Tillage Research* 29, 277–304.

Håkansson, I., Stenberg, M. and Rydberg, T. (1998) Long-term experiments with different depths of mouldboard ploughing in Sweden. *Soil and Tillage Research* 46, 209–223.

Håkansson, S. (1963a) *Allium vineale* L. as a weed – with special reference to the conditions in south-eastern Sweden. *Växtodling [Plant Husbandry]* 19. Uppsala, 208 pp.

Håkansson, S. (1963b) Inverkan av utsädesmängd och gödsling på höstsädens utveckling i två konkurrensförsök på Öland. [Summary:

Influence of seed rate and fertilizer application on the development of winter cereals in two competition trials on Öland] *Lantbrukshögskolan Meddelanden, Serie A* 3. Uppsala, 22 pp.

Håkansson, S. (1967) Experiments with *Agropyron repens* (L.) Beauv. I. Development and growth, and the response to burial at different developmental stages. *Annals of the Agricultural College of Sweden* 33, 823–867.

Håkansson, S. (1968a) Experiments with *Agropyron repens* (L.) Beauv. II. Production from rhizome pieces of different sizes and from seeds. Various environmental conditions compared. *Annals of the Agricultural College of Sweden* 34, 3–29.

Håkansson, S. (1968b) Experiments with *Agropyron repens* (L.) Beauv. III. Production of aerial and underground shoots after planting rhizome pieces of different lengths at varying depths. *Annals of the Agricultural College of Sweden* 34, 31–51.

Håkansson, S. (1968c) *Introductory Agro-Botanical Investigations in Grazed Areas in the Chilalo Awraja, Ethiopia.* Available in the Main Library, Swedish University of Agricultural Sciences, Uppsala, Sweden, 69 pp.

Håkansson, S. (1969a) Experiments with *Agropyron repens* (L.) Beauv. IV. Response to burial and defoliation repeated with different intervals. *Annals of the Agricultural College of Sweden* 35, 61–78.

Håkansson, S. (1969b) Experiments with *Agropyron repens* (L.) Beauv. V. Effects of TCA and amitrole applied at different developmental stages. *Annals of the Agricultural College of Sweden* 35, 79–97.

Håkansson, S. (1969c) Experiments with *Agropyron repens* (L.) Beauv. VI. Rhizome orientation and life length of broken rhizomes in the soil, and reproductive capacity of different underground shoot parts. *Annals of the Agricultural College of Sweden* 35, 869–894.

Håkansson, S. (1969d) Experiments with *Agropyron repens* (L.) Beauv. VII. Temperature and light effects on development and growth. *Annals of the Agricultural College of Sweden* 35, 953–978.

Håkansson, S. (1969e) Experiments with *Sonchus arvensis* L. I. Development and growth and the response to burial and defoliation in different developmental stages. *Annals of the Agricultural College of Sweden* 35, 989–1030.

Håkansson, S. (1970) Experiments with *Agropyron repens* (L.) Beauv. IX. Seedlings and their response to burial and to TCA in the soil. *Annals of the Agricultural College of Sweden* 36, 351–359.

Håkansson, S. (1972) Konkurrens kulturväxtogräs. Betydelsen av inbördes förskjutningar i uppkomsttid. [In Swedish: Competition between crops and weeds. Effect of changes in their relative time of emergence.) *Lantbrukshögskolan. 13:e svenska ogräskonferensen Ogräs och ogräsbekämpning. 1. Föredrag*, F 1–F 4.

Håkansson, S. (1974) Kvickrot och kvickrotsbekämpning på åker. [In Swedish: Couch grass and its control in arable fields.] *Lantbrukshögskolan. Meddelande B* 21. Uppsala, 82 pp.

Håkansson, S. (1975a) Grödan-växtodlingssystemet och ogräset på åkern. *Lantbrukshögskolan. Ogräs och ogräsbekämpning. 16:e svenska ogräskonferensen*, K 1–K 20.

Håkansson, S. (1975b) Grundläggande växtodlingsfrågor. I. Inflytande av utsädesmängden och utsädets horisontella fördelning på utveckling och produktion i kortvariga växtbestånd. [In Swedish: Basic research in crop production. I. Influence of seed rate and horizontal seed distribution on development and production in short-lived crop–weed stands.] *Lantbrukshögskolan. Institutionen för växtodling. Rapporter och avhandlingar* 33. Uppsala, 192 pp.

Håkansson, S. (1977) Växtodlingen och ogräset på åkern. 5. Villkor för växter av olika karaktär på åkern och i landskapet. *Lantmannen* 98(2), 14–16.

Håkansson, S. (1979) Grundläggande växtodlingsfrågor. II. Faktorer av betydelse för plantetablering, konkurrens och produktion i åkerns växtbestånd. [In Swedish: Basic research in crop production. II. Factors of importance for plant establishment, competition and production in crop–weed stands.] *Sveriges lantbruksuniversitetersitet, Inst. för växtodling. Rapporter och avhandlingar* 72. Uppsala, 88 pp.

Håkansson, S. (1981) Utsäde–såbädd–sådd. Några grundläggande sammanhang. [In Swedish: Seed–seedbed–sowing. Some fundamental contexts.] *Sveriges lantbruksuniversitetersitet. Kompendium i växtodling.* Uppsala, 26 pp.

Håkansson, S. (1982) Multiplication, growth and persistence of perennial weeds. In: Holzner, W. and Numata, N. (eds) *Biology and Ecology of Weeds*, Dr W. Junk Publishers, The Hague, pp. 123–135.

Håkansson, S. (1983a) Seasonal variation in the emergence of annual weeds – an introductory investigation in Sweden. *Weed Research* 23, 313–324.

Håkansson, S. (1983b) Competition and production in short-lived crop-weed stands. Density effects. *Swedish University of Agricultural Sciences. Department of Crop Production Science Report 127.* Uppsala, 85 pp.

Håkansson, S. (1984a) Seed rate and amount of weeds – influence on production in stands of cereals. *Swedish University of Agricultural Sciences. 25th Swedish Weed Conference: Reports*, 1–16.

Håkansson, S. (1984b) Row spacing, seed distribution in the row, amount of weeds – influence on production in stands of cereals. *Swedish University of Agricultural Sciences. 25th Swedish Weed Conference: Reports*, 17–34.

Håkansson, S. (1986) Competition between crops and weeds – influencing factors, experimental methods and research needs. *Proceedings of the European Weed Research Society Symposium 1986, Economic Weed Control*, pp. 49–60.

Håkansson, S. (1988a) Growth in plant stands of different density. *VIIIème Colloque International sur la Biologie, l'Ecologie et la Systematique des Mauvaises Herbes*, pp. 631–639. [European Weed Research Society Symposium, Dijon, September 1988.]

Håkansson, S. (1988b) Competition in stands of short-lived plants. Density effects measured in three-component stands. *Crop Production Science 3.* Swedish University of Agricultural Sciences, Uppsala, 181 pp.

Håkansson, S. (1989) Åkerogräs – egenskaper och förekomst. *Aktuellt från lantbruksuniversitetet 382. Mark/Växter*, pp. 3–10.

Håkansson, S. (1991) Growth and competition in plant stands. *Crop Production Science 12.* Swedish University of Agricultural Sciences, Uppsala, 241 pp.

Håkansson, S. (1992) Germination of weed seeds in different seasons. *IXth International Symposium on the Biology of Weeds*, Paris, pp. 45–54.

Håkansson, S. (1993) *Modellexperiment gällande växtodlingsfrågor.* I–XI. 95 pp. [Experiments used in teaching, illustrating basic biological questions in crop production: dormancy and germination, plant emergence and establishment, competition, effects and decomposition of herbicides, etc.] Copy available at the Department of Ecology and Crop Production Science, PO Box 7043, SE–750 07 Uppsala, Sweden.

Håkansson, S. (1995a) Weeds in agricultural crops. 1. Life-forms and occurrence under

Swedish conditions. *Swedish Journal of Agricultural Research* 25, 143–154.

Håkansson, S. (1995b) Weeds in agricultural crops. 2. Life-forms and occurrence in a European perspective. *Swedish Journal of Agricultural Research* 25, 155–161.

Håkansson, S. (1995c) Weeds in agricultural crops. 3. Life-forms, C_3 and C_4 photosynthesis and plant families in a global perspective. *Swedish Journal of Agricultural Research* 25, 163–171.

Håkansson, S. (1995d) Ogräs och odling på åker. *Aktuellt från lantbruksuniversitetetet. Mark/Växter* 437/438. Uppsala, 70 pp.

Håkansson, S. (1997a) Competitive effects and competitiveness in annual plant stands. 1. Measurement methods and problems related to plant density. *Swedish Journal of Agricultural Research* 27, 53–73.

Håkansson, S. (1997b) Competitive effects and competitiveness in annual plant stands. 2. Measurements of plant growth as influenced by density and relative time of emergence. *Swedish Journal of Agricultural Research* 27, 75–94.

Håkansson, S. and Jendeberg, H. (1972) Konkurrens kulturväxt-ogräs. Ogräsens utveckling vid olika grad av luckighet i första årets vall. [In Swedish: Competition between crops and weeds. Growth of weeds in gaps of different widths in ley stands.] *Lantbrukshögskolan. 13:e svenska ogräskonferensen Ogräs och ogräsbekämpning. 1. Föredrag*, F 12–F 16.

Håkansson, S. and Jonsson, E. (1970) Experiments with *Agropyron repens* (L.) Beauv. VIII. Responses of the plant to TCA and low moisture contents in the soil. *Annals of the Agricultural College of Sweden* 36, 135–151.

Håkansson, S. and Larsson, R. (1966) Ojämn uppkomst i vårsäd. [In Swedish: Uneven emergence in spring cereals.] *Aktuellt från Lantbrukshögskolan* 102, 36–42.

Håkansson, S. and Wallgren, B. (1972a) Experiments with *Sonchus arvensis* L. II. Reproduction, plant development and response to mechanical disturbance. *Swedish Journal of Agricultural Research* 2, 3–14.

Håkansson, S. and Wallgren, B. (1972b) Experiments with *Sonchus arvensis* L. III. The development from reproductive roots cut into different lengths and planted at different depths, with and without competition from barley. *Swedish Journal of Agricultural Research* 2, 15–26.

Håkansson, S. and Wallgren, B. (1976) *Agropyron repens* (L.) Beauv., *Holcus mollis* L. and *Agrostis gigantea* Roth as weeds – some

properties. *Swedish Journal of Agricultural Research* 6, 109–120.

Hallgren, E. (1972) Undersökning av slåttervallar år 1970. *Lantbrukshögskolan. 13:e svenska ogräskonferensen Ogräs och ogräsbekämpning. 1. Föredrag*, F 17–F 22.

Hallgren, E. (1974a) Utveckling och konkurrens i vallbestånd med ogräs. *Sveriges Lantbruksuniversitetersitet, Institutionen för växtodling. Rapporter och Avhandlingar* 9. Uppsala, 85 pp.

Hallgren, E. (1974b) Groning och plantutveckling hos vallväxter vid olika sådjup och vid olika markfuktighet. *Sveriges lantbruksuniversitetersitet, Institutionen för växtodling. Rapporter och Avhandlingar* 18. Uppsala, 34 pp.

Hallgren, E. (1976) Development and competition in stands of ley plants and weeds. 1. Influence of missing rows or gaps in rows. *Swedish Journal of Agricultural Research* 6, 247–253.

Hallgren, E. (1990a) Influence of yield level and amount of weeds in different crops on the yield increase after spraying. *Swedish University of Agricultural Sciences. 31st Swedish Weed Conference*, Uppsala, pp. 20–30.

Hallgren, E. (1990b) Olika faktorers inflytande på effekten av kemisk bekämpning i våroljeväxter. *Sveriges Lantbruksuniversitet. Växtodling* 16. Uppsala, 73 pp.

Hallgren, E. (1991a) Frekvensfördelningar av några variabler i fältförsök med kemisk ogräsbekämpning i höstsäd. *Sveriges Lantbruksuniversitet. 32:a svenska växtskyddskonferensen Ogräs och ogräsbekämpning. Rapporter*, Uppsala, pp. 9–21.

Hallgren, E. (1991b) Frekvensfördelningar av några variabler i fältförsök med kemisk ogräsbekämpning i vårsäd. *Sveriges Lantbruksuniversitet. 32:a svenska växtskyddskonferensen Ogräs och ogräsbekämpning. Rapporter*, Uppsala, pp. 23–37.

Hallgren, E. (1991c) Olika faktorers inflytande på effekten av kemisk ogräsbekämpning i höstoljeväxter. *Sveriges Lantbruksuniversitet. Växtodling* 34. Uppsala, 90 pp.

Hallgren, E. (1993) Do the weed flora and effect of a herbicide (Oxitril 4) change with time? *Swedish University of Agricultural Sciences. 34th Swedish Plant Protection Conference*, Uppsala, pp. 7–24.

Hallgren, E. (1997) Vilka örtogräsarter utgör störst andel av den totala ogräsmängden i olika grödor och på olika jordarter? *Sveriges Lantbruksuniversitetersitet. Växtodling* 53, 5–21.

Hallgren, E. (2000) Vilka ogräsarter förekommer mest frekvent och och utgör störst andel av den totala ogräsmängden i olika grödor? *Sveriges Lantbruksuniversitetersitet. Rapport från Fältforskningsenheten* 1, 5–31.

Hallgren, E. and Fischer, A. (1992) Olika faktorers inflytande på effekten av Roundup mot kvickrot i stubbåker, i vall före vallbrott och på träda. *Sveriges Lantbruksuniversitetersitet. 33:e svenska växtskyddskonferensen. Ogräs och ogräsbekämpning. Rapporter*, pp. 121–141.

Hamdoun, A. (1967) A study of the life history, regenerative capacity, and response to herbicide of *Cirsium arvense*. Master's thesis, University of Reading, UK, 226 pp.

Hammerton, J.L. (1970) Crop seed-rate and MCPA effectiveness. *Weed Research* 10, 178–180.

Hanf, M. (1982) *Ackerunkräuter Europas mit ihren Keimlingen und Samen.* BASF Aktiengesellschaft, Ludwigshafen, 496 pp.

Hansson, M.L. and Fogelfors, H. (1998) Management of permanent set-aside on arable land in Sweden. *Journal of Applied Ecology* 35, 758–771.

Hansson, M.L., Boström, U. and Fogelfors, H. (2001) Influence of harvest time and stubble height on weed seedling recruitment in a green cereal. *Acta Agriculturae Scandinavica, Section B, Soil and Plant Science* 51, 2–9.

Harper, J.L. (1977) *Population Biology of Plants.* Academic Press, London, 892 pp.

Hashimoto, G. (1982) Brazil. In: Holzner, W. and Numata, N. (eds) *Biology and Ecology of Weeds.* Dr W. Junk Publishers, The Hague, pp. 333–337.

Hayden, A. (1934) Distribution and reproduction of Canada thistle in Iowa. *American Journal of Botany* 21, 355–372.

Healy, A. (1953) Control of docks. *New Zealand Journal of Science and Technology* (Section A) 34, 473–475.

Heege, H.J. (1977) Technik der Getreidebestellung. *Landtechnik. DLG-Mitteilungen* 18/1977, 1004–1007.

Heimer, A. and Olrog, L. (1997) *Redovisning av resultat från projektet 'Metodutveckling inom ekologiskt lantbruk i västra Sverige'. Slutrapport inom programmet 'Ekologiskt lantbruk'.* Jordbruksverket, Jönköping, Sweden.

Heinonen, R. (1982) Jordens igenslamning och förhårdnande. [In Swedish: Slaking and hardening of soils.] *Sveriges Lantbruksuniversitetersitet. Speciella Skrifter* 12. Uppsala, 24 pp.

Henke, F. (1929) Untersuchungen über den Einfluss von Saatmenge und Reihenweite

aur Entwicklung und Ertrag des Roggens. Dissertation. Institut Pflanzenbau der Landwirtschaftlichen Hochschule Hohenheim. Buchdr. J. Särchen, Baruth/Mark-Berlin.

Henson, I.E. (1969) Studies on the regeneration of perennial weeds in the glasshouse. I. Temperate species. *Agricultural Research Council Weed Research Organization Oxford, Technical Report* 12. 23 pp.

Hillel, D. (1972) Soil moisture and seed germination. In: Kotzlowski, P.P. (ed.) *Water Deficit and Plant Growth, Vol. III.* Academic Press, New York, pp. 65–89.

Hodgson, J. (1964) Variations in ecotypes of Canada thistle. *Weeds* 12, 167–171.

Hodgson, J. (1968) The nature, ecology and control of Canada thistle. *US Department of Agriculture Technical Bulletin* 1386. Washington, DC.

Hodgson, J. (1971) Canada thistle and its control. *US Department of Agriculture Leaflet* 523. Washington, DC, 8 pp.

Hoffmann, J. (1994) Spontan wachsende C_4-Pflanzen in Deutschland und Schweden – eine Übersicht unter Berücksichtigung möglicher Klimaänderungen. *Angewandte Botanik* 68, 65–70.

Holliday, R. (1960a) Plant population and crop yield. *Nature (London)* 186, 22–24.

Holliday, R. (1960b) Plant population and crop yield. *Field Crop Abstracts* 13, 159–167.

Holm, L.G., Plucknett, D.L., Pancho, J.V. and Herberger, J.P. (1977) *The World's Worst Weeds – Distribution and Biology.* University Press of Hawaii, Honolulu, 609 pp.

Holm, L.G., Doll, J., Holm, E., Pancho, J. and Herberger, J. (1997) *World Weeds. Natural Histories and Distribution.* John Wiley & Sons, New York, 1129 pp.

Holmøy, R. and Netland, J. (1994) Band spraying, selective flame weeding and hoeing in late white cabbage. Part I. *Acta Horticulturae* 372, 223–234.

Holzner, W. (1982) Concepts, categories and characteristics of weeds. In: Holzner, W. and Numata, N. (eds) *Biology and Ecology of Weeds.* Dr W. Junk Publishers, The Hague, pp. 3–20.

Holzner, W., Hayashi, I. and Glauninger, J. (1982) Reproductive strategy of annual agrestals. In: Holzner, W. and Numata, N. (eds) *Biology and Ecology of Weeds.* Dr W. Junk Publishers, The Hague, pp. 111–121.

Holzner, W. and Numata, N. (eds) (1982) *Biology and Ecology of Weeds.* Dr W. Junk Publishers, The Hague, 461 pp.

Hoogerkamp, M. (1975) *Elytrigia repens* and its control in leys. *Proceedings of the Research Society Symposium European Weed, Paris,* pp. 322–327.

Hubbard, C.E. (1954) *Grasses. A Guide to Their Structure, Identification, Uses, and Distribution in the British Isles.* Penguin Books, Harmondsworth, 428 pp.

Hudson, H.J. (1941) Population studies with wheat. II. Propinquity. *Journal of Agriculturel Science* 31, 116–137.

Hudson, J. (1955) Propagation of plants by root cuttings. *Journal of Horticultural Science* 30, 242–251.

Huhtapalo, Å. (1982) Scandinavian principles for fertilizer placement. Utilization of fertilizer-N. *9th Conference of the International Soil Tillage Organization, ISTRO, Osijek, Yugoslavia,* pp. 669–674.

Hultén, E. (1950) *Atlas över växternas utbredning i Norden. Fanerogamer och ormbunksväxter.* Stockholm, 512 pp.

Hume, L. (1987) Long-term effects of 2,4-D application on plants. I. Effects on the weed community in a wheat crop. *Canadian Journal of Botany* 65, 2530–2536.

Inouye, R.S. and Schaffer, W.M. (1981) On the ecological meaning of ratio (de Wit) diagrams in plant ecology. *Ecology* 62 (5), 1679–1681.

Isaacson, D.L. and Charudattan, R. (1999) Biological approaches to weed management. In: Ruberson, J.R. (ed.) *Handbook of Pest Management.* Marcel Dekker, New York, pp. 593–625.

Jensen, H.A. (1969) Content of buried seeds in arable soil in Denmark and its relation to the weed population. *Dansk Botanisk Arkiv* 27 (2), 1–56.

Jensen, H.A. and Kjellsson, G. (1995) Frøpuljens størrelse og dynamik i moderne landbrug. I. Aendringer af frøindholdet i agerjord 1964–1989. *Bekaempelsesmiddelforskning fra Miljøstyrelsen* 13, 1–141.

Jensen, P.K. (1995) Effect of light environment during soil disturbance on germination and emergence pattern of weeds. *Annals of Applied Biology* 127, 561–571.

Johansson, D. (1997) Radhackning. *Sveriges Lantbruksuniversitet. Rapporter från Jordbearbetningsavdelningen* 91, 45–57.

Johnson, B.G. and Buchholtz, K.P. (1962) The natural dormancy of vegetative buds on the rhizomes of quackgrass. *Weeds* 10, 53–57.

Johnson, G.A., Defelice, M.S. and Helsel, Z.R. (1993) Cover crop management and weed control in corn (*Zea mays*). *Weed Technology* 7, 425–430.

Jordan, N.R., Zhang, J. and Huerd, S. (2000) Arbuscular-mycorrhizal fungi: potential roles in weed management. *Weed Research* 40, 397–410.

Jørnsgård, B., Rasmussen, K., Hill, J. and Christiansen, J.L. (1996) Influence of nitrogen on competition between cereals and their natural weed populations. *Weed Research* 36, 461–470.

Julien, M.H. and Griffiths, M.W. (1998) *Biological Control of Weeds. A World Catalogue of Agents and their Target Weeds,* 4th edn. CAB International, Wallingford, UK.

Känkänen, H.J., Mikkola, H.J. and Eriksson, C.I. (2001) Effect of sowing technique on growth of undersown crop and yield of spring barley. *Journal of Agronomy and Crop Science* 187, 127–136.

Karssen, C.M. (1982) Seasonal patterns of dormancy in weed seeds. In: Khan, A.A. (ed.) *The Physiology and Biochemistry of Seed Development, Dormancy and Germination.* Elsevier, Amsterdam, pp. 243–270.

Kaufmann, M.R. (1969) Effect of water potential on germination of lettuce, sunflower, and citrus seeds. *Canadian Journal of Botany* 47, 1761–1764.

Kees, H. (1962) Untersuchungen zur Unkrautbekämpfung durch Netzegge und Stoppelbearbeitungsmassnahmen unter besonderer Berücksichtigung des leichten Bodens. Dissertation. Arbeit aus dem Institut für Pflanzenschutz der Landwirtschaftlichen Hochschule, Universität Stuttgart-Hohenheim. 103 pp.

Kephart, L.W. (1923) Quack grass. *US Department of Agriculture, Farmer's Bulletin* 1307, 29 pp.

Kim, D.S., Brain, P., Marshall, E.J.P. and Caseley, J.C. (2002) Modelling herbicide dose and weed density effects on crop:weed competition. *Weed Research* 42, 1–13.

King, L.J. (1966) *Weeds of the World.* Interscience Publications, New York, 526 pp.

Kira, T., Ogawa, H. and Sakazaki, N. (1953) Intraspecific competition among higher plants. I. Competition-yield-density interrelationship in regularly dispersed populations. *Osaka City University Journal of the Institute of Polytechnics* 4, 1–16.

von Kirchner, O., Loew, E. and Schröter, C. (1912–1913) *Lebensgeschichte der Blütenpflanzen Mitteleuropas I:* 3. Stuttgart.

Knouwenhoven, J.K. (1981) Inter- and intra-row weed control with show-ridger. *Journal of Agricultural Engineering Research* 26, 127–134.

Knouwenhoven, J.K. (1997) Intra-row mechanical weed control – possibilities and problems. *Soil and Tillage Research* 41, 87–104.

Koch, W. (1964) Unkrautbekämpfung durch Eggen, Hacken und Meisseln in Getreide. 1. Wirkungsweise und Einsatzzeitpunkt von Egge, Hacke und Bodenmeissel. *Zeitschrift für Acker- und Pflanzenbau* 120, 369–382.

Koch, W. (1969) Einfluss von Umweltfaktoren auf die Samenphase annueller Unkräuter insbesondere unter dem Gesichtspunkt der Unkrautbekämpfung. Arbeiten der Universität Hohenheim (Landwirtschaftliche Hochschule) 50. Verlag Eugen Ulmer, Stuttgart, 204 pp.

Kolk, H. (1962) Viability and dormancy of dry stored weed seeds. *Växtodling (Plant Husbandry)* 18. Uppsala, 192 pp.

Kollar, D. (1968) The germination and viability of weed seeds matured in winter wheat at different soil depths. *Acta Fytotechnica Nitra* 17, 103–110.

Korsmo, E. (1930) *Unkräuter im Ackerbau der Neuzeit.* Berlin, 580 pp.

Korsmo, E. (1954) *Anatomy of Weeds.* Grøndahl & Søns Forlag, Oslo, 413 pp.

Kritz, G. (1983) Såbäddar för vårstråsäd. En stickprovsundersökning. [Summary: Physical conditions in cereal seedbeds. A sampling investigation in Swedish spring-sown fields.] *Swedish University of Agricultural Sciences, Division of Soil Management, Report 65.* Uppsala, 187 pp.

Kropff, M.J. (1988) Modelling the effects of weeds on crop production. *Weed Research* 28, 465–471.

Kropff, M.J. and Spitters, C.T.J. (1991) A simple model of crop loss by weed competition from early observations on relative leaf area of weeds. *Weed Research* 31, 97–105.

Kropff, M.J. and van Laar, H.H. (eds) (1993) *Modelling Crop–Weed Interactions.* CAB International, Wallingford and International Rice Research Institute, Manilla, 274 pp.

Krumbiegel, A. (1998) Growth forms of annual vascular plants in central Europe. *Nordic Journal of Botany* 18, 563–575.

Kvist, M. (1992) Catch crops undersown in spring barley – competitive effects and cropping methods. *Crop Production Science* 15, Swedish University of Agricultural Sciences, Uppsala, 210 pp.

Kvist, M. and Håkansson, S. (1985) Rytm och viloperioder i vegetativ utveckling och tillväxt hos några fleråriga ogräs. [English summary: Rhythm and periods of dormancy in the vegetative development and growth of some perennial weeds.] *Swedish University of Agricultural Sciences, Department of Plant Husbandry. Report 156.* Uppsala, 110 pp.

Landström, S. (1983) Control of *Agropyron repens* in different cropping systems. *Swedish University of Agricultural Sciences. 24th Swedish Weed Conference: Reports.* Uppsala, pp. 246–256.

Lang, A.L., Pendleton, J.W. and Dungan, G.H. (1956) Influence of population and nitrogen levels on yield and protein and oil contents of nine corn hybrids. *Agronomy Journal* 48, 284–289.

Larson, R.E., Klingman, D.L. and Fletchall, O.H. (1958) The effect of smoothing devices on the action of pre-emergence herbicides in soybeans and corn. *Weeds* 6, 126–132.

Larsson, R. (1974) Övervintringsproblem i höstvete och höstråg. *Institutionen för växtodling. Rapporter och Avhandlingar* 8, 29–40. Uppsala, Sweden.

Laursen, F. (1967) De almindeligste ukrudtsarters forekomst i danske afgrøder med saerlig henblik på fordelingen efter jordbundsforhold, Lt- og pH-vaerdier. Licentiataf-handling. Den Kongelige Landbo- og Veterinaerhøjskole, Copenhagen, 57 pp.

Law, R., Bradshaw, A.D. and Putwain, P.D. (1977) Life-history variation in *Poa annua. Evolution* 31, 233–246.

Lazenby, A. (1961a) Studies on *Allium vineale* L. I. The effects of soils, fertilizers and competition on establishment and growth of plants from aerial bulbils. *Journal of Ecology* 49, 519–541.

Lazenby, A. (1961b) Studies on *Allium vineale* L. II. Establishment and growth in different intensities of competition. *Journal of Ecology* 49, 543–548.

Lazenby, A. (1962a) Studies on *Allium vineale* L. III. Effect of depth of planting. *Journal of Ecology* 50, 97–109.

Lazenby, A. (1962b) Studies on *Allium vineale* L. IV. Effect of cultivations. *Journal of Ecology* 50, 411–428.

Leakey, R.R.B., Chancellor, R.J. and Vince-Prue, D. (1978a) Regeneration from rhizome fragments of *Agropyron repens* (L.) Beauv. I. The seasonality of shoot growth and rhizome reserves in single-node fragments. *Annals of Applied Biology* 87, 423–431.

Leakey, R.R.B., Chancellor, R.J. and Vince-Prue, D. (1978b) Regeneration from rhizome fragments of *Agropyron repens* (L.) Beauv. II. The breaking of 'late spring dormancy' and the influence of chilling and node position on growth from single-node fragments. *Annals of Applied Biology* 87, 433–441.

Leakey, R.R.B., Chancellor, R.J. and Vince-Prue, D. (1978c) Regeneration from rhizome fragments of *Agropyron repens* (L.) Beauv. III. Effects of nitrogen and temperature on the development of dominance amongst shoots on multi-node fragments. *Annals of Botany* 42, 197–204.

Leakey, R.R.B., Chancellor, R.J. and Vince-Prue, D. (1978d) Regeneration from rhizome fragments of *Agropyron repens* (L.) Beauv. IV. Effects of light on bud dormancy and the development of dominance amongst shoots on multi-node fragments. *Annals of Botany* 42, 205–212.

LeBaron, H.M. and Gressel, J. (eds) (1982) *Herbicide Resistance in Plants.* John Wiley & Sons, New York, 401 pp.

Légère, A. and Samson, N. (1999) Relative influence of crop rotation, tillage, and weed management on weed associations in spring barley cropping systems. *Weed Science* 47, 112–122.

Légère, A., Samson, N., Lemieux, C. and Rioux, R. (1990) Effects of weed management and reduced tillage on weed populations and barley yields. *Proceedings of the European Weed Research Society Symposium, Integrated Weed Management in Cereals*, pp. 111–118.

Leiss, K.A. (2001) Phenotic plasticity and genetic differentiation in ruderal and agricultural populations of the weed *Senecio vulgaris* L.: implication for its biological control. *European Weed Research Society. Newsletter 76: January 2001*, pp. 15–17.

Lemerle, D., Verbeek, B. and Coombes, N.E. (1996) Interaction between wheat (*Triticum aestivum*) and diclofop to reduce the cost of annual ryegrass (*Lolium rigidum*) control. *Weed Science* 44, 634–639.

Leuchovius, J.-O. (1972) Inventering gällande förekomsten av olika gräsarter som ogräs i stråsäd och oljeväxter i Skaraborgs län. *Sveriges Lantbruksuniversitetersitet, Institutionen för Växtodling. Examensarbeten 584.* Extended student work. Uppsala, 13 pp.

Liebl, R.A. and Worsham, A.D. (1983) Tillage and mulch effects on morning glory (*Ipomoea* spp.) and certain other weed species. *Proceedings of the Southern Weed Science Society, 36th Annual Meeting*, pp. 405–414.

Liebl, R.A., Simmons, F.W., Wax, L.M. and Stoller, E.W. (1992) Effect of rye (*Secale cereale*) mulch on weed control and soil moisture in soybeen (*Glycine max*). *Weed Technology* 6, 838–846.

Liebman, M., Mohler, C.L. and Staver, C.P. (2001) *Ecological Management of Agricultural Weeds.* Cambridge University Press, Cambridge, 532 pp.

Lotz, A.P., Christensen, C., Cloutier, D., Fernandez Quintanilla, C., Légère, A., Lemieux, C., Lutman, P.J.W., Pardo Eglesias, A., Salonen, J., Sattin, M., Stigliani, L. and Tei, F. (1996) Prediction of the competitive effects of weeds on crop yields based on the relative leaf area of weeds. *Weed Research* 36, 93–101.

Lund, S. and Rostrup, E. (1901) Marktidselen, *Cirsium arvense.* En monografi. *Den Kongelige Danske Videnskabernes Selskabs Skrifter, Raekke 6, Naturvidenskabelig og Mathematisk Afdenling* 10(3), 149–316.

Lundkvist, A. and Fogelfors, H. (1999) *Ogräsreglering på åkermark.* Institutionen för Ekologi och Växtproduktionslära. Rapport 1. Uppsala, 266 pp.

Mahn, E.G. (1984) The influence of different nitrogen levels on the productivity and structural changes of weed communities in agroecosystems. *VIIe Colloque International sur l'Ecologie et la Systematique des Mauvaises Herbes*, pp. 421–429.

Mann, H.H. and Barnes, T.W. (1947) The competition between barley and certain weeds under controlled conditions. II. Competition with *Holcus mollis. Annals of Applied Biology* 34, 252–266.

Manuhar, M.S. and Heydecker, W. (1964) Water requirements for seed germination. *University of Nottingham, School of Agriculture, Report 1964*, pp. 55–62.

Martin, J. (1943) Germination studies of the seeds of some common weeds. *Proceedings of the Iowa Academy of Science* 50, 221–228.

Martin, R.J. and Felton, W.L. (1993) Effect of crop rotation, tillage practice, and herbicides on the population dynamics of wild oats in wheat. *Australian Journal of Experimental Agriculture* 33, 159–165.

Matsubara, H. and Sugiyama, T. (1965) Influence of soil moisture content on germination and emergence of seeds. [In Japanese with English summary.] *Journal of the Japanese Society for Horticultural Science* 34, 105–112.

Mattsson, B., Nylander, C. and Ascard, J. (1990) Comparison of seven inter-row weeders. *3rd International Federation of Organic Agriculture Movements (IFOAM) Conference on Non-Chemical Weed Control. Linz, Oct. 1989.* 20, Bundesanstalt für Agrobiologie Linz/Donau. 20, 91–107.

Mayer, A.M. and Poljakoff-Mayber, A. (1978) *The Germination of Seeds*, 2nd edn. Pergamon Press, Oxford, 192 pp.

McIntyre, G.I. (1971) Apical dominance in the rhizome of *Agropyron repens*. Some factors affecting the degree of dominance in isolated rhizomes. *Canadian Journal of Botany* 49, 99–109.

McIntyre, G.I. (1972) Studies on bud development in the rhizome of *Agropyron repens*. II. The effect of nitrogen supply. *Canadian Journal of Botany* 50, 393–401.

Melander, B. (1988) Spredning og etablering af alm. kvik (*Elymus repens*) gennem frø. 5. *Danske Planteværnskonference/Ukrudt*, pp. 122–130.

Melander, B. (1995) Pre-harvest assessments of *Elymus repens* (L.) Gould. interference in five arable crops. *Acta Agricultrae Scandinavica Section B, Soil and Plant Science* 45, 188–196.

Melander, B. (1997) Optimization of the adjustment of a vertical axis rotary brush weeder for intra-row weed control in row crops. *Journal of Agricultural Engineering Research* 68, 39–50.

Melander, B. (1998) Interaction between soil cultivation in darkness, flaming and brush weeding when used for in-row weed control in vegetables. *Biological Agriculture and Horticulture* 16, 1–14.

Melander, B., Rasmussen, J. and Rasmussen, K. (1995) Icke-kemisk ukrudtsbekaempelse – muligheter og begraensninger i vinterraps og majs. *Statens Planteavlsforsøg. SP-rapport 3*, Tjele, Denmark, pp. 123–137.

Meyler, J.F. and Rühling, W. (1966) Mechanische Unkrautbekämpfung bei höheren Geschwindigkeiten. *Landtechnische Forschung* 16, 79–85.

Mikkelsen, G. (1997) Mekanisk renholdelse i majs og kartofler. *Statens Planteavlsforsøg. SP-rapport 7*, Tjele, Denmark, pp. 183–192.

Mikkelsen, V.M. and Laursen, F. (1966) Markukrudtet i Danmark omkring 1960. [English summary.] *Botanisk Tidsskrift* 62, 1–26.

Milberg, P. and Andersson, L. (1997) Seasonal variation in dormancy and light sensitivity in buried seeds of eight annual weed species. *Canadian Journal of Botany* 75, 665–672.

Milberg, P., Hallgren, E. and Palmer, M.W. (2000) Interannual variation in weed biomass on arable land in Sweden. *Weed Research* 40, 311–321.

Miles, J. (1987) Vegetation succession: past and present perceptions. In: Gray, A.J., Crawley, M.J. and Edwards, P.J. (eds) *Colonization, Succession and Stability*, Blackwell Scientific Publications, Oxford, pp. 1–29.

Miller, P.H.C. (1993) Spray drift and its measurement. In: Matthews, G.A. and Hislop, E.C. (eds) *Application Technology for Crop Protection*, CAB International, Wallingford, UK, pp. 101–122.

Mittnacht, A., Eberhardt, C. and Koch, W. (1979) Wandel in der Getreideunkrautflora seit 1948, untersucht an einem Beispiel in Südwestdeutschland. *Proceedings of the European Weed Research Society Symposium on the Influence of Different Factors on the Development and Control of Weeds*, pp. 209–216.

Mohler, C.L. and Galford, A.E. (1997) Weed seedling emergence and seed survival: separating the effects of seed position and soil modification by tillage. *Weed Research* 37, 147–155.

Morin, C., Blanc, D. and Darmency, H. (1993) Limits of a simple model to predict yield losses in maize. *Weed Research* 33, 261–268.

Morrow, G.E. and Hunt, T.F. (1891) Field experiments with corn, 1890. *Illinois Agricultural Experiment Station Bulletin* 13, 389–451.

Mortensen, K. (1998) Biological control of weeds using microorganisms. In: Boland, G.J. and Kuykendall, L.D. (eds) *Plant–Microbe Interactions and Biological Control*. Marcel Dekker, New York, pp. 223–248.

Mortimer, A.M. (1987) Contributions of plant population dynamics to understanding early succession. In: Gray, A.J., Crawley, M.J. and Edwards, P.J. (eds) *Colonization, Succession and Stability*. Blackwell Scientific Publications, Oxford, pp. 1–29.

Mossberg, B., Stenberg, L. and Ericsson, S. (1992) *Den Nordiska Floran*. Wahlström and Widstrand, Stockholm, 696 pp.

Mukula, J., Raatikainen, M., Lalluka, R. and Raatikainen, T. (1969) Composition of weed flora in spring cereals in Finland. *Annales Agriculturae Fenniae* 8, 59–110.

Müller, C.H. (1953) The association of desert annuals with shrubs. *American Journal of Botany* 40, 53–60.

Müller, C.H. (1969) Allelopathy as a factor in ecological processes. *Vegetatio* 18, 348–357.

Müller-Schärer, H. and Frantzen, J. (1996) An emerging system management approach for biological weed control in crops: *Senecio vulgaris* as a research model. *Weed Research* 36, 483–491.

Müller-Schärer, H. and Vogelgsang, S. (eds) (2000) Finding solutions for biological control of weeds in European crop systems. *European*

Commission. *Agriculture and Biotechnology. COST Action 816.* 79 pp.

Mulugeta, D. and Stoltenberg, D.E. (1997) Weed and seed bank management with integrated methods as influenced by tillage. *Weed Science* 45, 706–715.

Netland, J. (1985) Studium av tunrapp (*Poa annua* L.). Veksemåte, formering, konkurranseevne. [English summary: Studies on annual medow-grass, *Poa annua* L. – growth, reproduction, competitive ability.] Dissertation. Norsk Institute for Planteforskning. Avdeling Ugras. N-1432 Ås, Norway, 49 pp.

Nielsen, H.J. and Pinnerup, S.P. (1982) Reduced cultivation and weeds. *Swedish University of Agricultural Sciences. 23rd Swedish Weed Conference: Reports*, Uppsala, pp. 370–384.

Nieto, J. (1960) Elimine las hierbas a tiempo. *Agricultura Technica en Mexico* 9, 16–19.

Nieto, J., Brando, M.A. and Gonzalez, J.T. (1968) Critical periods of the crop growth cycle for competition from weeds. *PANS(C)* 14, 159–166.

Nikolaeva, M.G. (1977) Factors controlling the seed dormancy pattern. In: Khan, A.A. (ed.) *The Physiology and Biochemistry of Seed Dormancy and Germination.* Elsevier/North-Holland Biomedical Press, Amsterdam, pp. 52–74.

Numata, M. (1982) A methodology for the study of weed vegetation. In: Holzner, W. and Numata, N. (eds) *Biology and Ecology of Weeds.* Dr W. Junk Publishers, The Hague, pp. 21–34.

O'Donovan, J.T. (1996) Computerised decision support systems: aids to rational and sustainable weed management. *Canadian Journal of Plant Science* 76, 1–3.

O'Donovan, J.T., de St. Remy, E.A., O'Sullivan, P.A., Dew, D.A. and Sharma, A.K. (1985) Influence of the relative time of emergence of wild oat (*Avena fatua*) on yield loss of barley (*Hordeum vulgare*) and wheat (*Triticum aestivum*). *Weed Science* 33, 498–503.

O'Donovan, J.T., McAndrew, D.W. and Thomas, A.G. (1997) Tillage and nitrogen influence on weed population dynamics in barley (*Hordeum vulgare*). *Weed Technology* 11, 502–509.

Ohlander, L., Bergkvist, G., Stendahl, F. and Kvist, M. (1996) Yield of catch crops and spring barley as affected by time of undersowing. *Acta Agriculturae Scandinavica Section B, Soil and Plant Science* 46, 161–168.

Ohlsson, I. (1975) Utveckling, avkastning och kvalitet i olika bestånd av vårsådda oljeväxter. *Institutionen för Växtodling. Rapporter och Avhandlingar* 24, G1–G8.

Ohlsson, I. (1976) Sådd av vårraps och vårrybs med olika radavstånd – inverkan på avkastning och kvalitet. *Lantbrukshögskolan Meddelanden A* 262, Uppsala, 52 pp.

Olofsdotter, M. (2001) Rice – a step toward use of allelopathy. *Agronomy Journal* 93, 3–8.

Olofsdotter, M. and Mallik, A.U. (2001) Allelopathy symposium. Introduction. *Agronomy Journal* 93, 1–2.

Olofsdotter, M., Navarez, D., Rebulanan, M. and Streibig, J.C. (1999) Weed-suppressing rice cultivars – does allelopathy play a role? *Weed Research* 39, 441–454.

Olofsson, S. (1981) Flyghavreinventering i Vingåker. Sveriges Lantbruksuniversitetersitet, Institutionen för växtodlingslära. Examensarbete 705. Student work. Uppsala, 12 pp.

Olsson, E.G. (1987) Effects of dispersal mechanisms on the initial pattern of old-field forest succession. *Acta Oecologia* 8, 379–390.

Osvald, H. (1950) On antagonism between plants. *7th Proceedings of the International Congress of Botany*, Stockholm, pp. 167–171.

Palaveeva-Kovachevska, M. (1966) Certain biological peculiarities of *Cirsium arvense. Pastbiscmye Nauki* 2(6), 57–68.

Palmer, J.H. (1962) Studies on the behaviour of the rhizome of *Agropyron repens* (L.) Beauv. II. Effect of soil factors on the orientation of the rhizome. *Physiologia Plantarum* 15, 445–451.

Palmer, J.H. and Sagar, G.R. (1963) Biological flora of the British Isles. *Agropyron repens* (L.) Beauv. (*Triticum repens* L.; *Elytrigia repens* (L.) Nevski). *Journal of Ecology* 51, 783–794.

Parker, C. and Riches, C.R. (1993) *Parasitic Weeds of the World: Biology and Control.* CAB International, Wallingford, UK, 332 pp.

Pehkonen, A. and Sipilä, I. (1984) Improving the emergence in band sowing with wing coulters. *Helsingin Yliopisto Maatalousteknologian Laitos 43*, 48 pp.

Permin, O. (1961a) Bekaempelse af kvik ved jordbearbejdning efter høst. *Ugeskrift for Landmaend* 106, 594–596.

Permin, O. (1961b) Jordbearbejdningens betydning for bekaempelse af rodukrudt. *Tidsskrift for Planteavl* 64, 875–888.

Permin, O. (1973a) Virkningen af tør luft på spireevnen hos underjordiske udløbere af alm. kvik (*Agropyron repens*). *Tidsskrift for Planteavl* 77, 89–102.

Permin, O. (1973b) Virkning på den vegetative formering hos alm. kvik (*Agropyron repens*) når de underjordiske udløbere bliver skåret i stycker. *Statens Forsøgsvirksomhed i Plantekultur* 77, 357–369.

Peters, N.C.B. (1984) Time of onset of competition and effects of various fractions of an *Avena fatua* L. population on spring barley. *Weed Research* 24, 305–315.

Petersen, H.I. (1944) Ukrudtsundersøgelser ved Statens Ukrudtsforsøg i Aarene 1918–1928. *Tidsskrift for Planteavl* 48, 655–688.

Pfeiffer, R.K. and Holmes, H.M. (1961) A study of the competition between barley and oats as influenced by barley seed-rate, nitrogen level and barban treatment. *Weed Research* 1, 5–18.

Pipal, F.J. (1914) Wild garlic and its eradication. *Purdue University of Agriculture Experiment Station Bulletin* 176(18). 43 pp.

Pohjonen, V. and Paatela, J. (1974) Effect of planting interval and seed tuber size on the gross and net potato yield. *Acta Agriculturae Scandinavica* 24, 126–130.

von Polgár, J. (1984) Vältning efter vårsådd. *Sveriges Lantbruksuniversitetersitet. Rapporter från Jordbearbetningsavdelningen 69.* Uppsala, 16 pp.

Pollard, F. and Cussans, G.W. (1976) The influence of tillage on the weed flora of four sites sown to successive crops of spring barley. *Proceedings 1976 British Crop Protection Conference – Weeds*, pp. 1019–1028.

Pollard, F. and Cussans, G.W. (1981) The influence of tillage on the weed flora in a succession of winter cereal crops on a sandy loam soil. *Weed Research* 21, 185–190.

Pollard, F., Moss, S.R., Cussans, G.W. and Froud-Williams, R.J. (1982) The influence of tillage on the weed flora in a succession of winter wheat crops on a clay loam soil and silt loam soil. *Weed Research* 22, 129–136.

Popay, A.I. and Ivens, G.W. (1982) East Africa. In: Holzner, W. and Numata, N. (eds) *Biology and Ecology of Weeds.* Dr W. Junk Publishers, The Hague, pp. 345–372.

Powles, S.B. and Holtum, J.A.M. (eds) (1994) *Herbicide Resistance in Plants.* Lewis Publishers, Boca Raton, Florida, 353 pp.

Press, M.C., Scholes, J.D. and Riches, C.R. (2001) Current status and future prospects for managements of parasitic weeds (*Striga* and *Orobanche*). *2001 BCPC Symposium (in Brighton, UK). Proceedings 77: The World's Worst Weeds*, pp. 71–88.

Pülschen, L. and Koch, W. (1990) The altitudinal distribution of agrestal C_3 and C_4 Poaceae of the Shewa Province, Ethiopia. *Angewandte Botanik* 64, 281–288.

Putnam, A.R. (1988) Allelopathy: problems and opportunities in weed management. In: Altieri, M.A. and Liebman, M. (eds) *Weed Management in Agroecosystems: Ecological Approaches.* CRC Press, Boca Raton, Florida, pp. 77–88.

Raatikainen, M. and Raatikainen, T. (1975) Heinänurmien sato, kasvilajikoostumus ja sen muutokset. [Summary: Yield, composition and dynamics of flora in grasslands for hay in Finland.] *Annales Agriculturae Fenniae* 14, 57–191.

Raatikainen, M., Raatikainen, T. and Mukula, J. (1978) Weed species, frequencies and densities in winter cereals in Finland. *Annales Agriculturae Fenniae* 17, 115–142.

Rademacher, B. (1948) Gedanken über Begriff und Wesen des 'Unkrauts'. *Zeitschrift für Pflanzenkrankheiten (Pflanzenpathologie) und Pflanzenschutz* 55, 3–10.

Radosevich, S.R., Holt, J. and Ghersa, C. (1996) *Weed Ecology – Implication for Management.* John Wiley & Sons, New York, 589 pp.

Raghavendra, A.S. and Das, V.S.R. (1978) The occurrence of C_4-photosynthesis: a supplementary list of C_4 plants reported during late 1974–mid 1977. *Photosynthetica* 12, 200–208.

Rasmussen, J. (1991a) A model for prediction of yield response in weed harrowing. *Weed Research* 31, 401–408.

Rasmussen, J. (1991b) Reduceret jordbearbejdning og italiensk rajgraes som efterafgrøde. I. Vaextbetingelser, udbytter, afgrødeanalyser og ukrudt. *Tidsskrift for Planteavl* 95, 119–138.

Rasmussen, J. (1992) Testing harrows for mechanical control of annual weeds in agricultural crops. *Weed Research* 32, 267–274.

Rasmussen, J. (1998) Ukrudtsharvning i vinterhvete. *Danmarks Jordbrugsforskning. DJF-rapport Markbrug* 2, 179–189.

Rasmussen, J. and Pedersen, B.T. (1990) Forsøg med radrensning i korn – raeckeafstand og udsaedsmaengde. 7. *Danske Plantevaernskonference, Ukrudt*, pp. 187–189.

Rasmussen, J. and Svenningsen, T. (1995) Selective weed harrowing in cereals. *Biological Agriculture and Horticulture* 12, 29–46.

Raunkiær, C. (1934) *The Life Forms of Plants and Statistical Plant Geography.* Oxford University Press, Oxford, 632 pp.

Regnier, E.E. (1995) Teaching seed bank ecology in an undergraduate laboratory exercise. *Weed Technology* 9, 5–16.

Rice, E.L. (1984) *Allelopathy*, 2nd edn. Academic Press, Orlando, Florida, 422 pp.

Richens, R.H. (1947) Biological flora of the British Isles. No. 1862. *Allium vineale* L. *Journal of Ecology* 34, 209–226.

Roberts, H.A. and Dawkins, P.A. (1967) Effect of cultivation on the numbers of viable weed seeds in soil. *Weed Research* 7, 290–301.

Roberts, H.A. and Feast, P.M. (1970) Seasonal distribution of emergence in some annual weeds. *Experimental Horticulture* 21, 36–41.

Roberts, H.A. and Feast, P.M. (1972) Fate of some annual weeds in different depths of cultivated and undisturbed soil. *Weed Research* 12, 313–324.

Roberts, H.A. and Hewson, R.T. (1971) Herbicide performance and soil surface conditions. *Weed Research* 11, 69–73.

Roberts, H.A. and Neilson, J.E. (1980) Seed survival and periodicity of seedling emergence in some species *Atriplex, Chenopodium, Polygonum* and *Rumex. Annals of Applied Biology* 94, 111–120.

Roberts, H.A. and Potter, M.E. (1980) Emergence patterns of weed seedlings in relation to cultivation and rainfall. *Weed Research* 20, 377–386.

Roberts, H.A. and Ricketts, M. (1979) Quantitative relationship between the weed flora after cultivation and the seed population in the soil. *Weed Research* 19, 269–275.

Rodskjer, N. and Tuvesson, M. (1976) Observations on soil temperature under short grass cover and in variously managed bare soils at Ultuna, Sweden, 1968–1972. *Swedish Journal of Agricultural Research* 6, 243–246.

Röhrig, M. and Stützel, H. (2001a) Dry matter production and partitioning of *Chenopodium album* in contrasting competitive environments. *Weed Research* 41, 111–128.

Röhrig, M. and Stützel, H. (2001b) Canopy development of *Chenopodium album* in pure and mixed stands. *Weed Research* 41, 129–142.

Rola, J. (1979) Das Zusammenwirken von Kulturpflanzen und der angewendeten Herbizide auf die Verunkrautung. *Proceedings of the European Weed Research Society Symposium on the Influence of Different Factors on the Development and Control of Weeds*, pp. 281–285.

Rösiö, G. (1964) Utsädesmängdens inflytande på plantornas och beståndens utveckling i vårsäd. Dissertation. Kungliga Lantbrukshögskolan, Institutionen för Växtodling, Uppsala.

Rydberg, N.T. (1985) Ogräsharvning – en litteraturstudie. *Sveriges Lantbruksuniversitet., Institutionen för Växtodling. Rapport 154.* Uppsala, 25 pp.

Rydberg, N.T. (1993) Weed harrowing – driving speed at different stages of development. *Swedish Journal of Agricultural Research* 23, 107–113.

Rydberg, N.T. (1994) Weed harrowing – the influence of driving speed and driving direction on the degree of soil covering and the growth of weed and crop plants. *Biological Agriculture and Horticulture* 10, 197–205.

Rydberg, N.T. (1995) Weed harrowing in growing cereals. Dissertation. Swedish University of Agricultural Sciences, Department of Crop Production Science. 18 pp. + appended papers.

Rydberg, N.T. and Milberg, B. (2000) A survey of weeds in organic farming in Sweden. *Biological Agriculture and Horticulture* 18, 175–185.

Rydberg, T. (1992) Ploughless tillage in Sweden. Results and experiences from 15 years of field trials. *Soil and Tillage Research* 22, 253–264.

Sagar, G. and Rawson, H. (1964) The biology of *Cirsium arvense. Proceedings of the 7th British Weed Control Conference*, Brighton, UK, pp. 553–562.

Salisbury, E.J. (1961) *Weeds and Aliens.* 2nd edn. Collins, London, 384 pp.

Salonen, J. (1992a) Efficacy of reduced herbicide doses in spring cereals of different competitive ability. *Weed Research* 32, 483–491.

Salonen, J. (1992b) Yield responses of spring cereals to reduced herbicide doses. *Weed Research* 32, 493–499.

Salonen, J. (1993) Weed infestation and factors affecting weed incidence in spring cereals in Finland – a multivariate approach. *Agricultural Science in Finland* 2, 525–536.

Salonen, J., Hyvönen, T. and Jalli, H. (2001a) Weed flora in organically grown spring cereals in Finland. *Agricultural and Food Science in Finland* 10, 231–242.

Salonen, J., Hyvönen, T. and Jalli, H. (2001b) Weeds in spring cereal fields in Finland – a third survey. *Agricultural and Food Science in Finland* 10, 347–364.

Schnieders, B.J. (1999) A quantitative analysis of inter-specific competition in crops with a row structure. Thesis, Agricultural University Wageningen. The Netherlands, 165 pp.

Schroeder, D., Mueller-Schaerer, H. and Stinson, C.A.S. (1993) A European weed survey in 10 major crop systems to identify targets for biological control. *Weed Research* 33, 449–458.

Seavers, G.P. and Wright, K.J. (1999) Crop canopy development and structure influence weed suppression. *Weed Research* 39, 319–328.

Semb Tørresen, K. (1998) Emergence and longevity of weed seeds in soil with different tillage treatments. *Aspects of Applied Biology* 51, 197–204.

Semb Tørresen, K., Skuterud, R., Weiseth, L., Tandsaether, H.J. and Haugan Jonsen, S. (1999) Plant protection in spring cereal production with reduced tillage. I. Grain yield and weed development. *Crop Protection* 18, 595–603.

Sherwood, L.V. and Fuelleman, R.F. (1948) Experiments in eradicating field bindweed. *Illinois Agricultural Experiment Station Bulletin* 525, 473–506.

Shinosaki, K. and Kira, T. (1956) Intraspecific competition among higher plants. VII. *Osaka City University Journal of the Institute of Polytechnics, Series* D 7, 35–72.

Sjörs, H. (1971) *Ekologisk botanik*. Almqvist and Wiksell, Stockholm, 296 pp.

Skuterud, R. (1984) Growth of *Elymus repens* (L.) Gould and *Agrostis gigantea* Roth. at different light intensities. *Weed Research* 24, 51–57.

Skuterud, R., Semb, K., Saur, J. and Mygland, S. (1996) Impact of reduced tillage on the weed flora in spring cereals. *Norwegian Journal of Agricultural Sciences* 10, 519–532.

Spitters, C.J.T. (1983a) An alternative approach to the analysis of mixed cropping experiments. 1. Estimation of competition effects. *Netherlands Journal of Agricultural Science* 31, 1–11.

Spitters, C.J.T. (1983b) An alternative approach to the analysis of mixed cropping experiments. 2. Marketable yield. *Netherlands Journal of Agricultural Science* 31, 143–155.

Stenberg, M. (2000) Influences of seedbed bottom properies on seed germination and water relations. *Acta Agriculturae Scandinavica, Section B, Soil and Plant Science* 50, 114–121.

Stenberg, M., Aronsson, H., Lindén, B., Rydberg, T. and Gustfson, A. (1999) Soil mineral nitrogen and nitrate leaching losses in soil tillage systems combined with a catch crop. *Soil and Tillage Research* 50, 115–125.

Stenberg, M., Håkansson, I., von Polgár, J. and Heinonen, R. (1994) Sealing, crusting and hardsetting soils in Sweden – occurrence, problems and research. *Proceedings of the 2nd International Conference on Sealing, Crusting and Hardening Soils*. Brisbane, Australia, pp. 287–292.

Stoller, E.W. (1973) Effect of minimum soil temperature on differential distribution of *Cyperus rotundus* and *C. esculentus* in the United States. *Weed Research* 13, 209–217.

Strand, E. (1968) Radavstand ved såing av korn, engvekster m. v. En sammenstilling og bearbeidelse av inn- og utenlandske forsøksresultater. *Nordisk Jordbrugsforskning* 50, 429–445.

Streibig, J.C. (1979) Numerical methods illustrating the phytosociology of crops in relation to weed flora. *Journal of Applied Ecology* 16, 577–587.

Streibig, J.C. (1984) Measurement of phytotoxicity of commercial and unformulated soil-applied herbicides. *Weed Research* 24, 327–331.

Streibig, J.C. (1988) Herbicide bioassay. *Weed Research* 28, 479–484.

Streibig, J.C. (1992) Quantitative assessment of herbicide phytotoxicity with dilution assay. Dissertation. Department of Agricultural Sciences, Royal Veterinary and Agricultural University, Copenhagen, 98 pp.

Sundelin, G. and Gustafsson, H. (1946) Ogräsbekämpning. Sammanfattning av under ledning av Jordbruksförsöksanstalten utförda ogräsbekämpningsförsök åren 1932–1944. *Jordbruksförsöksanstalten. Meddelande 15*. Uppsala, 51 pp.

Suomela, H. and Paatela, J. (1962) The influence of irrigation, fertilizing and MCPA on the competition between spring cereals and weeds. *Weed Research* 2, 90–99.

Svensson, B. and Carlsson, H. (1974) Potatisbestånd och kvävegödsling – inverkan på avkastning och kvalitet. *Institutionen for växtodling. Rapporter och Avhandlingar* 6. Uppsala, 41 pp.

Swan, D.G. and Chancellor, R.J. (1974) The regeneration of field bindweed root fragments. *Proceedings of the Society of Weed Science* 27, 20–21.

Teräsvuori, K. (1929) Die Quecke als Kulturpflanze und Unkraut. *Acta Agralia Fennica* 18, 33–58.

Terbøl, M. and Petersen, P.H. (1999) Forsøg og erfaringer med mekanisk ukrudtsbekaempelse. *16th Danske Plantevaernskonference. Plantebeskyttelse i økologisk jordbrug. DJF Rapport Markbrug* 10, 71–83.

Thind, K.B. (1975) The anatomy, physiology and biochemical factors controlling regeneration from roots of *Cirsium arvense* (L.) Scop. PhD thesis. Brunel University, UK, 234 pp.

Thompson, B.K., Weiner, J. and Warwick, S.I. (1991) Size-dependent reproductive output in agricultural weeds. *Canadian Journal of Botany* 69, 442–446.

Thompson, K., Bakker, J.P. and Bekker, R.M. (1997) *The Soil Seed Banks of North West Europe – Methodology, Density and Longevity*. Cambridge University Press, Cambridge, 276 pp.

Thörn, K.-G. (1967) Norrländsk vallodling. II. Resultat av vallinventering 1959–1961. *Skogs- och Lantbruksakademiens Tidskrift 1967*. Supplement 7. Stockholm, 57 pp.

Thorup, S. and Pinnerup, S.P. (1981) Jordbehandlingens indflydelse på ukrudtsbestanden. I. V. Nielsen: Reduceret jordbehandling. Ørritsgård 1972–80. *Statens Jordbrugstekniske Forsøg, Rapport 1*. Copenhagen.

Thurston, J.M. (1982) Wild oats as successful weeds. In: Holzner, W. and Numata, N. (eds) *Biology and Ecology of Weeds*. Dr W. Junk Publishers, The Hague, pp. 191–199.

Thurston, J.M. and Williams, E.D. (1968) Growth of perennial grass weeds in relation to the cereal crop. *Proceedings of the 9th British Weed Control Conference*, pp. 1115–1123.

Tilman, D. (1986) Resources, competition and the dynamics of plant communities. In: Crawley, M.J. (ed.) *Plant Ecology*. Blackwell Scientific Publications, Oxford, pp. 51–75.

Tilman, D. (1987) Secondary succession and the pattern of plant dominance along experimental nitrogen gradients. *Ecological Monographs* 57, 189–214.

Tilman, D. (1990) Constraints and tradeoffs: toward a predictive theory of competition and succession. *Oikos* 58, 3–15.

Timmons, F.L. and Bruns, V.F. (1951) Frequency and depth of shoot-cutting in eradication of certain creeping perennial weeds. *Agronomy Journal* 43, 371–375.

Torstensson, L. (2000) Experiences of biobeds in practical use in Sweden. *Pesticide Outlook* 11(5), 206–211.

Torstensson, L. and Castillo, M.dP. (1997) Use of biobeds in Sweden to minimize environmental spillages from agricultural spraying equipment. *Pesticide Outlook* 8(3), 24–27.

Tutin, T.G., Heywood, V.H., Burges, N.A., Moore, D.M., Vakentine, D.H., Walters, S.M. and Webb, D.A. (eds) (1964–1980) *Flora Europaea*. Cambridge University Press, Cambridge.

Vanhala, P. and Pitkänen, J. (1998) Long-term effects of primary tillage on above-ground weed flora and on the weed seedbank. *Aspects of Weed Biology* 51, 99–104.

Vengris, J. (1962) The effect of rhizome length and depth of planting on the mechanical and chemical control of quackgrass. *Weeds* 10, 71–74.

Vleeshouwers, L.M. and Bouwmeester, H.J. (2001) A simulation model for seasonal changes in dormancy and germination of weed seeds. *Seed Science Research* 11, 77–92.

Vleeshouwers, L.M., Bouwmeester, H.J. and Karssen, C.M. (1995) Redefining seed dormancy: an attempt to integrate physiology and ecology. *Journal of Ecology* 83, 1031–1037.

Vleeshouwers, L.M., Streibig, J.C. and Skovgaard, I. (1989) Assessment of competition between crops and weeds. *Weed Research* 29, 273–280.

Watkinson, A.R. (1981) Interference in pure and mixed populations of *Agrostemma githago*. *Jouranl of Applied Ecology* 18, 967–976.

Wedin, D.A. and Tilman, D. (1996) Influence of nitrogen loading and species composition on the carbon balance of grasslands. *Science (Washington)* 274, 1720–1723.

Weissmann, R. (2002) Plant growth-affecting bacteria as weed suppressing agents. *Acta Universitatis Agriculturae Sueciae. Agraria* 312, 1–60.

Wells, M.J. and Stirton, C.H. (1982) South Africa. In: Holzner, W. and Numata, N. (eds) *Biology and Ecology of Weeds*, Dr W. Junk Publishers, The Hague, pp. 339–343.

Werner, P.A. and Rioux, R. (1977) *Agropyron repens* (L.) Beauv. No. 24. *Canadian Journal of Plant Science* 57, 905–919.

Whybrew, Y.E. (1958) Seed rates and nitrogen for winter wheat. Experiments at the Norfolk Agricultural Station. *Experimental Husbandry* 3, 35–39.

Wicks, G.A., Ramsel, R.E., Nordquist, P.T. and Schmidt, J.W. (1986) Impact of wheat cultivars on establishment and suppression of summer annual weeds. *Agronomy Journal* 78, 59–62.

Wiggans, R.G. (1939) The influence of space and arrangement on the production of soybean plants. *Journal of the American Society of Agronomy* 31, 314–321.

Wikander, G. (1988) *Ogräsharvning. Harvningsdjupets betydelse för effekten på ogräs och gröda*. Sveriges Lantbruksuniversitet. Institutionen för Växtodling. Seminarier och Examensarbeten 813. Uppsala, 37 pp.

Willey, R.W. and Heath, S.B. (1969) The quantitative relationships between plant population and crop yield. *Advances in Agronomy* 21, 281–321.

Willey, R.W. and Rao, M.R. (1980) A competitive ratio for quantifying competition between intercrops. *Experimental Agriculture* 16, 117–125.

Williams, E.D. (1968) Preliminary studies of germination and seedling behaviour in *Agropyron repens* (L.) Beauv. and *Agrostis gigantea* Roth. *Proceedings of the 9th British Weed Control Conference*, pp. 119–124.

Williams, E.D. (1970a) Studies on the growth of seedlings of *Agropyron repens* (L.) Beauv. and *Agrostis gigantea* Roth. *Weed Research* 10, 321–330.

Williams, E.D. (1970b) Effects of decreasing the light intensity on the growth of *Agropyron repens* (L.) Beauv. in the field. *Weed Research* 10, 360–366.

Williams, E.D. (1977) Growth of seedlings of *Agropyron repens* L. Beauv. and *Agrostis gigantea* Roth. in wheat and barley: effect of time of emergence, nitrogen supply and cereal seed rate. *Weed Research* 17, 69–76.

Williams, E.D. and Attwood, P.J. (1971) Seed production of *Agropyron repens* (L.) Beauv. in England and Wales in 1969. *Weed Research* 11, 22–30.

Williamson, G.B. (1990) Allelopathy, Koch's postulates, and the neck riddle. In: Grace, J.B. and Tilman, D. (eds) *Perspectives in Plant Competition*. Academic Press, New York, pp. 143–162.

Wilson, B.J. (1986) Yield responses of winter cereals to the control of broad-leaved weeds. *Proceedings of the European Weed Research Society Symposium 1986, Economic Weed Control*, pp. 75–82.

Wilson, B.J. and Lawson, H.M. (1992) Seed-bank persistence and seedling emergence of seven weed species in autumn-sown crops following a single year's seeding. *Annals of Applied Biology* 120, 105–116.

Wilson, B.J. and Wright, K.J. (1990) Predicting the growth and competitive effects of annual weeds in wheat. *Weed Research* 30, 201–211.

Wilson, B.J., Wright, K.J., Brain, P., Clements, M. and Stephens, E. (1995) Predicting the competitive effects of weed and crop density on weed biomass, weed seed production and crop yield in wheat. *Weed Research* 35, 265–278.

de Wit, C.T. (1960) On competition. *Verslagen van landbouwkundige onderzoekingen* 66. 8. Wageningen, 82 pp.

de Wit, C.T. and van den Bergh, J.P. (1965) Competition between herbage plants. *Netherlands Journal of Agricultural Science* 13, 212–221.

Yenish, J.P., Doll, J.D. and Buhler, D.D. (1992) Effects of tillage on vertical-distribution and viability of weed seed in soil. *Weed Science* 40, 429–433.

Yoshida, S. (1972) Physiological aspects of grain yield. *Annual Review of Plant Physiology* 23, 437–464.

Zimdahl, R.L. (1980) *Weed–Crop Competition*. International Plant Protection Center, Oregon State University Corvallis, Oregon, 196 pp.

Zimdahl, R.L. (1988) The concept and application of the critical weed-free period. In: Altieri, M.A. and Liebman, M. (eds) *Weed Management in Agroecosystems: Ecological Approaches*, CRC Press, Boca Raton, Florida, pp. 145–155.

Species Index

Numbers preceded by 'F' or 'T' are numbers of figures and tables, respectively; others are page numbers.

Subject Index

Numbers preceded by 'F' or 'T' are numbers of figures and tables, respectively; others are page numbers.